Electric Vehicles and the Future of Energy Efficient Transportation

Umashankar Subramaniam
Prince Sultan University, Saudi Arabia

Sheldon S. Williamson
Ontario Tech University, Canada

Mohan Krishna S.
Alliance University, India

Febin Daya J. L.
Vellore Institute of Technology, India

IGI Global
PUBLISHER of TIMELY KNOWLEDGE

A volume in the Advances in Mechatronics and Mechanical Engineering (AMME) Book Series

Published in the United States of America by
IGI Global
Engineering Science Reference (an imprint of IGI Global)
701 E. Chocolate Avenue
Hershey PA, USA 17033
Tel: 717-533-8845
Fax: 717-533-8661
E-mail: cust@igi-global.com
Web site: http://www.igi-global.com

Library of Congress Cataloging-in-Publication Data

Names: Subramaniam, Umashankar, editor. | Williamson, Sheldon S., editor. | Krishna, Mohan, 1987- editor. | Febin Daya, J.L., 1980- editor.
Title: Electric vehicles and the future of energy efficient transportation / Umashankar Subramaniam, Sheldon Williamson, Mohan Krishna, and J.L. Febin Daya, editors.
Description: Hershey, PA : Engineering Science Reference, [2021] | Includes bibliographical references and index. | Summary: "This book creates awareness and enlighten the readers about the growth and adoption of electric vehicles for the purpose of sustainable transportation offering critical analysis in terms of the economics, technology and environment perspectives"-- Provided by publisher.
Identifiers: LCCN 2020049420 (print) | LCCN 2020049421 (ebook) | ISBN 9781799876267 (hardcover) | ISBN 9781799876274 (paperback) | ISBN 9781799876281 (ebook)
Subjects: LCSH: Electric vehicles. | Transportation--Environmental aspects. | Electric vehicles--Technological innovations. | Sustainable engineering--Research.
Classification: LCC TL220 .E4475 2021 (print) | LCC TL220 (ebook) | DDC 629.22/93--dc23
LC record available at https://lccn.loc.gov/2020049420
LC ebook record available at https://lccn.loc.gov/2020049421

This book is published in the IGI Global book series Advances in Mechatronics and Mechanical Engineering (AMME) (ISSN: 2328-8205; eISSN: 2328-823X)

British Cataloguing in Publication Data
A Cataloguing in Publication record for this book is available from the British Library.

All work contributed to this book is new, previously-unpublished material. The views expressed in this book are those of the authors, but not necessarily of the publisher.

For electronic access to this publication, please contact: eresources@igi-global.com.

Advances in Mechatronics and Mechanical Engineering (AMME) Book Series

J. Paulo Davim
University of Aveiro, Portugal

ISSN:2328-8205
EISSN:2328-823X

Mission

With its aid in the creation of smartphones, cars, medical imaging devices, and manufacturing tools, the mechatronics engineering field is in high demand. Mechatronics aims to combine the principles of mechanical, computer, and electrical engineering together to bridge the gap of communication between the different disciplines.

The **Advances in Mechatronics and Mechanical Engineering (AMME) Book Series** provides innovative research and practical developments in the field of mechatronics and mechanical engineering. This series covers a wide variety of application areas in electrical engineering, mechanical engineering, computer and software engineering; essential for academics, practitioners, researchers, and industry leaders.

Coverage

- Control Methodologies
- Computer-Based Manufacturing
- Computational Mechanics
- Intelligent Sensing
- Tribology and surface engineering
- Intelligent Navigation
- Manufacturing Methodologies
- Sustainable and green manufacturing
- Autonomous Systems
- Nanomaterials and nanomanufacturing

IGI Global is currently accepting manuscripts for publication within this series. To submit a proposal for a volume in this series, please contact our Acquisition Editors at Acquisitions@igi-global.com or visit: http://www.igi-global.com/publish/.

The Advances in Mechatronics and Mechanical Engineering (AMME) Book Series (ISSN 2328-8205) is published by IGI Global, 701 E. Chocolate Avenue, Hershey, PA 17033-1240, USA, www.igi-global.com. This series is composed of titles available for purchase individually; each title is edited to be contextually exclusive from any other title within the series. For pricing and ordering information please visit http://www.igi-global.com/book-series/advances-mechatronics-mechanical-engineering/73808. Postmaster: Send all address changes to above address.

Titles in this Series

For a list of additional titles in this series, please visit:
http://www.igi-global.com/book-series/advances-mechatronics-mechanical-engineering/73808

Modeling and Optimization of Solar Thermal Systems Emerging Research and Opprtunities
Jagadish (National Institute of Technology, Raipur, India) and Agnimitra Biswas (National Institute of Technology, Silchar, India)
Engineering Science Reference • © 2021 • 215pp • H/C (ISBN: 9781799835233) • US $185.00

Recent Trends on Electromagnetic Environmental Effects for Aeronautics and Space Aplications
Christos D. Nikolopoulos (Department of Electronic Engineering, School of Engineering, Hellenic Mediterranean University, Greece)
Engineering Science Reference • © 2021 • 285pp • H/C (ISBN: 9781799848790) • US $265.00

Modern Machining Processes and Techniques for Aerospace Materials
Anand Pandey (Manipal University Jaipur, India)
Engineering Science Reference • © 2021 • 300pp • H/C (ISBN: 9781799845102) • US $215.00

Evaluation and Applications for Ultrasonic Methods of Non-Destructive Testing
Ashish Khaira (Samrat Ashok Technological Institute, India)
Engineering Science Reference • © 2021 • 215pp • H/C (ISBN: 9781799845645) • US $195.00

Handbook of Research on Advancements in Manufacturing, Materials, and Mechanical Engineering
Leonid Burstein (Independent Researcher, Israel)
Engineering Science Reference • © 2021 • 462pp • H/C (ISBN: 9781799849391) • US $295.00

Genetic Algorithms and Remote Sensing Technology for Tracking Flight Debris
Maged Marghany (Syiah Kuala University, Indonesia)
Engineering Science Reference • © 2020 • 334pp • H/C (ISBN: 9781799819202) • US $195.00

Humanizing Cities Through Car-Free City Development and Transformation
Rahma M. Doheim (Assiut University, Egypt & University of Business and Technology, Saudi Arabia) Alshimaa Aboelmakarem Farag (Zagazig University, Egypt) and Ehab Kamel (University of Central Lancashire, UK)
Engineering Science Reference • © 2020 • 305pp • H/C (ISBN: 9781799835073) • US $215.00

Airline Green Operations Strategies Emerging Research and Opportunities
Yazan Khalid Abed-Allah Migdadi (Qatar University, Qatar)
Engineering Science Reference • © 2020 • 231pp • H/C (ISBN: 9781799842552) • US $185.00

IGI Global
PUBLISHER of TIMELY KNOWLEDGE

701 East Chocolate Avenue, Hershey, PA 17033, USA
Tel: 717-533-8845 x100 • Fax: 717-533-8661
E-Mail: cust@igi-global.com • www.igi-global.com

Table of Contents

Detailed Table of Contents

The mode of transit in the current trend is gradually shifting from internal combustion engine operated vehicle to battery operated electric vehicle. The need of electric vehicle began the revolution from traditional gasoline-powered vehicles to electric vehicles (EVs). An electric vehicle uses electric traction motors for propulsion. It could also be powered through a collector system by electricity from off-vehicle sources or could also be self-contained with a battery, solar panels, or an electrical generator to convert fuel to electricity.

In the last two decades the pollution problems and the increase of the cost of fossil energy (oil, gas) have become planetary problems. The automobile manufacturers started to react to the urban pollution problems in the nineties by commercializing the electric vehicle. But the battery weight and cost problems were not solved. The batteries must provide energy and peak power during the transient states. These conditions are severe for the batteries. To decrease these severe conditions, the supercapacitors and batteries associated with a good power management present a promising solution. Supercapacitors are storage devices that enable to supply the peaks of power to hybrid vehicles during the transient states. During the steady states, batteries will provide the energy requested. This methodology enables the decrease of the weight and increase of the lifespan of the batteries. Hybridization using batteries and supercapacitors for transport applications is needed when energy and power management are requested during the transient states and steady states.

An efficient state estimator is critical for the development of an autonomous plug-in hybrid electric vehicle (PHEV). To achieve effective autonomous regulation of the powertrain, the latency period and estimation error should be minimum. In this work, a novel error state extended kalman filter (ES-EKF)-based state estimator is developed to perform sensor fusion of data from light detection and ranging sensor (LIDAR), the inertial measurement unit sensor (IMU), and the global positioning system (GPS) sensors, and the estimation error is minimized to reduce latency. The estimator will provide information to an intelligent energy management system (IEMS) to regulate the powertrain for effective load sharing in the PHEV. The integration of the sensor fusion data with the vehicle model is simulated in MATLAB environment. The PHEV model is fed with the proposed state estimator output, and the response parameters of the PHEV are monitored.

A non-sustainable wellspring of energy is for the most part non-plentiful in nature and can be utilized until they exist in plenitude. As and when the acknowledgment of the decrease of the bountiful assets and natural risks hit, an auxiliary idea of use of non-pollutable-sustainable power sources shows up into place. Different vulnerabilities as for the plan and supplies of a vehicle emerge with less or weakened association of the sustainable assets. A few vulnerabilities will emerge from the capital venture part and extents from the general population and private segments. Another challenge is the contribution of the innovation to travel the flow ordinary vehicles to electrical vehicles (EV) and to coordinate the renewable energy sources (RES), for example, wind-vitality, sunlight-based photovoltaics, and differing assortments of bio-energies to constantly work according to the prerequisite to reduce greenhouse gas (GHG). This chapter deals with the integration and charging challenges and strategies, coordination of vehicles, demand integration, and solutions to them.

In the present work, three different methods for obtaining the DC response for modeling practical DC-DC buck and boost converters operating in continuous conduction mode (CCM) are demonstrated using MATLAB/Simulink. The method of selection for inductor, diode, and MOSFET for a DC-DC converter is discussed in detail. The governing equations for the non-ideal converters were derived using volt-sec and amp-sec balance equations. Mathematical modeling of basic converters was carried out using 'commonly used blocks' reducing the dependence on SimPower System tool box in Simulink. The non-ideal parameters in the converter caused a drastic variation in the duty cycle and output voltage which in turn had an adverse effect on the efficiency. The transients in output voltages and inductor currents were observed. In addition, a Li ion polymer battery was mathematically modeled. Accurate battery modeling for pulse charging was proposed. A comparative analysis of 1, 2, …, 5 RC pair/s modeling of the battery was presented.

This chapter focuses on developing a sustainable architecture for public electric motorbike charging stations. Electric motorbikes or electric bicycles (both referred to as e-bikes) are compact electric vehicles which are primarily battery-powered and driven solely by electric motors. This work conceptualizes a microgrid architecture which utilizes the integration of distributed generation energy resources providing the charging station nodes with sustainable power and increased fault tolerance. The charging stations proposed in the study increase the long-time energy savings of the infrastructure maintenance authorities while also reducing reliance on the public grid during peak hours. The photovoltaic-based DC microgrid is integrated with e-bike charging infrastructure, moving towards a future of eco-friendly and power-efficient technology.

Hybrid electric vehicles have great potential to reduce fuel consumption and palliate harmful emissions for better population health and climate change. These power trains are essential to meet current emissions legislation, contribute to more environmentally friendly mobility and improve air quality in big cities. In the past two decades, due to the advent growth of modern electric drives, battery technology, and intelligent charging methodology, the overall performance of the electric vehicles have improved. Many research works have been conducted to improve the efficiency and reliability. It increases the applications of electric vehicles. Apart from grid power, the use of renewable sources makes zero-emission rate and zero fuel cost. This chapter deals with the role of power converters in hybrid electric vehicles (PHEVs). It deals with power converters used in hybrid electric vehicles, types of battery chargers, emerging power electronic devices, and thermal management of HEV power electronics, which are presented with the simulation using MATLAB software.

In the present era, electric vehicles (EV) have revolutionized the world with their dominant features like cleanliness and high efficiency compared to that of the internal combustion (IC) engine-based vehicles. To crave for the higher efficiency of the EV during the braking, the kinetic energy of the EV is converted into electrical energy, which is harvested into storage system, called regenerative braking. Various techniques such as artificial neural network (ANN) and fuzzy-based controllers consider factors like state of charge of the battery and supercapacitor and brake demand for calculating the regenerative braking energy. A force distribution curve is designed to ensure that the braking force is distributed and applied on the four wheels simultaneously. In real-time optimization, an operating area is formed for maximizing the regenerative force which is evaluated by linear programming. It is proved that the drive range of the vehicle is increased by 25.7% compared to the one with non-RBS. In this work, RTO-based control loop for regenerative braking system is simulated in MATLAB/Simulink.

Solar powered wireless electric vehicle charging technology functions independently without interface with the utility grid. Wireless power transfer (WPT) technology is incorporated for wireless charging, which brings the benefits of safe operation, less pollution, and little maintenance cost. WPT technology necessitates no physical connection between the charging device and vehicle, thus hazards and inconvenience produced by conventional charging methods have been minimized. WPT in electric vehicle can be used to reduce the charging time, range, and cost. In this chapter, the various configurations of WPT like inductive, capacitive, resonant, and roadway power transfer techniques have been presented. The small-scale prototype of wireless charging has been developed in the laboratory by incorporating inductive power transfer technique. The experimental results have been presented to validate the feasibility of the system in real time.

As an experienced automotive manufacturer, Turkey is following the global trend of the EV transition and on an enthusiastic start. A government-led action group for EV production called TOGG was initiated in 2018. Automotive sector in Turkey yields (but is not limited to) the following titles at the same time: a strength, a public revenue source, a trade dilemma, an (national) ambition. To tackle power grid problems due to increasing demand, renewable energy usage seems both challenging and necessary. There is a potentially favorable demand for EVs. Promoting the EVs for the customers requires strong infrastructure which seems lacking in Turkey. The EVCSs have been scarce against large land area of the country. On the contrary, consumers in Turkey buy too much BEVs against PHEVs compared to the EU which needs to be directed by the government accordingly to make a healthy transition in the future.

Preface

Electric vehicles have been seen as a promising technology option and the use of electric cars, trucks, and buses is on the rise. Several national governments have successfully implemented policies to promote the technology. The rationale is simple: electric vehicles have lower cost of ownership than their conventionally powered peers, they emit less pollution, they are energy efficient and they enable emerging mobility technologies and business models. There is a consensus among auto industry executives and analysts that a tipping point is approaching where mass adoption will become unavoidable because of falling battery costs, pressure from regulators and generous government subsidies. These factors have come together to force the traditional industry to take electrification faster than expected around the globe.

Electric vehicles are complex electro-mechanical systems. They require holistic consideration of vehicle and tire dynamics, powertrain, electric motors and batteries, and control and estimation modules that are integrated through hardware- software interface. One of the core elements of the Electric Vehicle is an Electric motor. The rapid development in the field of Power electronics and control techniques has created a space for various types of electric motors to be used in Electric Vehicles. The other important element as far as Electric Vehicle is concerned is the Battery and its management system. All Electric Vehicles employ rechargeable batteries. Hence, proper care of these batteries is critical to the operation of the vehicle. In order to understand the impact of vehicle design in the electrical grid, we have to look at the characteristics of these batteries, which are usually li-ion. Electric vehicle battery packs are made up of multiple cell modules arranged in series and in parallel. The battery pack has a battery management system (BMS) that monitors components close to the battery cells themselves. During the charging and discharging of an Electric vehicle battery, each cell within the battery pack must be closely and accurately monitored. Any kind of unbalanced cell voltage conditions can impact battery performance. Intelligent battery management and charging of the battery is necessary to protect the safety of the vehicle. Overvoltage or under-voltage conditions can cause a battery failure that affects vehicle operation. Charging EV batteries could pose a challenge to the nation's electricity grid as more Electric vehicles appear on the roads. New behavioral patterns of Electric Vehicle create big implications for the electricity grid. This may require substantial development in the grid with costly investments in the nation's electricity system.

This book would tend to address all these major sticking points by eliciting research inputs which would aid the stakeholders in all the critical areas mentioned. The significant impacting factors in the power electronics for electric vehicle market include surge in demand for energy-efficient battery-powered devices, stringent emission regulations to reduce vehicle weight and emission, and government initiatives to balance environmental pollution and vehicle emission. In addition, power electronics supports high input impedance and improved parallel current sharing, which increases the adoption of power electronic components in Electric vehicle. The recent advancements in power electronics used for Electric vehicle applications are addressed. Major challenges in charging infrastructure required for electric vehicles, current evolution in the field of charging infrastructure and the ongoing standardization efforts in the field are highlighted. It also addresses the techno-economic feasibility of different EV technologies being developed.

OBJECTIVE OF THE BOOK

The objective of this book is to create awareness and enlighten the readers about the growth and adoption of EV for the purpose of sustainable transportation. This would ensure critical analysis in terms of the economics, technology and environment perspectives. Also, the impact on the society would be multi frontal as it would educate the consumer about the benefits and limitations of the same. This book holds great contemporary relevance as there is increased attention to the EV technology by all the stakeholders and would be of great value to them.

TARGET AUDIENCE

The proposed book would play a significant role for a wide umbrella of researchers and policy makers involved in the technology, economics, policy and environmental aspects of EV. The book includes chapters offering overview of Electric vehicle technology, charging techniques, battery management systems and integration of renewable system for Electric vehicle applications, which would be useful for audience from any background. The details for specific charging technique such as wireless charging and their integration with renewable system are presented. The book includes 10 chapters, presenting recent advancements on Electric vehicle and their future for efficient energy transportation.

Chapter 1 enlightens about some of the challenging issues, regarding the battery management system (BMS) from a control theoretic perspective. The basic functionality, battery charging methods, state of charge (SoC) estimation methods, state of health (SoH) determination and battery communication interfaces are described in detail. The chapter also highlights selected open areas of research in the discussed areas.

In Chapter 2, the authors design and develop a bidirectional DC-DC converter using battery /Super-capacitor for Electric vehicle application, specifically for electric assisted bicycle using state of the art hub motor technology. The design of a power converter is presented based on the energy requirements of the system. The implementation procedures are given in detail. The simulation and experimental results are presented for various cases. The amelioration of the proposed work is presented with clarity.

Chapter 3 is on another important domain, the regulation of an autonomous PHEV using an improved Error State Extended Kalman Filter as the State Estimator to perform sensor fusion. The author introduces the modeling and development of a split parallel plug-in hybrid Electric vehicle. Further, the selection of the sensors and its calibration is presented. The implementation process is explained with focus on the Error State Extended Kalman Filter State Estimator modeling. The chapter also discusses about the Intelligent Energy Management System and its role in regulation of the plug-in hybrid Electric vehicle model. Finally the results are presented and the merits are discussed considering the effects of sensor calibration error on the vehicle state estimation.

In Chapter 4, the important issues of integrating renewable sources in the Electric vehicle, their challenges and advantages are presented. Usage of renewable energy systems as input source for battery charging infrastructure creates a sustainable power system. The chapter aims to address various challenges such as energy conversion, storage of energy into the batteries, integration of renewable sources, vehicle to grid interface and effective utilization of the stored energy and battery charging. The author highlights the various solution techniques available with their merits.

In Chapter 5, the author focuses on mathematical modeling of DC-DC Converters and Li ion Battery using MATLAB / Simulink. Buck and Boost converters were modeled in three different methods using MATLAB/ Simulink and its DC response was analyzed. The design and procedure for selection of every component in the DC-DC converter are discussed in this chapter. Further, the battery modeling for pulse charging is presented with simulation results.

Chapter 6 focuses on developing a sustainable architecture for public electric motorbike charging stations. The Micro grid architecture of the proposed charging station is presented, considering the real-world implementation and constraints. The chapter also expounds the use of photovoltaic arrays as the renewable energy resource used as supply to the Micro grid. Further, the input date and the evaluation parameters are validated using simulation and the results are presented. As the proposed work focuses on developing a market deployable model, the economic constraints are studied and the environmental impact of the proposed architecture is presented comparing with conventional charging stations.

Chapter 7 deals about the role of power converters in hybrid electric vehicle applications. The current research in this area mainly concentrates on improvement of the power semiconductor device, fault-tolerant converter topologies, novel control techniques and thermal management for power semiconductor devices. This chapter gives a glimpse of some of the recent converter/ inverter topologies used for hybrid electric vehicle applications. A detailed comparison of various topologies with respect to cost, power density and efficiency is presented by the author.

In Chapter 8, the author presents detailed analysis of the regenerative braking system and effective optimization of the process to estimate the regenerative braking force. To estimate the regenerative braking force, there is requirement of complete understanding of the brake system used in electric vehicles. To ensure the vehicle safety and stability, the applied braking force is distributed among the front and rear wheels according to the ideal braking force distribution curve which is discussed in this chapter. Some of the important factors such as coefficient of adhesion which influence the braking performance and various scenarios of braking and the problems faced by the vehicle at different scenarios are presented. The optimization process is implemented using MATLAB/Simulink and the results are evaluated for an urban drive cycle to evaluate the braking force scenarios at different speeds. The results are compared with other optimization techniques and presented to validate the robustness of the proposed technique.

Chapter 9 is about wireless power transfer which is gaining momentum in the recent years. With the advent of power converters, power in kW range can be transferred in wireless mode between coils. Wireless power transfer has wide range of applications and one such is for battery charging of Electric vehicles. The author mainly concentrates on solar powered Electric vehicle, and gives a brief review on various wireless power transfer technologies used. The basic principle of wireless power transfer, various types of wireless power transfer systems and the testing methods are presented. The comparison of various wireless power transfer technologies and the challenges in implementation are highlighted in this chapter.

In Chapter 10, a comprehensive review about the current outlook and possibilities regarding Turkish automotive industry and the impact of Electric vehicle adaptation. The chapter presents a clear report on the work published by Turkish researchers and industries in electric vehicle research. Turkey is one of the countries that has prioritized the Electric vehicle promotion and looks to further boost the sales of hybrid and electric automobiles. It is focusing on developing and using low-consumption and environmentally friendly automobile technologies. The author also presents the challenges, trends and scope of Electric vehicle in revolutionizing the Turkish automation sector.

We strive to include a rich range of recent developments in Electric vehicle technology. A wide range of topics in the area of Electric vehicle research is discussed in every chapter. We assume that the book would be a valuable tool for a wide audience, including researchers on Electric vehicle technology.

Umashankar Subramaniam
Sheldon S. Williamson
Mohan Krishna S.
Febin Daya J. L.

Chapter 1
Battery Management Systems (BMS) for EV:
Electric Vehicles and the Future of Energy-Efficient Transportation

Rakshitha Ravi
RVITM, India

Usha Surendra
CHRIST University, India

ABSTRACT

The mode of transit in the current trend is gradually shifting from internal combustion engine operated vehicle to battery operated electric vehicle. The need of electric vehicle began the revolution from traditional gasoline-powered vehicles to electric vehicles (EVs). An electric vehicle uses electric traction motors for propulsion. It could also be powered through a collector system by electricity from off-vehicle sources or could also be self-contained with a battery, solar panels, or an electrical generator to convert fuel to electricity.

INTRODUCTION

Electric vehicle companies are enabling modern technologies like Artificial Intelligence and IOT to improve efficiency. Nowadays, vehicles are essential in the day to day life and for industrial use as well. Sufficient effort is being done to withdraw the combustion engines by electric motors (Liu et al., 2016).A system with IOT will definitely streamline the performance of EV charging and looks the impacts. IOT will improve the city planning and makes the city life easy. Automotive/transportation sector is undergoing a phenomenal course of rapid, change. This transformation is enabled by similar factors that are driving advancements in the Internet of Things including increased computing resources and ubiquitous connectivity. They're easier on the environment. EV's don't even have an exhaust system, meaning they have zero emissions. And since gas-powered vehicles are large contributors to greenhouse-gas buildup

DOI: 10.4018/978-1-7998-7626-7.ch001

in the earth's atmosphere, making the switch to an electric car can help contribute to cleaner air and a healthier planet. Electric vehicles (EVs) burn no gasoline and have no tailpipe emissions, but producing the electricity used to charge them does generate global warming emissions.

Benefits of Electric Vehicles

There are a number of great benefits to electric vehicles (EVs) over conventional petrol/diesel cars.

- Cheaper to run. Owners of an EV have the advantage of much lower running costs.
- Cheaper to maintain. ...
- Other savings. ...
- Better for the environment.

ADVANTAGES

Better for the Environment

There are a ton of people who choose to purchase an electric vehicle solely because these vehicles are better for the environment than the alternatives. These cars do not contain an exhaust system at all, which means they aren't creating emissions like other cars. Vehicles that are powered by gas are major contributors to the buildup of greenhouse gases, which stay in the earth's atmosphere. As such, having an electric car helps keep air cleaner and the planet healthier.

Electricity is Less Expensive than Gas

When you break everything down, the average American pays about 15 cents a mile to drive a gas-powered vehicle. This might seem negligible but that changes when you look at the cost to drive an electric car. Rather than 15 cents a mile, many electric cars run on five cents a mile due to the fact that electricity is largely less expensive than gasoline. Most people will also be charging their cars in the garage, which means installing a few solar panels can cut that price even further while offering savings on powering your entire home.

Less Maintenance at a Lower Cost

With a gasoline-powered car, there's a ton of maintenance. It starts with oil changes and only goes on from there. Electric cars don't have oil and don't require that maintenance task or others associated with a gas engine. Anything related to the combustion engine in a traditional car isn't needed with an electric car. In addition, the brakes on electric vehicles tend to last longer. That only adds to the cost savings of choosing an electric car rather than one that uses gasoline. It all starts to add up when you begin to think about it.

Figure 1. Importance of Electric vehicle

BATTERY MANAGEMENT SYSTEM IN ELECTRIC VEHICLES

The mode of transit in the current trend is gradually shifting from Internal combustion engine operated vehicle to Battery operated Electric Vehicle. The need of Electric Vehicle began the revolution from traditional gasoline-powered vehicles to Electric Vehicles (EVs). An Electric Vehicle uses electric traction motors for propulsion. It could also be powered through a collector system by electricity from off-vehicle sources, or could also be self-contained with a battery, solar panels or an electrical generator to convert fuel to electricity.

Battery Management System (BMS) for Electric Vehicles

The Lithium-ion batteries have proved to be the battery of interest for Electric Vehicle manufacturers because of its high charge density and low weight. Even though these batteries pack in a lot of punch for its size they are highly unstable in nature. It is very important that these batteries should never be over charged or under discharge at any circumstance which brings in the need to monitor its voltage and current. This process gets a bit tougher since there are a lot of cells put together to form a battery pack in EV and every cell should be individually monitored for its safety and efficient operation which requires a special dedicated system called the **Battery Management System**.

LITERATURE REVIEW

Since Battery electric vehicle is involved type of battery and its importance is included. Along with battery importance Types of electric vehicles and importance of battery in electric vehicle is discussed here.

LITHIUM-ION BATTERY

History

The idea of Lithium Ion battery was first coined by G.N Lewis in the 1912, but it became feasible only in the year 1970's and the first non-rechargeable lithium battery was put into commercial markets. Later in 1980's engineers attempted to make the first rechargeable battery using lithium as the anode material and were partially successful. They failed to notice that these types of lithium batteries were unstable during the charging process and it would create a short inside the battery increasing the temperature and causing a thermal runaway. In 1991, one such lithium battery used in mobile exploded over a man's face in Japan.

LI-ION BATTERY

Chemistry and Working

As the name obviously indicates, the Lithium Ion batteries use the Lithium ions to get the job done. Lithium is a very light metal with high energy density, this property enables the battery to be light in weight and provide high current with a small form factor. Energy density is the amount of energy that can be stored in per unit volume of the battery, the higher the energy density the smaller the battery will be. Despite the overwhelming properties of lithium metal, it cannot be used as an electrode directly in the batteries since lithium is highly unstable because of its metallic nature. Hence we use lithium-ions which more or less has the same property of a lithium metal but it is non-metallic and is comparatively safer to use.

Nominal Voltage: The nominal voltage is the actual voltage rating of an 18650 Cell. By default, it is 3.6V and will remain the same for all 18650 cells despite of its manufactures.

Full discharge voltage: An 18650 cell should never be allowed to discharge below 3.2V, failing to do so will alter the internal resistance of the battery which will damage the battery permanently and might also lead to explosion.

Full charge Voltage: The charging voltage for lithium ion cell is 4.2V. Care should be taken that the cell voltage does not increase 4.2V at any given time.

mAh Rating: The capacity of a cell is normally given in terms of mAh (Milli Ampere hour) rating. This value will vary based on the type of cell you have purchased. For example, let's assume our cell here is 2000mAh which is nothing but 2Ah (Ampere/hour). This means that if we draw 2A from this battery it will last for 1 hour and similarly if we draw 1A from this battery it will last for 2 hours. So, if you want to know how long the battery will power you project (Run-time) then you have to calculate it using the mAh Rating.

Run Time (in hours) = Current drawn / mAh Rating
Where, current drawn should be within the C rating limit.

C Rating: The C rating of the battery again changes for each battery, let's assume that the battery we have is a 2Ah battery with 3C rating. The value 3C means that the battery can output 3 times the rated Ah rating as its maximum current. In this case it can supply upto 6A (3*2 = 6) as the maximum current. Normally 18650 cells have a 1C rating only.

Maximum current drawn from battery = C Rating * Ah Rating

Charging Current: Normally it will be 0.5C, meaning half the value of the Ah rating. For a 2Ah rating battery the charging current will be 1A (0.5*2 = 1).

Charging time: The minimum charging time required for a single 18650 cell to charge to can be calculate by using the value of charge current and Ah rating of battery.

Internal Resistance (IR): The health and capacity of a battery can be predicted by measuring the internal resistance of the battery.

State of Charge (SOC) %: The state of charge is nothing but the capacity of the battery, similar to the ones shown in our mobile phone. The capacity of a battery cannot be plainly calculated with its voltage valve, it is normally calculated using current integration to determine the change in battery capacity over time.

Depth of Discharge (DOD) %: No battery will have 100% discharges since as we know it will damage the battery. Normally an 80% depth of discharge is set for all batteries.

Cell dimension: Another unique and interesting feature of the 18650 cell is its dimension. Every cell will have a diameter of 18mm and a height of 650mm which makes this cell gets its name 18650.

Literature Review

In this chapter a literature review of Electric vehicles is done.

BATTERY MANAGEMENT SYSTEM

Paper number : .... 1[2]

Paper type

(Journal/conference) : .... International journal of Energy [ENERGIES]
Name of the publisher : .... MDPI
Year : .... Energies 2019, 12, 446; doi:10.3390/en12030446
Title : .... Towards a Smarter Battery Management System for Electric Vehicle Applications: A Critical Review of Lithium-Ion Battery State of Charge Estimation
Authors : .... Muhammad Umair Ali, Amad Zafar, Sarvar Hussain Nengroo, Sadam Hussain, Muhammad Junaid Alvi and Hee-Je Kim,

Inferences : an efficient BMS also provides information on the battery states, such as the state of available power (SOP), state of charge (SOC), state of life (SOL), and state of health (SOH). The BMS can sense the battery voltage, battery current, and temperature to avoid overcharge and over discharge conditions.

conclusion : This review has given a focus on multiple performance features, such as output power, component count, switching frequency, electromagnetic interference (EMI), losses, effectiveness, cost and reliability which directly influence the selection of a particular DC-DC converter for respective BEV and PHEV powertrains.

limitation : the MEIBC is the best option for mainly high-power BEVs and PHEVs for having very low input current ripples and output voltage ripples, bidirectionality, low EMI, high efficiency as well as reliability, and moderate cost and compact size or volume.

Paper number : ………… 4[7]

Paper type

(Journal/conference) : ………… Technical article https://doi.org/10.1007/s41918-019-00060-4
Name of the publisher : ………… Center for Study of Science, Technology and Policy (CSETP) Electrochemical Energy Reviews (2020) 3:1–42
Year : ………… Received: 8 April 2019 / Revised: 13 June 2019 / Accepted: 5 December 2019 / Published online: 28 December 2019
Title : ………… Battery Management System: The key to Electric Vehicle adoption
Authors : ………… Ms. Epica MANDAL SARKAR, Research Consultant&Mr.TanmaySarkar, Senior Research Engineer, Center for Study of Science, Technology and Policy (CSETP)

conclusion : Lithium-ion batteries (LIBs), with relatively high energy density and power density, have been considered as a vital energy source in our daily life, especially in electric vehicles.

limitation : Battery Management System BMS, an electronic system, protects a battery from overcharging/over-discharging which ensures the safe operation of EVs.

Paper number : ………… 10[17]

Paper type

(Journal/conference) : ………… Universal Journal of Electrical and Electronic Engineering 4(1): 10-15, 2016
Name of the publisher : ………… Universal Journal of Electrical and Electronic Engineering
Year : ………… Volume 10, Number 2, 2016
Title : ………… Algorithm for based Battery Management System
Authors : ………… Lan-Rong Dung, Yu-Syuan Peng

Inferences : Importance of lithium ion battery and cell balancing technique is discussed .In recent years, as energy-saving and environmental protection were widely concerned around the world, the lithium iron phosphate (LFP) battery has been widely used in hybrid electric vehicle (HEV) and electric vehicle (EV). Generally the battery pack for HEV is composition of the number of cells in series.

conclusion : In this case, imbalance among cells due to the difference of degradation and temperature will be accelerated by the cycles of charge and discharge without an appropriate battery equalization management system. However, the balancing system only intended to reduce the difference of cell voltage

limitation : Electric Vehicles is more advantageous because of its features.Within that Battery Management System plays a key role in Battery Electric Vehicles. Lithium Ion battery is used among all other batteries.

Paper number : .... 11[1]

Paper type

(Journal/conference) : .... INTERNATIONAL JOURNAL OF ENERGY
Name of the publisher : .... SAE INTERNATIONAL
Year : .... Volume 4, Issue 2 july 2015
Title : .... Model-Based Parameter Identification of Healthy and Aged Li-ion Batteries for Electric Vehicle Applications
Authors : .... Ryan Ahmed, McMaster Univ.

Inferences : The battery management system (BMS) is responsible for monitoring the battery state-of-charge (SOC), state-of-health (SOH), state-ofpower (SOP), and remaining useful life . The BMS also performs thermal management, cell balancing, and attempts to avoid battery overcharge or under-discharge that might affect driver safety and shorten battery life.

conclusion : The battery management system (BMS) is responsible for monitoring the battery state-of-charge (SOC), state-of-health (SOH), state-ofpower (SOP), and remaining useful life.The BMS also performs thermal management, cell balancing, and attempts to avoid battery overcharge or under-discharge that might affect driver safety and shorten battery life.

limitation : installed charging stations are not able to meet the increasing charging demand of Electric Vehicles. So, if that is overcome then effectively electric cars can be into effect. Among the configurations of electric vehicles battery Electric Vehicles is more advantageous because of its features.

Paper number : .... 12[10]

Paper type

(Journal/conference) : .... Journal of power source
Name of the publisher : .... SCIENCE DIRECT
Year : .... Volume 10, Number 8, 2013
Title : .... Online battery state of health estimation based on Genetic Algorithm for electric and hybrid vehicle applications
Authors : .... Zheng Chen, Chunting Chris Mi*, Yuhong Fu, Jun Xu, Xianzhi Gong

Inferences : A battery management system (BMS) is essential to manage, observe and protect the battery for safe and reliable operations of the vehicle. State of charge(SOC)and state of health (SOH) are very basic functions of the BMS.

conclusion : A battery management system (BMS) is essential to manage, observe and protect the battery for safe and reliable operations of the vehicle. State of charge(SOC)and state of health (SOH) are very basic functions of the BMS.

limitation : Limitations are the installed charging stations are not able to meet the increasing charging demand of Electric Vehicles.

TYPES OF ELECTRIC VEHICLES:

BEV, PHEV AND HEV

There are three main types of electric vehicles (EVs), classified by the degree that electricity is used as their energy source. BEVs, or battery electric vehicles, PHEVs of plug-in hybrid electric vehicles, and HEVs, or hybrid electric vehicles.

Electric Vehicles can be categorized as follows:

Figure 2. Types of Electric Vehicles

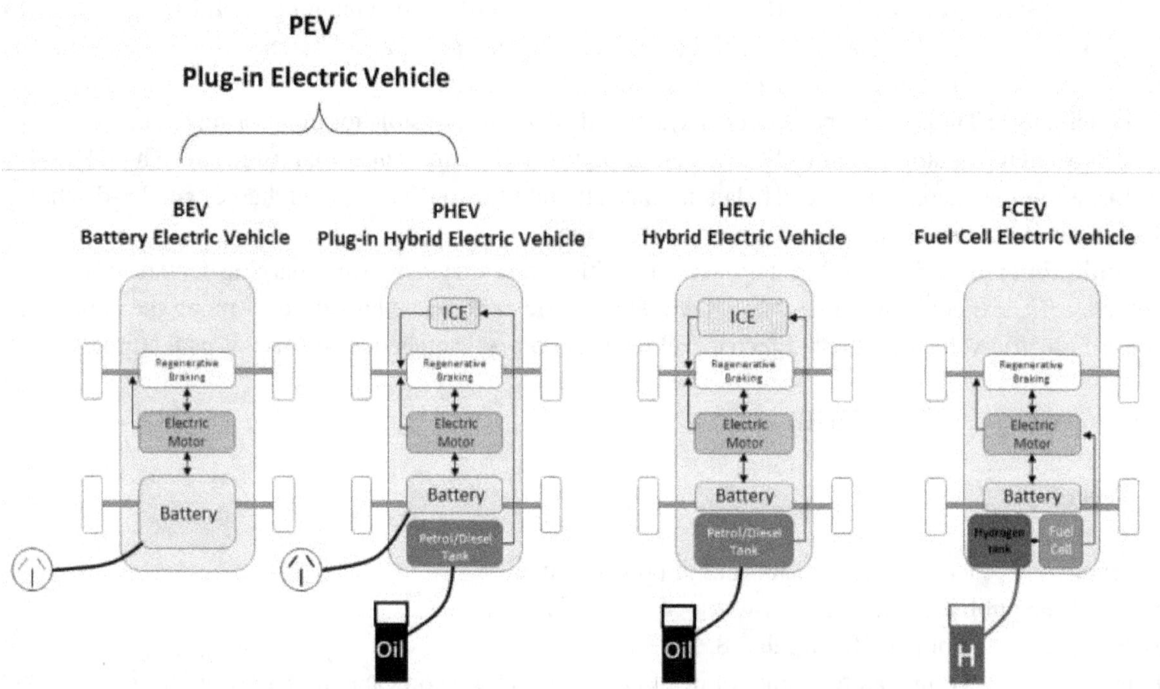

Battery Electric Vehicles (BEV)

Battery Electric Vehicles, also called BEVs, and more frequently called EVs, are fully-electric vehicles with rechargeable batteries and no gasoline engine. Battery electric vehicles store electricity onboard with high-capacity battery packs. Their battery power is used to run the electric motor and all onboard electronics. BEVs do not emit any harmful emissions and hazards caused by traditional gasoline-powered vehicles. BEVs are charged by electricity from an external source. Electric Vehicle (EV) chargers are classified according to the speed with which they recharge an EVs battery.

The classifications are Level 1, Level 2, and Level 3 or DC fast charging. Level 1 EV charging uses a standard household (120V) outlet to plug into the electric vehicle and takes over 8 hours to charge an EV for approximately 75-80 miles. Level one charging is typically done at home or at your workplace.

- Level 1 chargers have the capability to charge most EVs on the market.
- Level 2 charging requires a specialized station which provides power at 240v. Level 2 chargers are typically found at workplaces and public charging stations and will take about 4 hours to charge a battery to 75-80 miles of range.
- Level 3 charging, DC fast charging, or simply fast charging is currently the fastest charging solution in the EV market. DC fast chargers are found at dedicated EV charging stations and charge a battery up to 90 miles range in approximately 30 minutes.

Figure 3. Battery Electric Vehicles (BEV)

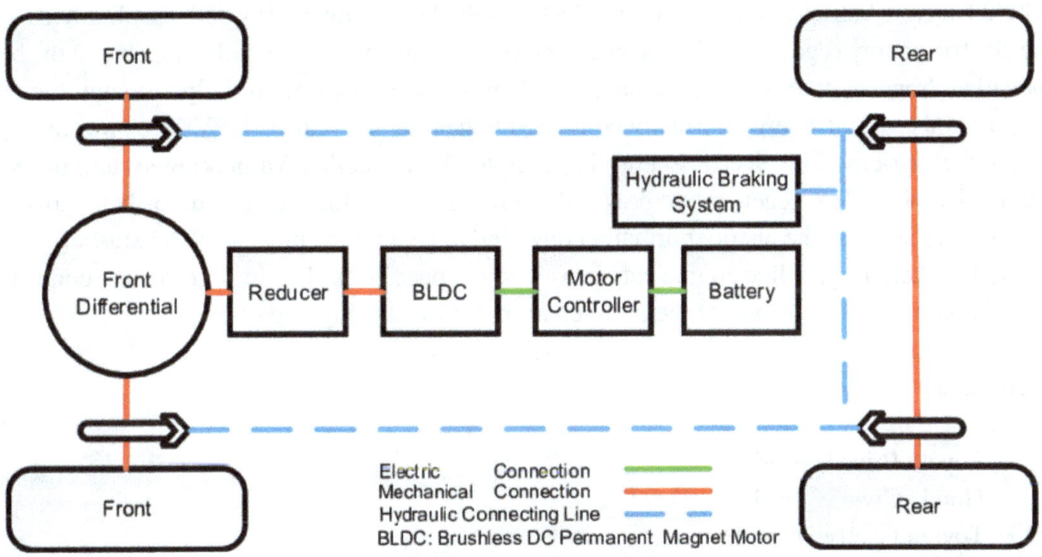

Plug-in Hybrid Electric Vehicle (PHEV): Plug in hybrid electric vehicles can recharge the battery through both regenerative braking and plugging in to an external source of electrical power. While standard hybrids can go about 1-2 miles before the gasoline engine turns on, plug in hybrid electric vehicle models can go anywhere from 10-40 miles before their gas engines provide assistance. Plug-in Hybrid Electric Vehicles or PHEVs can recharge the battery through both regenerative braking and "plugging in" to an external source of electrical power.

PHEV Examples

- Chevy Volt

Figure 4. Plug-in Hybrid Electric Vehicle

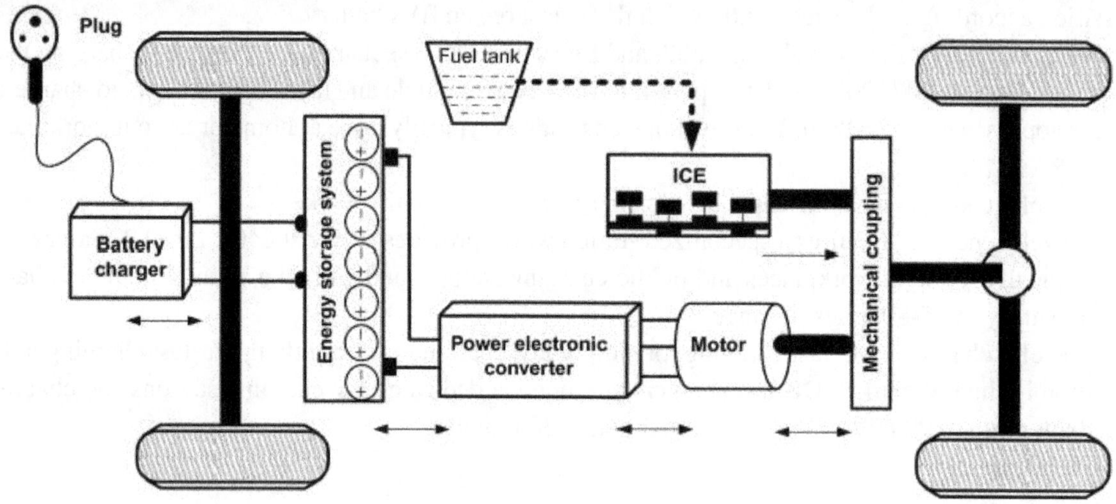

Hybrid Electric Vehicles (HEV): Hybrid Electric Vehicles are powered by both gasoline and electricity. The electric energy is generated by the braking system of the moving vehicle to recharge the battery. This is called "regenerative braking", a process where the electrical motor helps to slow the vehicle and uses a number of the energy normally converted to heat by the brakes. HEVs are powered by both gasoline and electricity. The electric energy is generated by the car's own braking system to recharge the battery. This is called 'regenerative braking', a process where the electric motor helps to slow the vehicle and uses some of the energy normally converted to heat by the brakes. HEVs start off using the electric motor, then the gasoline engine cuts in as load or speed rises. The two motors are controlled by an internal computer, which ensures the best economy for the driving conditions.

HEV Examples

- Toyota Prius Hybrid
- Honda Civic Hybrid
- Toyota Camry Hybrid

Fuel Cell Electric Vehicle (FCEV): A Fuel Cell Vehicle (FCV) or Cell Electric Vehicle (FCEV) may be a sort of electric vehicle which uses a cell, rather than a battery, or together with a battery or supercapacitor, to power its on-board motor.

Figure 5. Hybrid Electric Vehicle

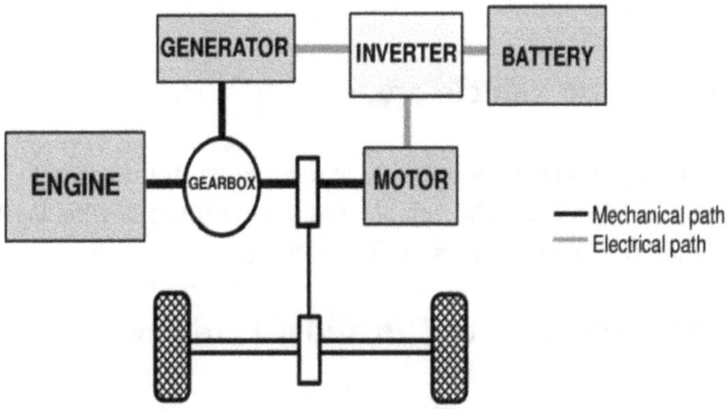

Figure 6. Fuel cell

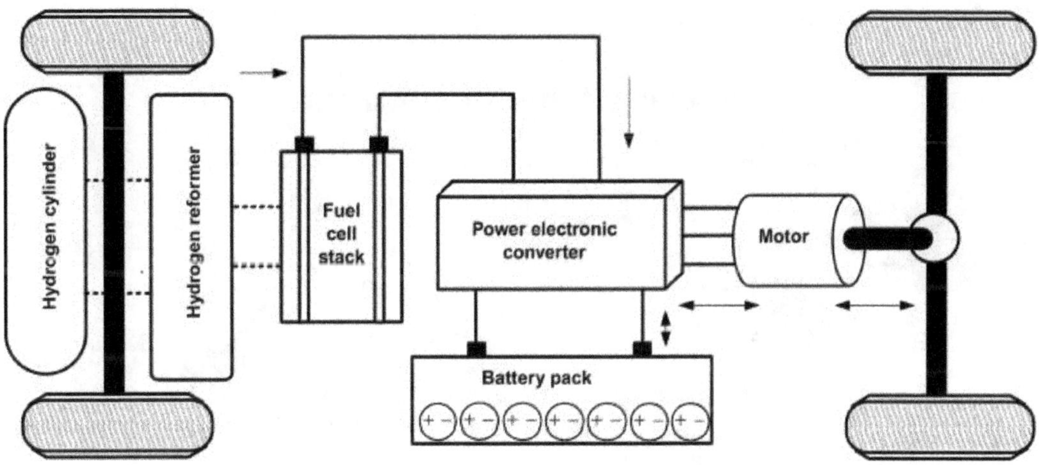

Table 1.

Types of EV	PEV	HEV	FCEV
Energy source	(i) Battery	(i) Battery/ultracapacitor (ii) Internal combustion engines	(i) Fuel cells
Propulsion technique	(i) Electric motor drives	(i) Electric motor drives (ii) Internal combustion engines	(i) Electric motor drives
Characteristics and feature	(i) Zero emission (ii) Short driving range (iii) Higher initial costs	(i) Low emission (ii) Longer range (iii) Complex	(i) Zero emission (ii) Highest initial costs (iii) Medium driving range
Major techniques	(i) Electric motor control (ii) Battery management (iii) Charging device	(i) Electric motor control (ii) Battery management (iii) Managing multiple energy sources and optimal system efficiency (iv) Components sizing	(i) Fuel processor (ii) Fueling system (iii) Fuel cell cost
Regenerative braking	(i) Yes	(i) Yes	(i) Yes

COMPARISON BETWEEN DIFFERENT TYPES OF EV

Components of the Electric Vehicle are,

EV Powertrain: power train provides power to the vehicle. Powertrain refers to the set of components that generate the power required to move the vehicle and deliver it to the wheels. An EV powertrain has 60% fewer components than the powertrain of an ICE vehicle.

Difference between Powertrains of EVs and ICE vehicles

The powertrain of an electric vehicle is a simpler system, comprising of far fewer components than a vehicle powered by an internal combustion engine.

Figure 7. Components of vehicle drive

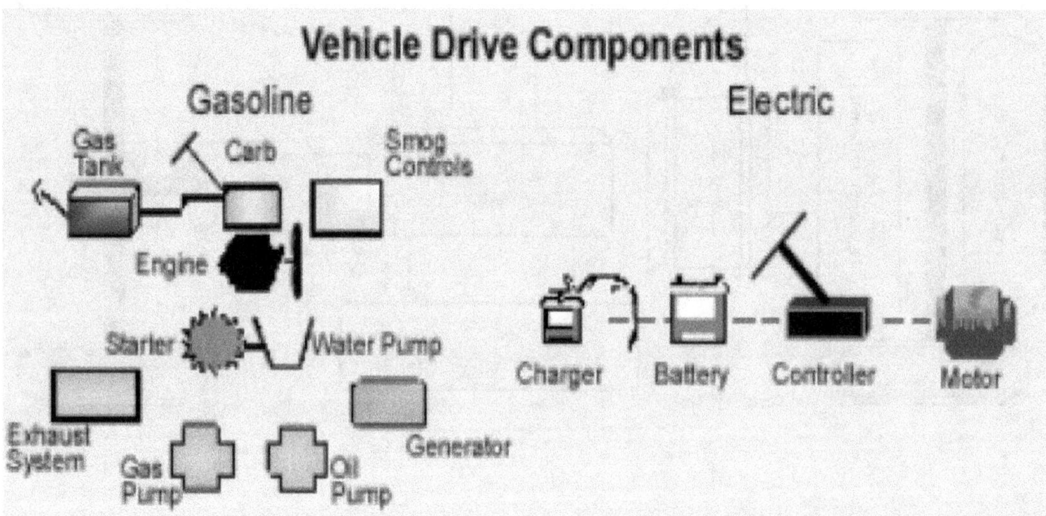

Main Components of an ICE vehicle powertrain – An ICE vehicle has hundreds of moving parts. Main components of its powertrain are Engine, Transmission and Driveshaft. Power is generated by the engine and transmitted to the driveshaft.

Battery Pack:The battery pack is made up of multiple Lithium-ion cells and stores the energy needed to run the vehicle. Battery packs provide direct current (DC) output.

DC-DC Converter: The DC supplied by battery pack is converted to AC and supplied to the electric motor. This power transfer is managed by a sophisticated **motor control mechanism** (also referred to as Powertrain Electronic Control Unit) that controls the frequency and magnitude of the voltage supplied to the electric motor in order to manage the speed and acceleration as per driver's instructions communicated via acceleration/brakes.

DC-AC Converter The DC supplied by battery pack is converted to AC and supplied to the electric motor. This power transfer is managed by a sophisticated motor control mechanism

Electric Motor: Converts electrical energy to mechanical energy, that is delivered to the wheels via single ratio transmission. Many EVs use motor generators that can perform regeneration as well.

On-board Charger: Converts AC received through charge port to DC and controls the amount of current flowing into the battery pack.

Body Control Module (BCM): The BCM is responsible for supervising and controlling the functions of electronic accessories such as power windows, mirrors, security and vehicle access control.

Figure 8. Block diagram of Battery Management System

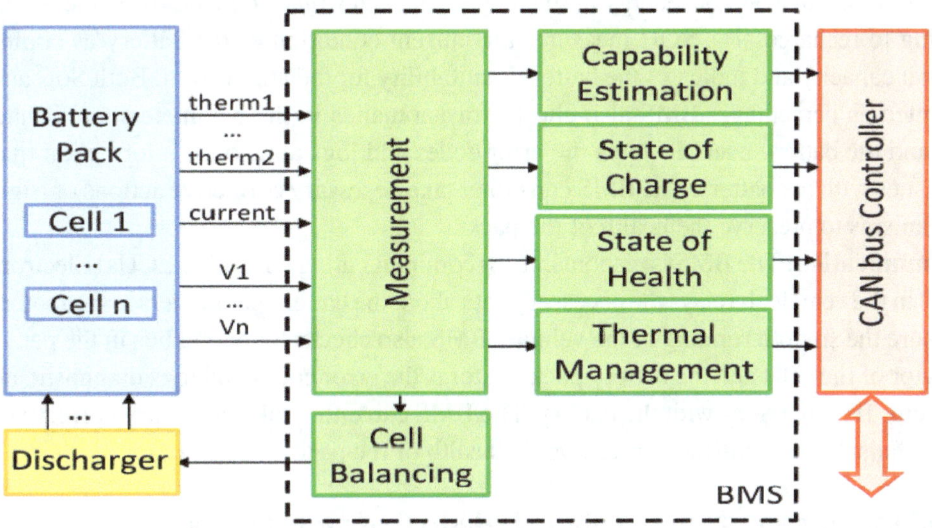

The Primary Functions of the BMS for an EV battery consists of

1. **Safety**: Electric vehicles run on high voltage Lithium-ion battery packs. Lithium-ion batteries have higher energy density (i.e. 100-265 Wh/kg) than other battery chemistries. These batteries come with a risk of catching fire under unusual circumstances. It is imperative to operate the EV batteries in pre-defined safe limits to ensure the safety of the user as well as the vehicle. The Battery Management System continuously monitors parameters such as temperature, voltage and current in and out of the pack to ensure it is being operated in safe conditions the entire time. BMS is responsible for thermal management of the battery and monitors its temperature continuously. If required, BMS can adjust cooling and trigger other safety mechanisms to cease operations and minimize the risk. e.g. in Hyundai Kona Electric, if overheating of the battery pack is detected by the BMS, the vehicle's power output is automatically limited and the car is put in fail-safe mode. Overcharging of lithium-ion cells can also lead to thermal runaway and potentially an explosion. BMS continuously monitors the voltage of the pack as well as individual battery cells and controls the supply of the current to avoid overcharging. BMS can enforce the limits of maximum charge or discharge current according to temperature.
2. **Performance Optimization:** BMS is responsible for optimizing the performance of the battery pack. Lithium-ion batteries perform best when their State of Charge (SOC) is maintained between

the minimum and maximum charge limits defined in the battery profile. Overcharging as well as deep discharging degrades the capacity of the battery, thereby shortening its life. At the time of charging, BMS determines how much current can safely go in and communicates the same to the EVSE (Electric Vehicle Supply Equipment or the Charger). During discharge of the battery, BMS would communicate with the motor controller to avoid the cell voltages reaching too low. The vehicles can show a corresponding alert to the user to charge the battery pack. The BMS also controls the recharging of the battery pack by energy generated through regenerative braking.

3. **Health Monitoring and Diagnostics:** The BMS uses the collected data points (temperature, voltage, current etc.) to estimate the State of Charge and State of Health (SOH) of the battery pack. The SoC refers to available energy in the battery and determines how far the vehicle can go before needing to recharge. The SOH measures the current condition of the battery as compared to its original capacity and indicates the battery's suitability for the application. Both SoC and SOH are presented as percentages.BMS also checks for anomalies in the parameters and behavior of the cells and the battery pack. It stores the error codes and logs diagnostic information that helps fix any issues with the battery. The BMS can either take necessary corrective actions or trigger failsafe mechanisms to preserve the health of the pack.
4. **Communication:** The BMS is responsible for communicating with other ECUs (Electronic Control Units) in the vehicle. It relays the necessary data about the battery parameters to the motor controller to ensure the smooth running of the vehicle. BMS also checks for anomalies in the parameters and behavior of the cells and the battery pack. It stores the error codes and logs diagnostic information that helps fix any issues with the battery. The BMS can either take necessary corrective actions or trigger failsafe mechanisms to preserve the health of the pack.

Battery Management system (BMS) Design Considerations

There are lot of factors that are to be considered while designing a BMS.

Discharging Control: The primary function of a BMS is to maintain the lithium cells within the safe operating region. For example, a typical Lithium 18650 cell will have an under-voltage rating of around 3V. It is the responsibility of the BMS to make sure that none of the cells in the pack get discharged below 3V.

Charging Control: Apart from the discharging the charging process should also be monitored by the BMS. Most batteries tend to get damaged or get reduced in lifespan when charged inappropriately. For lithium battery charger a 2-stage charger is used. The first stage is called the Constant Current (CC) during which the charger outputs a constant current to charge the battery. When the battery gets nearly full the second stage called the Constant Voltage (CV) stage is used during which a constant voltage is supplied to the battery at a very low current

State-of-Charge (SOC) Determination: SOC as the fuel indicator of the EV. It actually tells us the battery capacity of the pack in percentage. But it is not as easy as it sounds. The voltage and charge/discharge current of the pack should always be monitored to predict the capacity of the battery. Once the voltage and current are measured there are a lot of algorithms that can be used to calculate the SOC of the Battery pack.

State-of-Health (SOH) Determination: The capacity of the battery not only depends on its voltage and current profile but also on its age and operating temperature. The SOH measurement tells us about the age and expected life cycle of the battery based on its usage history. This way we can know how much the mileage (distance covered after full charge) of the EV reduces as the battery ages and also, we can know when the battery pack should be replaced. The SOH should also be calculated and kept in track by the BMS.

CELL BALANCING TECHNIQUE

Cell balancing is the process of equalizing the voltages and state of charge among the cells when they are at a full charge. Most typical battery chargers detect full charge by checking whether the voltage of the entire string of cells has reached the voltage regulation point. In passive balancing, energy is drawn from the most charged cell and dissipated as heat, usually through resistors. In active balancing, energy is drawn from the most charged cell and transferred to the least charged cells, usually through capacitor-based, inductor-based or DC-DC converters.

Active cell balancing is a more complex balancing technique that redistributes charge between battery cells during the charge and discharge cycles, thereby increasing system run time by increasing the total useable charge in the battery stack, decreasing charge time compared with passive balancing, and decreasing heat. Passive balancing allows the stack to look like every cell has the same capacity as the weakest cell.

Figure 9. Cell Balancing Technique

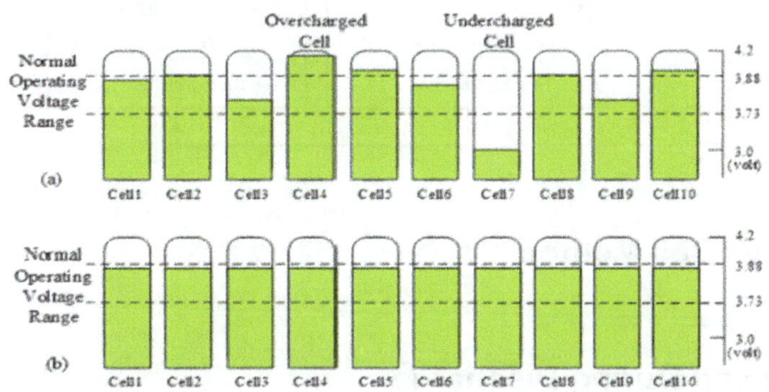

Active Cell Balancing Technique

With passive and active cell balancing, each cell in the battery stack is monitored to maintain a healthy battery state of charge (SoC). This extends battery cycle life and provides an added layer of protection by preventing damage to a battery cell due to deep discharging over overcharging. Passive balancing results in all battery cells having a similar SoC by simply dissipating excess charge in a bleed resistor; it does not however, extend system run .Active cell balancing is a more complex balancing technique that redistributes charge between battery cells during the charge and discharge cycles, thereby increasing system run time by increasing the total useable charge in the battery stack, decreasing charge time compared with passive balancing, and decreasing heat generated while balancing.

Figure 10.

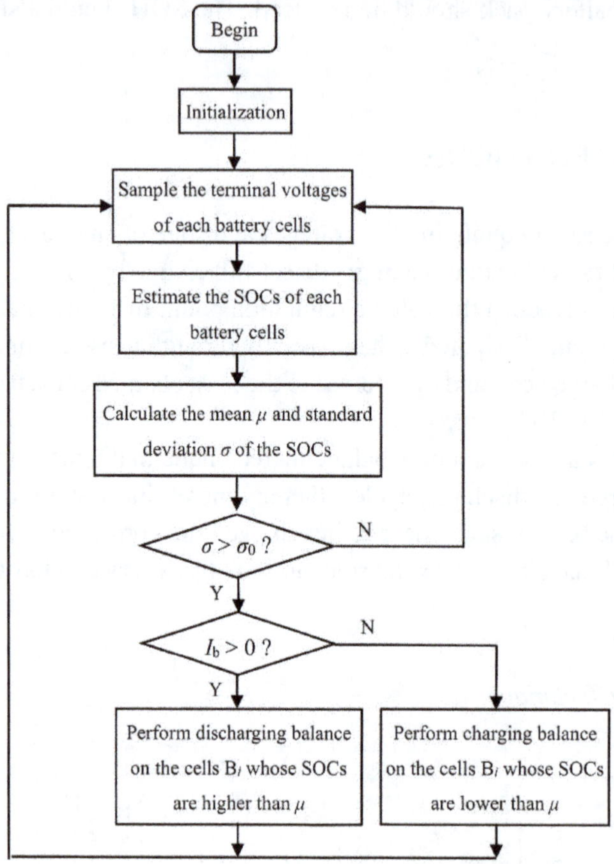

Flowchart of Active Cell Method

Active Cell Balancing During Discharge

The diagram below represents a typical battery stack with all cells starting at full capacity. In this example, full capacity is shown as 90% of charge because keeping a battery at or near its 100% capacity

Figure 11. Full capacity.

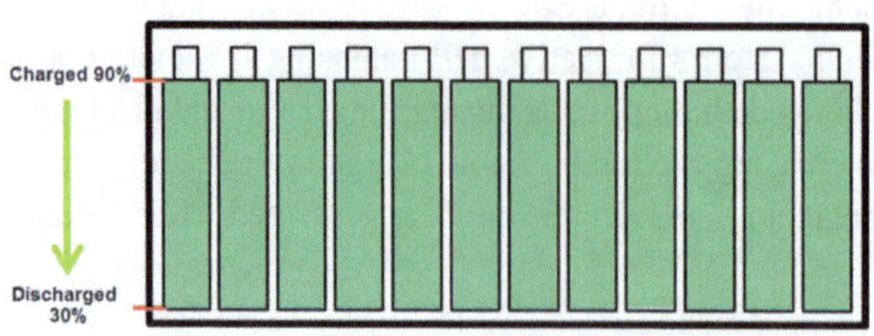

Figure 12. Mismatched discharge.

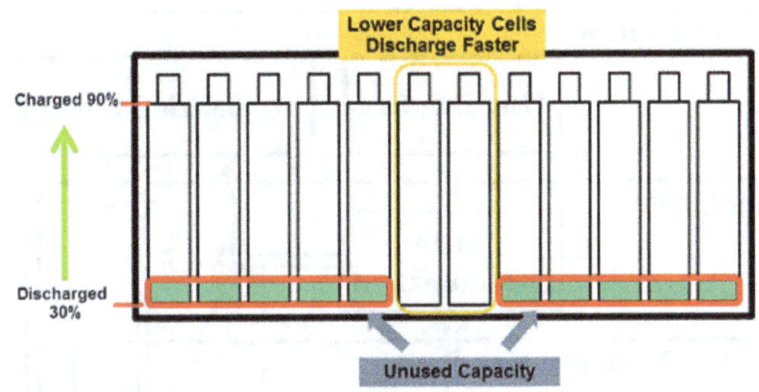

Figure 13. Full depletion with active balancing.

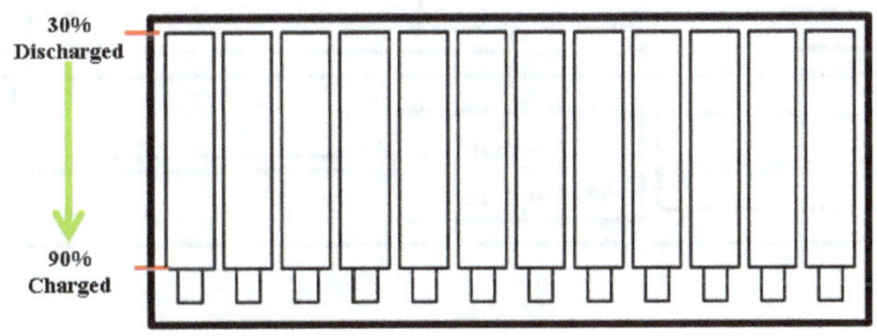

point for long periods of time degrades lifetime faster. 30% represents fully discharged to prevent deep discharge of the cells.

Over time, some cells will become weaker than others, resulting in a discharge profile represented by the figure below.

It can be seen that even though there may be quite a bit of capacity left in several batteries, the weak batteries limit the runtime of the system. A battery mismatch of 5% results in 5% of the capacity unused. With large batteries, this can be an excessive amount of energy left unused. This becomes critical in remote systems and systems that are difficult to access since it results in an increase in the number of battery charge and discharge cycles, which reduces the lifetime of the battery, leading to higher costs associated with more frequent battery replacement. With active balancing, charge is redistributed from the stronger cells to the weaker cells, resulting in a fully depleted battery stack profile.

Figure 14. Charging without balancing.

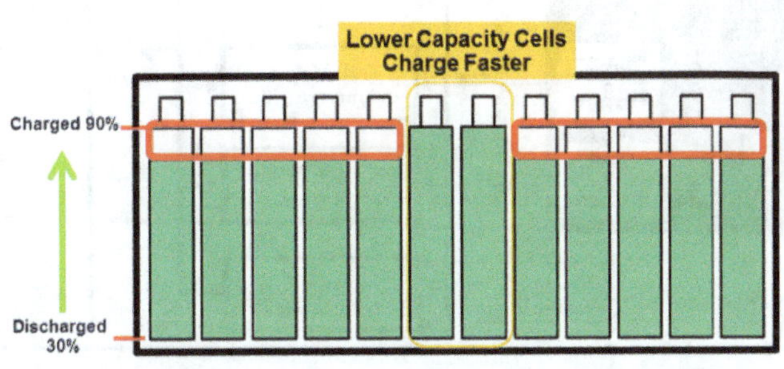

Figure 15. Block diagram of Battery Management System

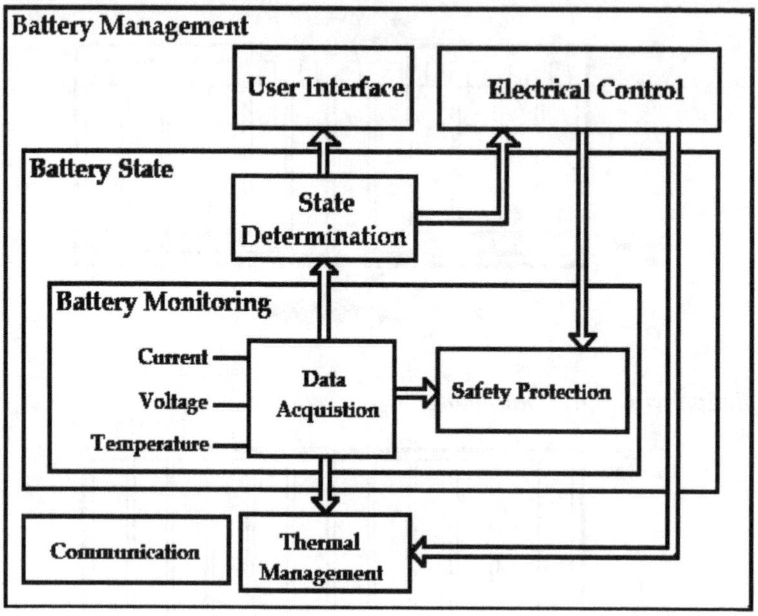

Active Cell Balancing While Charging: When charging the battery stack without balancing, the weak cells reach full capacity prior to the stronger batteries. Again, it is the weak cells that are the limiting factor; in this case they limit how much total charge our system can hold. The diagram below illustrates charging with this limitation.

Building Blocks of a BMS: From a hardware structure perspective there are only three types of BMS based on its topology they are Centralized BMS, distributed BMS and Modular BMS.

Figure 16. Block diagram of cell monitoring system

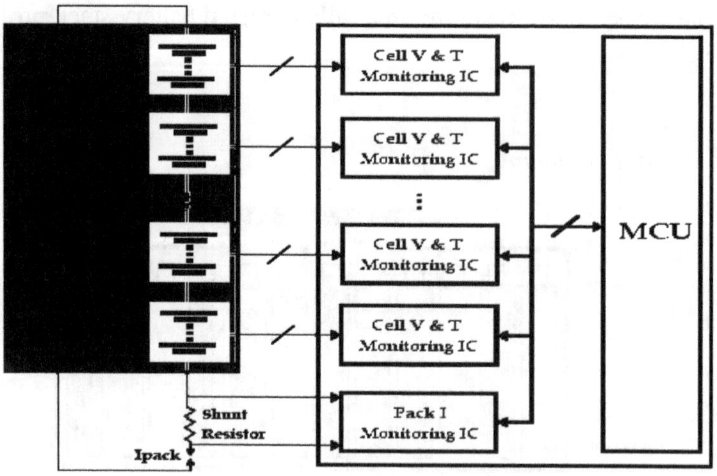

BMS Data Acquisition

The primary function of the BMS is to monitor the Battery for which it needs to measure three vital parameters such as the voltage, current and temperature from every cell in the battery pack. If a set of cells are connected in series then we have to measure voltage across each cell but current for the entire set will be same since current will be same in a series circuit.

Battery state Estimation: The major computational power of a BMS is dedicated to estimate the Battery state. This includes the measurement of SOC and SOH. SOC can be calculated using the cell voltage, current, charging profile and discharging profile. SOH can be calculated by using the number of charge cycle and performance of the battery.

Basic Methods

§ Coulomb Counting method
§ Ampere-hour (Ah) method
§ Open-Circuit Voltage (OCV) method
§ Impedance / IR Measurement Method

Machine Learning Based Algorithms

§ Neural Network Fuzzy Logic
§ Support Vector Machine

Advanced Method

§ State-Space Model Estimation using Kalman Filter

Coulomb Counting Technique: Coulomb counting Technique is the most used and easy to understand algorithm. It is based on the fact that the ratio between the Total charge Input and the Maximum capacity of the battery will give us the SOC value. The formula for the same is given below.

SOC = Total Charge Input / Maximum Capacity

But this method is not very reliable because the maximum capacity of the battery will get reduced as the battery ages. Hence many other algorithms were developed.

STATE OF CHARGE[SOC]:

The state of charge (SOC) of a cell denotes the capacity that is currently available as a function of the rated capacity. The value of the SOC varies between 0% and 100%. If the SOC is 100%, then the cell is said to be fully charged, whereas a SOC of 0% indicates that the cell is completely discharged. In practical applications, the SOC is not allowed to go beyond 50% and therefore the cell is recharged when the SOC reaches 50%. Similarly, as a cell starts aging, the maximum SOC starts decreasing. This means that for an aged cell, a 100% SOC would be equivalent to a 75%–80% SOC of a new cell. Battery SOC estimation is key component for battery management system. It helps in describing the actual energy level available at the battery. Battery cell current, cell voltages, and cell temperature are critical parameters for the estimation of battery SOC. It is common to estimate the SOC percentage using charge transfer in and out of the battery, and in this case typical coulomb-counting methods are used. In this case, we can use the rated capacity of the battery, Q_{rated}, which is not a function of temperature or aging (since it is usually estimated at a standard laboratory condition at a controlled temperature). However, to accommodate the SOC variation with aging, it is necessary to use a different term, $Q_{discharge}$. This is defined as the maximum total electrical charge, expressed in Ah that a battery can deliver from the fully charged state (SOC of 100%) to a fully discharged state (SOC of 0%) as indicated.

DESCRIPTION OF STATE OF CHARGE

State of Charge: State of charge (SOC) is the level of charge of an electric battery relative to its capacity as seen in fig 16. The units of SoC are percentage points (0% = empty; 100% = full). An alternative form of the same measure is the depth of discharge (DoD), the inverse of SoC (100% = empty; 0% = full).

Figure 17. State of charge calculation in an electric car

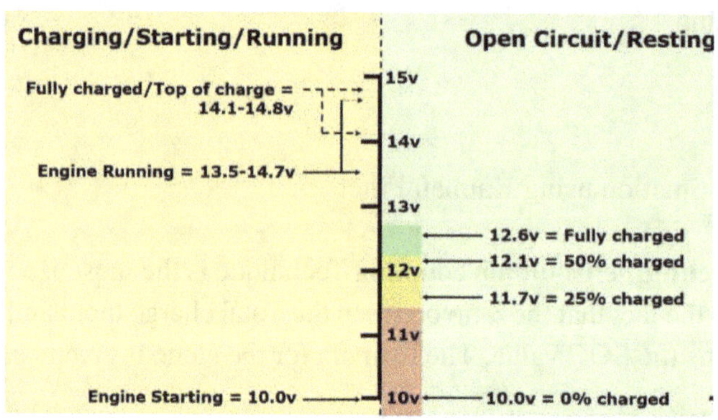

METHODOLOGY AND RESULTS

Comparative charge status analysis in a Battery Electric Vehicle using Model based Method has been carried out with suitable algorithms and simulation results. Among the various algorithms used to calculate state of charge, detail analysis of Kalman filter and Open Circuit Voltage is explained below:

Case a) Open Circuit Voltage: In this method RC model has been built and simulated in Scilab. Simulation Results are compared with experimental results (EU's Future Noise Policy, 2020) and it is found that it is same and it is mapped with experimental results.

Figure 18. Block diagram of State of charge calculation in an electric car

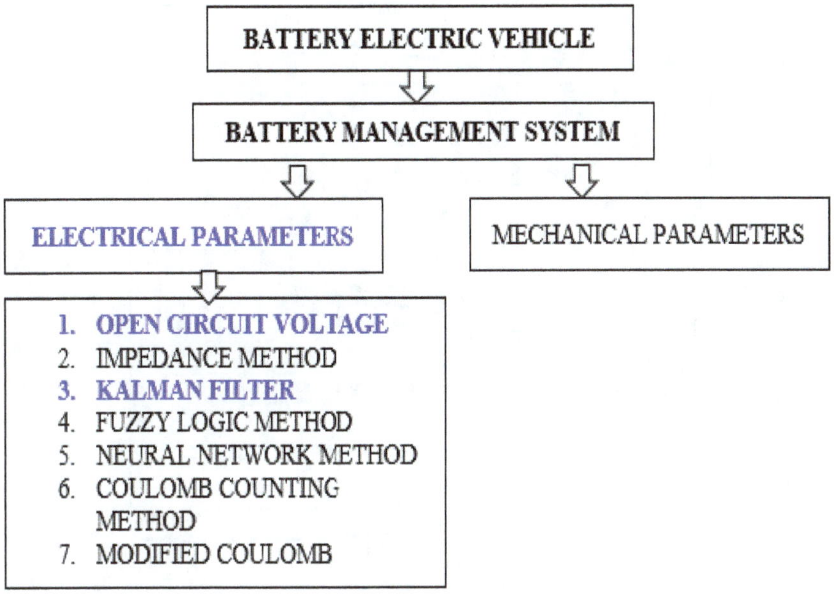

Open Circuit Voltage Method

Open circuit voltage (OCV) has a considerable influence on the accuracy of battery state of charge (SOC) estimation. The OCV curve can be locally reconstructed even when the accumulated data only cover a partial range of SOC, which is suitable for electric vehicle (EV) operation conditions. From the simulation model, consider RC network in addition with voltage source. State of charge is measured at output (Wei et al., 2019). Voltage is measured at terminal 2 and total resistance is measured at terminal 1. Voltage and Resistance of the RC Circuit model has been developed using Scilab and has been indicated in fig 18 Open circuit voltage, Voc, internal resistances during discharging, Rdis, and charging, Rcha, are involved. The dependent variables considered for State of charge estimation includes:

Open circuit voltage (Voc),

Figure 19. RC Circuit model of Open Circuit Voltage

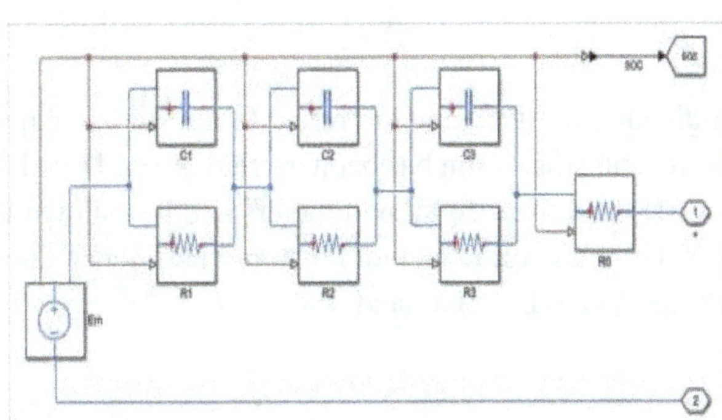

Figure 20. Battery open circuit voltage and internal resistance with respect to SOC.

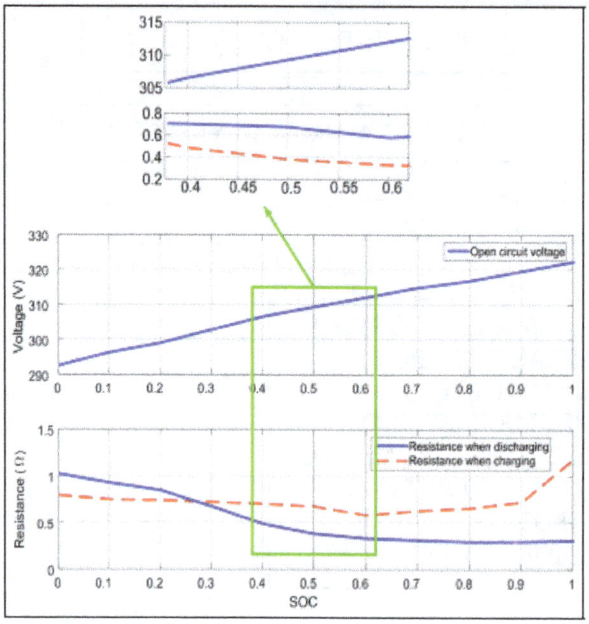

Internal Resistance during discharge -Rdis

The battery current and output voltage can be calculated as follows:

$$I = \frac{V_{oc} - V_b}{R_b} \qquad V_b = \frac{P_{ele,m}}{I}$$

where Rb is the internal resistance of the battery depending on discharge and charge mode, I is the circuit current, and Vb is the battery output voltage. SOC can be obtained by integrating the current. When the current is positive, it means that the battery is discharging. On the other hand, the battery is recharging if the current has a negative value.

$$SOC = SOC_{ind} - \int_0^t \frac{I}{Q_b} dr$$

Here, SOCini is the initial value of SOC, and Qb refers to the storage capacity of the battery. From the developed RC circuit model and simulation has been carried out and results has been compared with existing paper(Pallas et al., 2018) and mapped all parameters and it is shown in fig 18 Hence method is easy to implement. However the initial value cannot be stored and hence data retrieval is found out to be difficult when state of charge has to be calculated.

Kalman Filter Method

To overcome the shortcomings of the open circuit voltage method, Kalman filter method is used. The battery can be modeled with an electrical model which the Kalman filter will use to predict the over-voltage, due to the current. This method is complex but data can be stored and data retrieval is also possible but has few errors like

Integration of current sensor error: Integration of the current obtained by an erroneous sensor causes the results to drift.

1. Degradation of battery: Degradation of the battery results in a loss of battery capacity, which causes an estimation error.
2. Self-discharge: Self-discharge is the consumption of charge inside the battery. This cannot be observed by current measurement.

SOH: State of health (SOH) is a figure of merit of the condition of a battery (or a cell, or a battery pack), compared to its ideal conditions. The units of SOH are percent points (100% = the battery's conditions match the battery's specifications).

Coulomb Counting Method. The **Coulomb counting** method measures the discharging current of a battery and integrates the discharging current over time in order to **estimate SOC**. **Coulomb counting** method is done to **estimate** the, which is estimated from the discharging current, and previously estimated **SOC** values. In this book chapter, an efficient SOC estimation algorithm based on the Coulomb-counting algorithm which is a book-keeping approach is proposed. Using a piecewise linear relationship mapping between SOC and the Open Circuit Voltage (OCV) we try to overcome the drawbacks of this approach so as to improve its accuracy.

Motivation Behind the Proposed Work

- Electric vehicles is currently the research area where extensive research work is going on .
- If Electric Vehicle is used carbon –di-oxide emission can be reduced .
- Information regarding state of charge, state of health can be given to user with the help of IOT Based coulomb counting algorithm .
- Monitoring of an Electric Vehicle can be achieved using IOT architecture.
- Information regarding identification of charging network station can be given to user with the help of IOT.
- Hence Battery Electric vehicle is more advantageous when compared to Internal combustion engine.

Coulomb Counting

Advantages

- Linear system and nonlinear system
- Easy to implement
- Error propagation

- Battery voltages of up to 1kV.
- Charging time required is 30 MIN.
- DC CHARGING TYPE
- Error propagation: Can be reduced from 5% to 1%

Methodology of Coulomb Counting

Figure 21. State of charge controller.

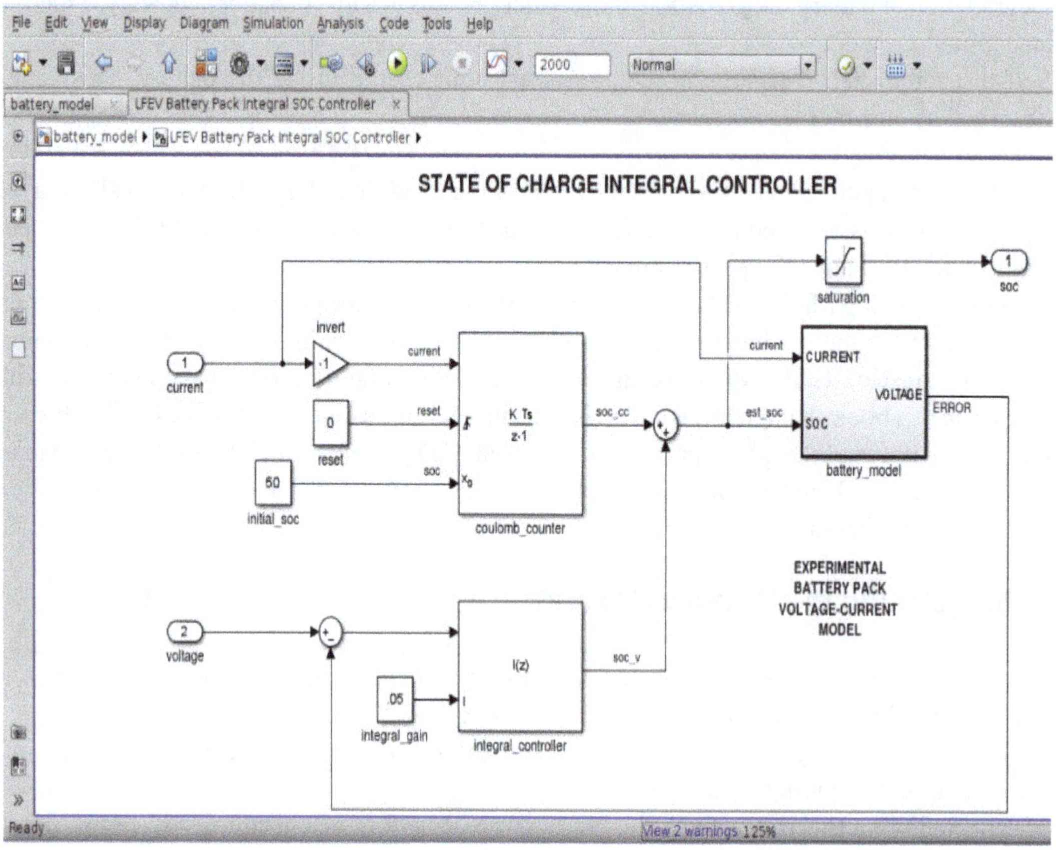

DESCRIPTION OF STATE OF CHARGE

State of Charge: State of charge (SOC) is the level of charge of an electric battery relative to its capacity as seen in fig 21. The units of SoC are percentage points (0% = empty; 100% = full). An alternative form of the same measure is the depth of discharge (DoD), the inverse of SoC (100% = empty; 0% = full). Fig 21 indicates the block diagram representation of various algorithms used to calculate State of charge.

METHODOLOGY AND RESULTS

Figure 22. State of charge controller - charging profile

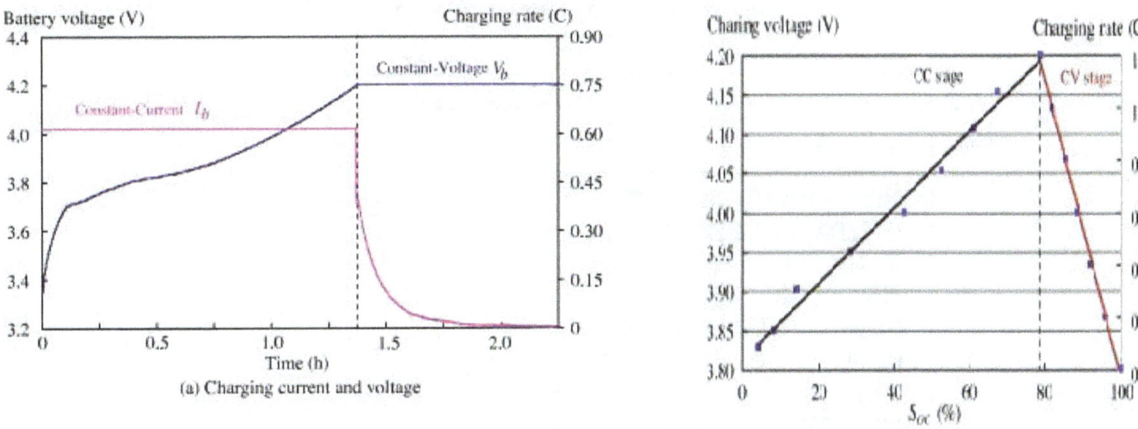

Comparative charge status analysis in a Battery Electric Vehicle using Model based Method has been carried out with suitable algorithms and simulation results. Among the various algorithms used to calculate state of charge, detail analysis of Kalman filter and Open Circuit Voltage is explained below:

Figure 23. State of charge calculation in an electric car

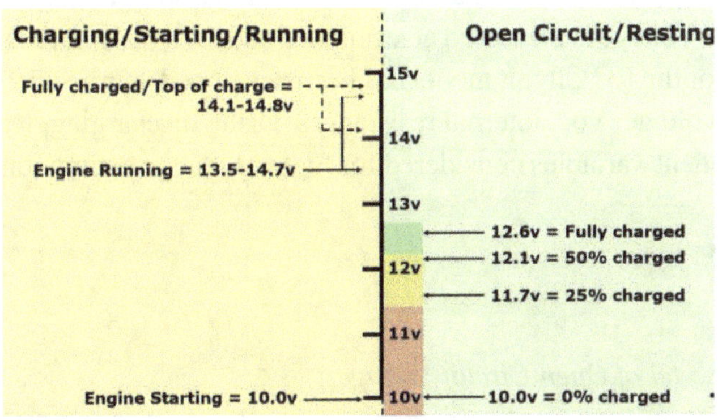

Case a) Open Circuit Voltage: In this method RC model has been built and simulated in Scilab. Simulation Results are compared with experimental results (EU's Future Noise Policy, 2020) and it is found that it is same and it is mapped with experimental results.

Figure 24. Block diagram of State of charge calculation in an electric car

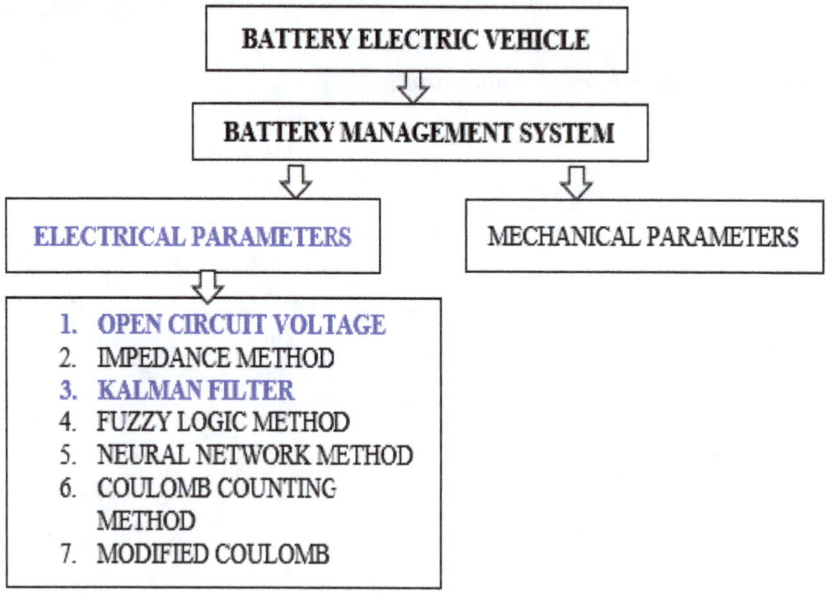

Open Circuit Voltage Method

Open circuit voltage (OCV) has a considerable influence on the accuracy of battery state of charge (SOC) estimation. The OCV curve can be locally reconstructed even when the accumulated data only cover a partial range of SOC, which is suitable for electric vehicle (EV) operation conditions. From the simulation model, consider RC network in addition with voltage source. State of charge is measured at output (Wei et al., 2019). Voltage is measured at terminal 2 and total resistance is measured at terminal 1. Voltage and Resistance of the RC Circuit model has been developed using Scilab and has been indicated in fig 3.1. Open circuit voltage, Voc, internal resistances during discharging, Rdis, and charging, Rcha, are involved. The dependent variables considered for State of charge estimation includes:

Open circuit voltage (Voc),

Figure 25. RC Circuit model of Open Circuit Voltage

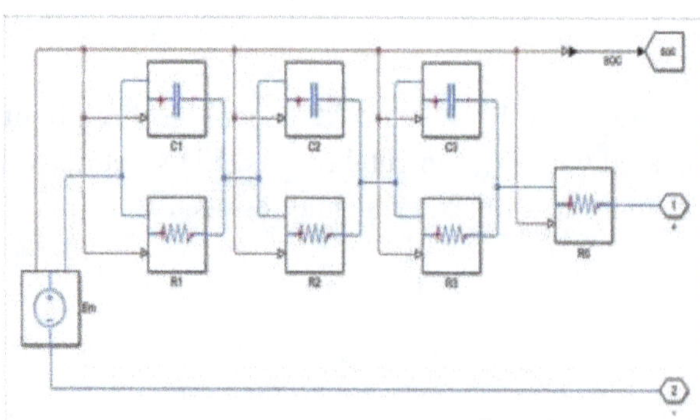

Table 2. Simulation result for a Nissan leaf electric car lithium ion battery identified RC mode

State of charge	12v battery	Voltage per cell (if comprised of separate 2v cells)	Comments
100%	12.70v+	2.12v	Cycling your battery in this zone will ensure a reasonable life expectancy
95%	12.60v	2.10v	
90%	12.50v	2.08v	
80%	12.40v	2.06v	
70%	12.30v	2.05v	
60%	12.20v	2.03v	Occasionally dropping into this zone is OK but not recommended
50%	12.05v	2.00v	
40%	11.90v	1.98v	
30%	11.75v	1.95v	
20%	11.55v	1.92v	Cycling a battery into this zone is certainly not recommended unless unavoidable. Battery life will be massively shortened.
10%	11.30v	1.88v	
Below 10% DEAD	10.5v Or less	1.75v Or less	

Internal Resistance during discharge -Rdis

The battery current and output voltage can be calculated as follows:

Figure 26. Battery open circuit voltage and internal resistance with respect to SOC

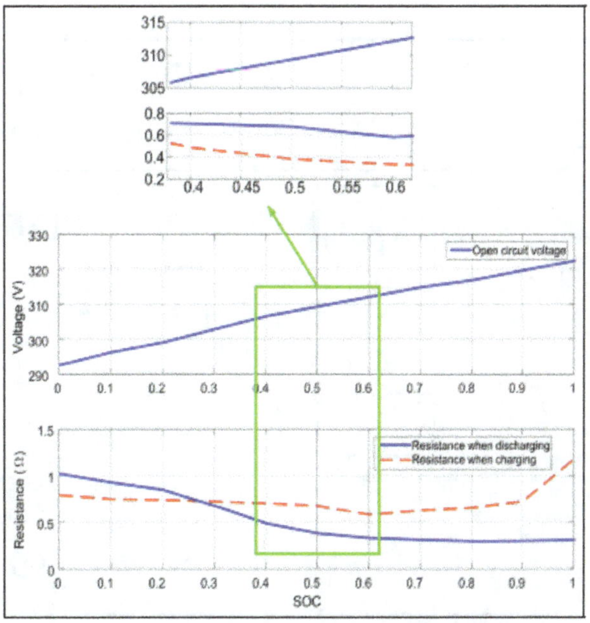

$$I = \frac{V_{oc} - V_b}{R_b} \qquad V_b = \frac{P_{ele,m}}{I}$$

where Rb is the internal resistance of the battery depending on discharge and charge mode, I is the circuit current, and Vb is the battery output voltage. SOC can be obtained by integrating the current. When the current is positive, it means that the battery is discharging. On the other hand, the battery is recharging if the current has a negative value.

$$SOC = SOC_{ind} - \int_0^t \frac{I}{Q_b} dr$$

Here, SOCini is the initial value of SOC, and Qb refers to the storage capacity of the battery. From the developed RC circuit model and simulation has been carried out and results has been compared with existing paper(Pallas et al., 2018) and mapped all parameters and it is shown in fig 24 Hence method is easy to implement. However the initial value cannot be stored and hence data retrieval is found out to be difficult when state of charge has to be calculated.

From Table 2 Simulation result for a Nissan leaf electric car lithium ion battery identified RC model is compared for a 12V battery. It is found that Nissan leaf electric car has used Level 2 charging and experimental and simulation results is same and it is validated.

Figure 27. Enhanced Coulomb Counting

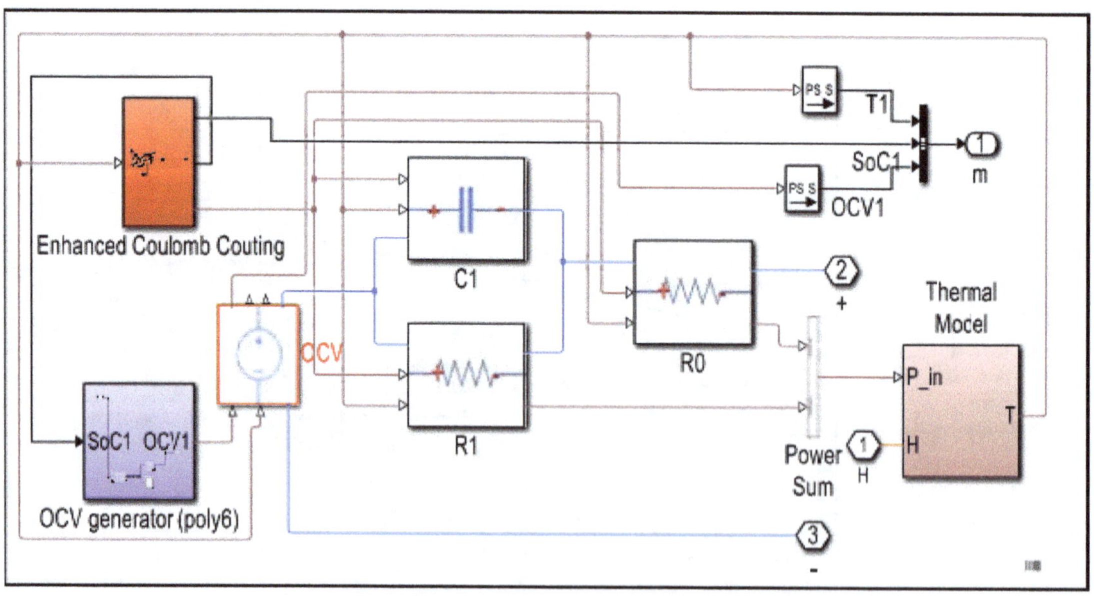

Kalman Filter Method

To overcome the shortcomings of the open circuit voltage method, Kalman filter method is used. The battery can be modeled with an electrical model which the Kalman filter will use to predict the over-voltage, due to the current. This method is complex but data can be stored and data retrieval is also possible but has few errors like

Integration of current sensor error: Integration of the current obtained by an erroneous sensor causes the results to drift.

1. Degradation of battery: Degradation of the battery results in a loss of battery capacity, which causes an estimation error.
2. Self-discharge: Self-discharge is the consumption of charge inside the battery. This cannot be observed by current measurement.

Battery Model and Parameter Identification of Kalman Filter

While the demand on energy increases day by day, the amount of available fossil fuel reserves deplete. Relying on the fossil fuel consumption causes damages to the environment and incremented greenhouse gas emissions as well. Consequently, the renewable energy technologies have gained more attention and derived the use of more electric vehicles. Lithium-ion batteries are getting popular in both renewable energy systems and electric vehicles thanks to their high power and energy density. Therefore, accurate

Figure 28. Graph related to enhanced coulomb counting

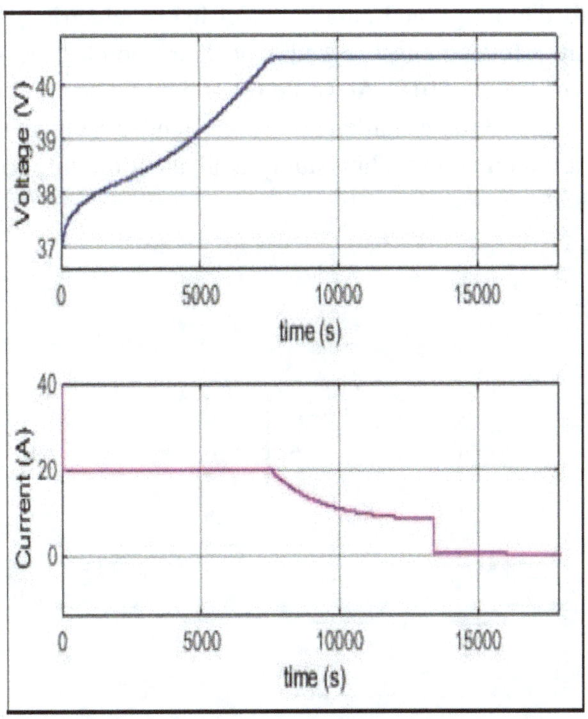

Figure 29. 2RC Li-ion battery improved physical model design and its Simscape library elements

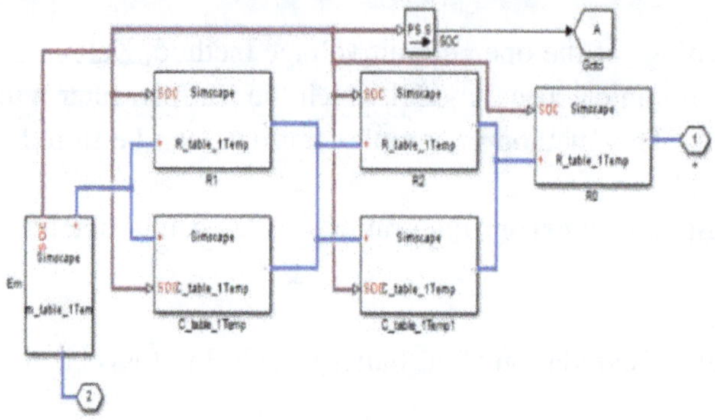

Figure 30. Two RC Branch Equivalent circuit model

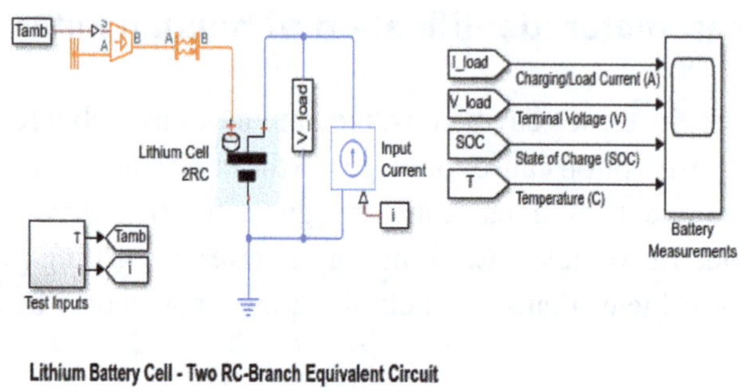

battery models are vital to the design and simulation of hybrid/electric vehicle propulsion systems. Modelling and batteries are a toilsome task because of their complex electrochemical structure and nonlinear characteristics. (Liu et al., 2018) Accurate real-time SOC estimation reporting to drivers is also difficult. This work addressed these challenges using extended Kalman filter (EKF) algorithm and a two-RC-block equivalent circuit shown This battery equivalent circuit model is designed in Matlab

Figure 31. RC Thevenin Model

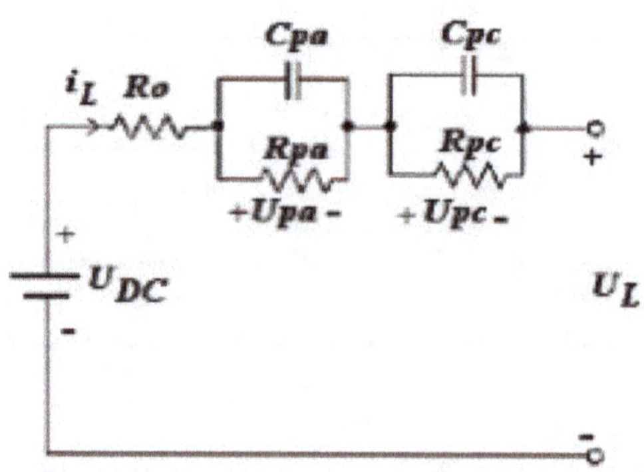

Simulink using the Simscape Language. Then, an algorithm with the EKF approach is developed to enhance the SOC estimation.

Li-Ion Battery Modeling using Kalman Filter

The Thevenin model is widely used to model the Lithium ion battery, but it is not accurate enough as all of its elements can change, depending on the state of the battery and its conditions. To improve the model reliability, unlike the general equivalent Thevenin model, extra RC branch is added .These blocks shown in the figure were created in Simulink Simscape Language to define the custom components as text files. The texts include complete parameterization, physical connections, and equations represented as a couple of causal implicit differential algebraic equations. The mathematical relation has been obtained as a result from experimental work as explained in (Wu et al., 2018) .

The Thevenin model shown in fig 26 includes three parts as open circuit voltage-Voc, internal resistances, and equivalent capacitances. The internal resistances include ohmic resistance-Ro, electrochemical polarization resistance- Rpa, and concentration polarization resistance- Rpc. The equivalent capacitances (Kong et al., 2018)which include electrochemical polarization capacitance-Cpa and concentration polarization capacitance- Cpc, Charging current iL and UL are the charging/discharging current and the terminal voltage respectively. the soc output voltage is explained .

The state of charge estimation using Kalman filter method and its equivalent battery model has been shown .The Extended Kalman filter is a method for predicting the future state of a system based on the previous ones and convenient form for online real time processing. It consists of two equations. The equations include A, B, C and D matrices that can be constituted using Ro, Rpa, Rpc, Cpa and Cpc. The expressions of matrices and vectors are the same as (Wu et al., 2018). xk is the system state matrix and one of the matrix values represents SoC. Therefore, xk captures the system dynamics. Input of the system is uk which is a control variable matrix and known or can be measured. However, the measurement could result in errors, assumed to be stochastic process noise, wk, which cannot be measured and affects the state of the system

$$xk + 1 = A \times xk + B \times uk + wk$$

Figure 32. Battery SOC and current (2RC model)

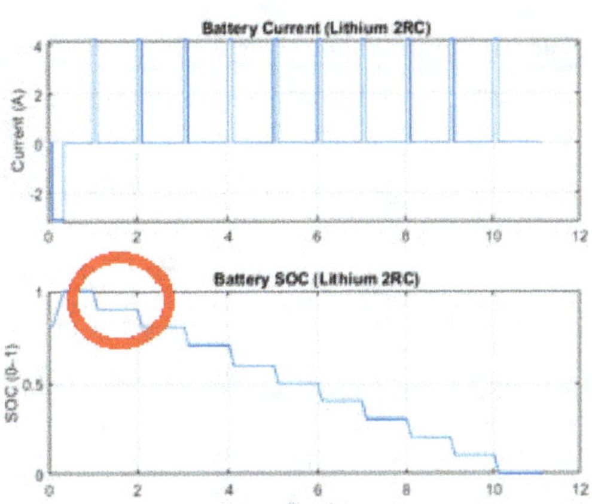

Figure 33. Equivalent battery model based SOC Kalman filter method [6]

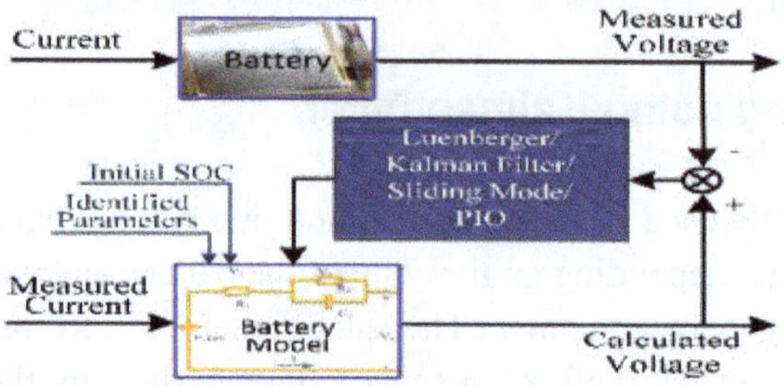

$$yk = C \times xk + D \times uk + vk$$

The second equation is the measurement equation and models the output voltage of the system yk, in terms of the input, the state vector and the noise in the measurement of the output, vk which is called measurement noise. For the Lithium ion battery module, some parameters for the Extended Kalman Filter code are specified as follows:

Figure 34. Experimental data results for Kalman filter

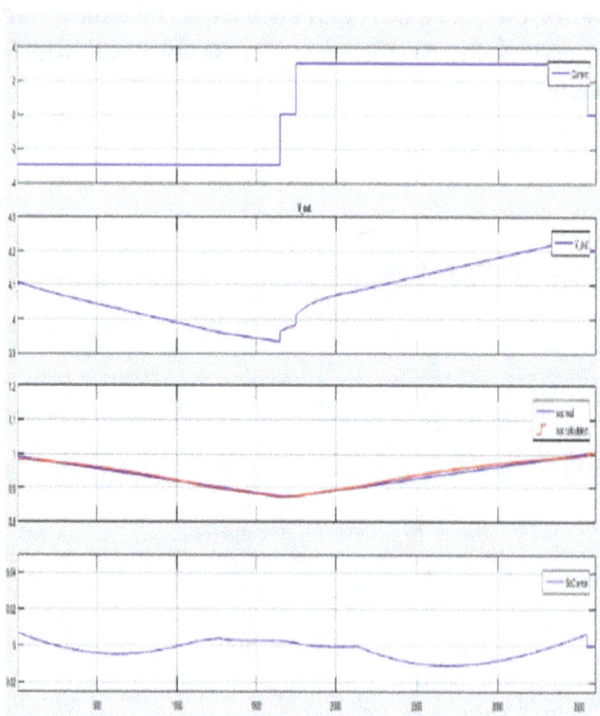

Figure 35. Theoretical results of Kalman filter

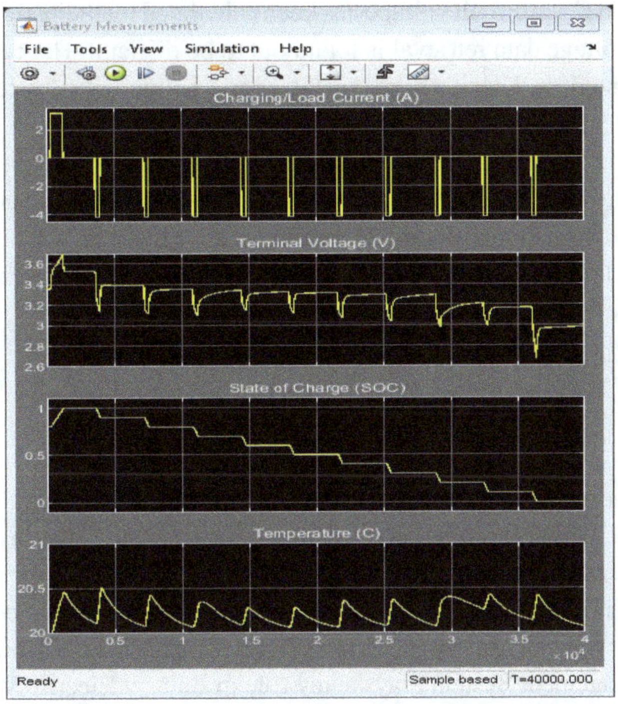

$$Q = \begin{bmatrix} 0.1 & 0 & 0 \\ 0.000001 & 0 & 0 \\ 0 & 0 & 10 \end{bmatrix}$$

$$R = \begin{bmatrix} 1 & 0 & 0 \\ 0 & 1 & 0 \\ 0 & 0 & 0.1 \end{bmatrix}$$

P = [10]

State of Charge Estimation

Using the Matlab Optimization Tool Box, parameters of the battery such as Cpa, Cpc, R0, Rpa, Rpc, Uoc were determined .The Charge and discharge input current signals as well as the real and estimated SOC curves are shown in Fig 33. The results are compared and found out to be mapped in all aspects (EU's Future Noise Policy, 2020).

Fig 34 and 35 indicate the Compare the simulation results with research paper results and found out to be same (EU's Future Noise Policy, 2020). Maximum SOC error is found out to be 1% from the simulation results as per fig 35 and it proved to be successful implementation of Kalman filter technique.

For monitoring the charge status suitable algorithms has to be implemented. In this paper, Comparison between Kalman filter and Open circuit voltage has been calculated mathematically with suitable battery model. In Open circuit voltage data retrieval is found out to be difficult. Hence to retrieve data Kalman filter has been implemented. In Kalman filter method for a Lithium ion battery model, the proposed improved Thevenin model has been implemented, and its parameter identification has been performed using the Extended Kalman Filter algorithm. Maximum SOC error is found out to be 1% from the simulation results. Kalman filter is applicable only for nonlinear systems and it is hard to implement in linear system as there is no feedback compensation mechanism. Hence to overcome this issue of Kalman filter, other suitable algorithm like coulomb counting, fuzzy logic, modified coulomb counting method may be considered for reliable operation of a battery in a battery electric vehicle.

CONCLUSION

Electric vehicles are trending nowadays compared to internal combustion engine because of its advantages and less limitations. Global warming is effective with electric vehicles because of no emissions of gases. Limitations are the installed charging stations are not able to meet the increasing charging demand of Electric Vehicles. So, if that is overcome then effectively electric cars can be into effect. Among the configurations of electric vehicles battery Electric Vehicles is more advantageous because of its features. Within that Battery Management System plays a key role in Battery Electric Vehicles.

REFERENCES

Bundy, K., Karlsson, M., Lindbergh, G., & Lundqvist, A. (1998). An electrochemical impedance spectroscopy method for prediction of the state of charge of a nickel-metal hydride battery at open circuit and during discharge. *Journal of Power Sources*, *72*(2), 118–125. doi:10.1016/S0378-7753(97)02695-5

EU's Future Noise Policy. (2020). *Position paper on dose response relationships between transportation noise and annoyance*. European Commission.

Feng, X., Sun, J., Ouyang, M., Wang, F., He, X., Lu, L., & Peng, H. (2015). Characterization of penetration induced thermal runaway propagation process within a large format lithium ion battery module. *Journal of Power Sources*, *275*, 261–273. doi:10.1016/j.jpowsour.2014.11.017

Huet, F. (1998). A review of impedance measurements for determination of the state-of-charge or state-of-health of secondary batteries. *Journal of Power Sources*, *70*(1), 59–69. doi:10.1016/S0378-7753(97)02665-7

Kong, L., Li, C., Jiang, J., & Pecht, M. (2018). Li-Ion Battery Fire Hazards and Safety Strategies. *Energies*, *11*(9), 2191. doi:10.3390/en11092191

Lelie, M., Braun, T., Knips, M., Nordmann, H., Ringbeck, F., Zappen, H., & Sauer, D. (2018). Battery Management System Hardware Concepts: An Overview. *Applied Sciences (Basel, Switzerland)*, *8*(4), 534. doi:10.3390/app8040534

Li, R., Wu, J. F., Wang, H. Y., & Li, G. C. (2010). Prediction of state of charge of lithium-ion rechargeable battery with electrochemical impedance spectroscopy theory. *Proceedings of the 5th IEEE Conference on Industrial Electronics and Applications (ICIEA '10)*, 684–688.

Liu, K., Liu, Y., Lin, D., Pei, A., & Cui, Y. (2018). Materials for lithium-ion battery safety. *Science Advances*, *4*(6), eaas9820. doi:10.1126ciadv.aas9820 PMID:29942858

Liu, Y., & Xie, J. (2015). Failure Study of Commercial LiFePO4 Cells in Overcharge Conditions Using Electrochemical Impedance Spectroscopy. *J. Electro Chem. State of Charge.*, *162*, A2208–A2217.

Liu, Z., Ahmed, Q., Zhang, J., Rizzoni, G., & He, H. (2016). Structural analysis-based sensors fault detection and isolation of cylindrical lithium-ion batteries in automotive applications. *Control Engineering Practice*, *52*, 46–58. doi:10.1016/j.conengprac.2016.03.015

Lu, L., Han, X., Li, J., Hua, J., & Ouyang, M. (2003). A review on the key issues for lithium-ion battery management in electric vehicles. *J. Power Sources*, *226*, 272–288.

Lyu, D., Ren, B., & Li, S. (2019). Failure modes and mechanisms for rechargeable Lithium-based batteries: A state-of-the-art review. *Acta Mechanica*, *230*(3), 701–727. doi:10.100700707-018-2327-8

Ng, K. S., Moo, C. S., Chen, Y. P., & Hsieh, Y. C. (2009). Enhanced Coulomb counting method for estimating state-of-charge and state-of health of lithium-ion batteries. *Applied Energy*, *86*(9), 1506–1511. doi:10.1016/j.apenergy.2008.11.021

Pallas, M. A., Kennedy, F. J., Walker, I. I. (2018). *Noise emission of electric and hybrid electric vehicles*. University of Bath.

Rodrigues, S., & Munichandraiah, N. (2000). A review of state-of-charge indication of batteries by means of A.C. impedance measurements. *Journal of Power Sources*, *87*(1-2), 12–20.

Sato, S., & Kawamura, A. (2002). A new estimation method of state of charge using terminal voltage and internal resistance for lead acid battery. *Proceedings of the Power Conversion Conference*, 565–570. 10.1109/PCC.2002.997578

Venkatasubramanian, VRengaswamy, RYin, KKavuri, S.N. (n.d.). A review of process fault detection and diagnosis. *Computers & Chemical Engineering*, *27*, 293–311.

Wei, J., Dong, G., & Chen, Z. (2019). Model based fault diagnosis of Lithium ion battery using strong tracking Extended Kalman Filter. *Energy Procedia*, *158*, 2500–2505. doi:10.1016/j.egypro.2019.01.391

Wu, C., Zhu, C., Ge, Y., & Zhao, Y. (2018). A Review on Fault Mechanism and Diagnosis Approach for Li-Ion Batteries. *Journal of Nanomaterials*, 1–9.

Xiong, R., Yu, Q., & Shen, W. (2018). Review on sensors fault diagnosis and fault-tolerant techniques for lithium ion batteries in electric vehicles. *Proceedings of the 2018 13th IEEE Conference on Industrial Electronics and Applications (ICIEA)*, 406–410. 10.1109/ICIEA.2018.8397751

Chapter 2
Design and Development of Bidirectional DC-DC Converters Using Battery/Supercapacitor for Electric Vehicle Applications

Bharathi Sankar Ammaiyappan
School of Electronics Engineering, VIT University Chennai Campus, Chennai, India

Seyezhai Ramalingam
Renewable Energy Conversion Lab, Department of EEE, SSN College of Engineering, Chennai, India

ABSTRACT

In the last two decades the pollution problems and the increase of the cost of fossil energy (oil, gas) have become planetary problems. The automobile manufacturers started to react to the urban pollution problems in the nineties by commercializing the electric vehicle. But the battery weight and cost problems were not solved. The batteries must provide energy and peak power during the transient states. These conditions are severe for the batteries. To decrease these severe conditions, the supercapacitors and batteries associated with a good power management present a promising solution. Supercapacitors are storage devices that enable to supply the peaks of power to hybrid vehicles during the transient states. During the steady states, batteries will provide the energy requested. This methodology enables the decrease of the weight and increase of the lifespan of the batteries. Hybridization using batteries and supercapacitors for transport applications is needed when energy and power management are requested during the transient states and steady states.

1. INTRODUCTION

The use of a hybrid energy storage system is therefore particularly convenient for a complete satisfaction of the electric vehicle demands, in terms of both energy density and power density. An interesting hybrid architecture for electric vehicles is the one based on the battery/supercapacitors combination as shown

DOI: 10.4018/978-1-7998-7626-7.ch002

in Figure.1. The battery should provide to the vehicle the average power level, while the supercapacitors should react to the peak power demands. A hybrid energy storage system requires the combination between an optimal sizing and a proper power management strategy, to precisely determine the size of the different energy sources and to control the power sharing between them according to various possible scenarios of mobility (Amari et al., 2015; Ates et al., 2010; Marzougui et al., 2016).

Figure 1. Block diagram of battery/supercapacitor hybrid energy storage system

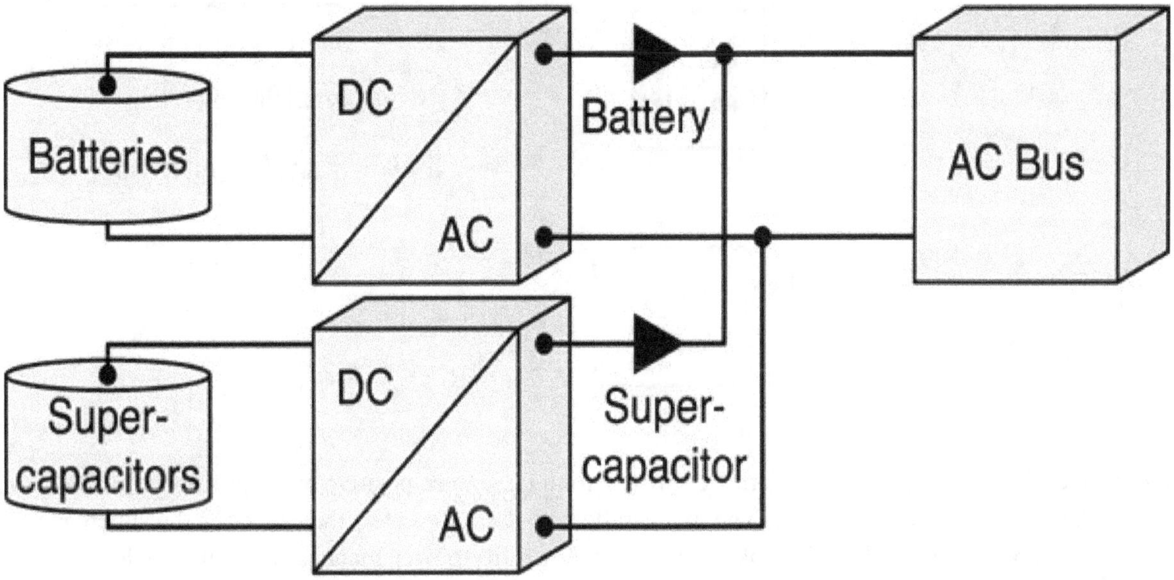

A new strategy of energy management between battery and supercapacitors for an urban electric vehicle is suggested. The battery/supercapacitors integration is able to provide notable advantages in the power management of an electric vehicle, in terms of capability of both high energy storage and readiness to deal with fast load variations. In order to manage the charge or discharge of the supercapacitors from or towards a DC voltage bus, a proper bi-directional converter is required. These two sources are connected in parallel to the DC bus through two bidirectional DC–DC converters enabling separate control over the power flow of each source.

A bi-directional DC-DC converter for the supercapacitors application is proposed. In Figure. 2 a possible architecture of the battery/supercapacitors system is shown. Two bi-directional DC-DC converters connect the battery and the supercapacitors to a common bus. The bus voltage is generated by means of the battery-connected DC-DC converter, which also regulates the average current required by the load; the supercapacitors-connected DC-DC converter deals with the peak power load demands: if the load requires a power peak, the supercapacitors are discharged; if the Supercapacitors require to be recharged, the energy is recovered from the bus (Cultura & Salameh, 2015; Payman et al., 2008; Stefan, 2009; Zhang et al., 2015).

Two bi-directional DC–DC converters are linked to the same DC-Link, one manages the battery power ñow and the other manages the supercapacitors power ñow, as shown in Figure. 2. Energy transfer is enabled between sources and load in the mentioned structure as well as from one source to another

Figure 2. Configuration of battery/supercapacitor hybrid energy management system.

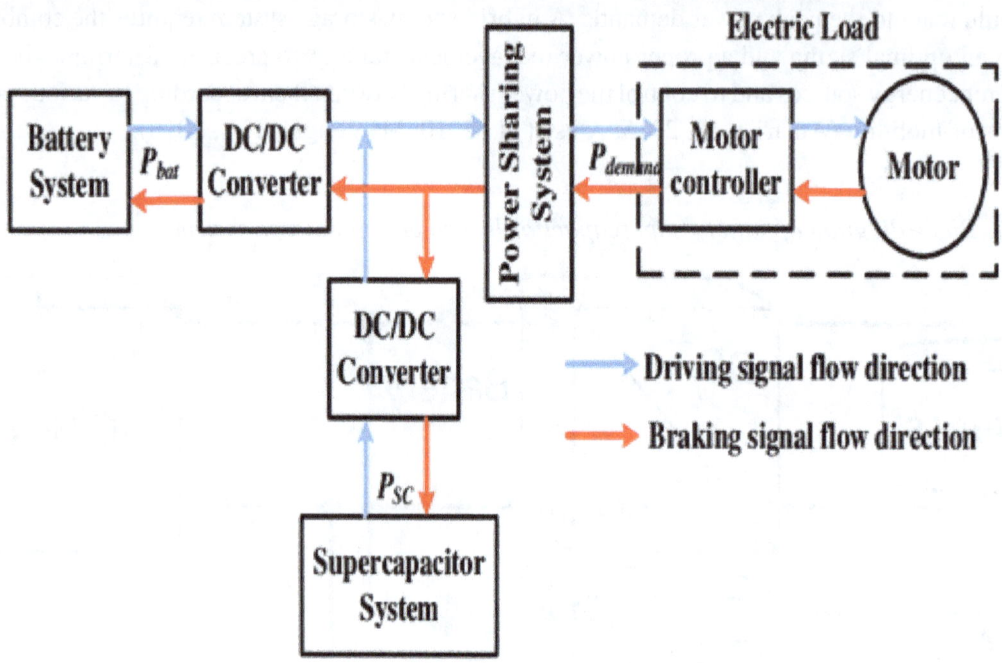

independently of energy ñow direction. DC-link current, I_{DC}, can be positive or negative whereas voltage across the DC bus is always positive. The two parallel DC–DC converters that interface the supercapacitors and the battery with the DC-Link give a good ñexibility power management either when energy ñow to DC-Link rising the voltage level which is boost behaviour, or when energy ñows from DC-Link to sources showing a typical buck behaviour (Ban et al., 2013; Shi & Crow, 2008; Zhang, Zhenpo, Sun et al, 2014).

2. HYBRID ENERGY MANAGEMENT SYSTEM:

Vehicle dynamics with load torque applied on the shaft motor is to be considered. This strategy of energy management permits dividing energy between the two sources depending on the state of charge of each source as well as on the vehicle displacement state such as stopping, acceleration, cruising down and uphill, and deceleration. The aim of the proposed strategy is the best use of energy through maximizing the use of Supercapacitors by transferring energy from batteries to Supercapacitors during the standstill phase or when the load applied to the vehicle is small supercapacitors will then be ready in critical situations such as rapid acceleration or in high hills climbing. In order to validate the control design and evaluate our energy management strategy performance, a simulation of an urban hybrid electric vehicle movement with the Matlab/Simulink software is implemented as shown in Figure.3 & 4.

In order to improve the efficiency of energy conversion and increase the driving range of electric vehicles, the regenerative energy captured during braking process is stored in the energy storage devices and then will be re-used. Due to the high power density of supercapacitors, they are employed to withstand high current in the short time and essentially capture more regenerative energy. The measuring

Figure 3. Matlab/Simulink hybrid energy management system.

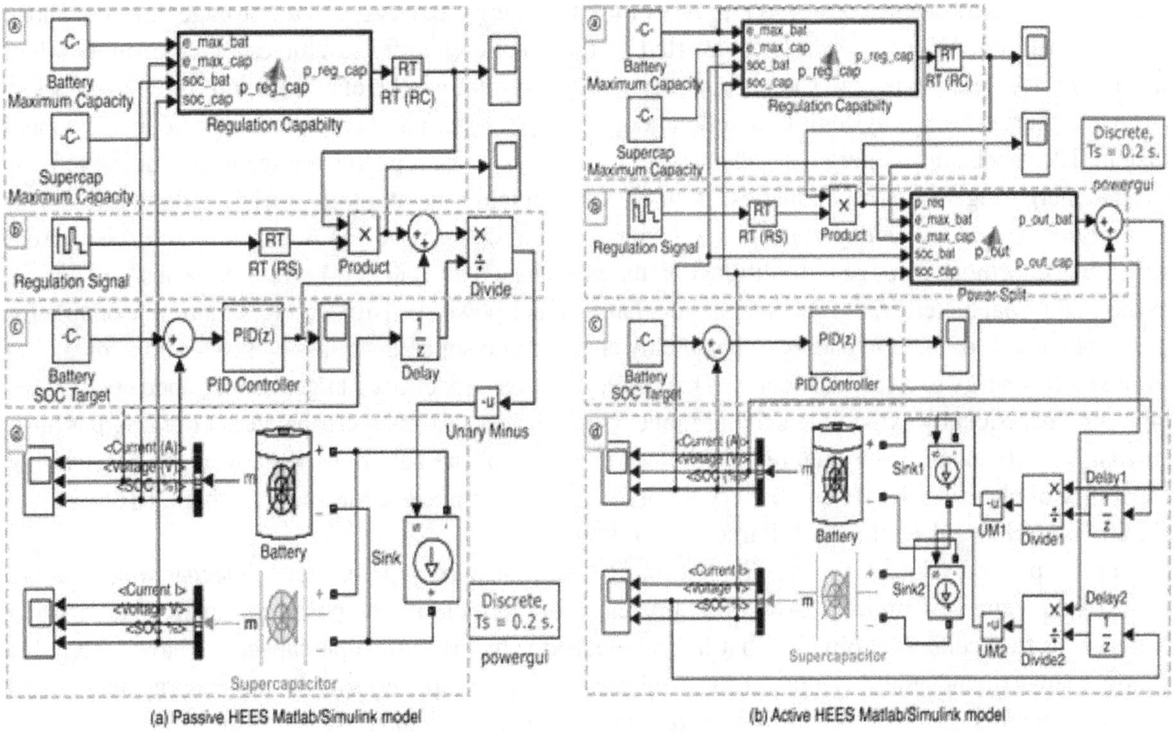

methods for regenerative energy should be investigated to estimate the energy conversion efficiency and performance of electric vehicles (Fletcher et al., 2017; Fuyuan et al., 2010; Hala et al., 2015; Jin et al., 2015; Zhang, Wang, Sun et al, 2014).

Figure 4. Proposed hybrid energy management system.

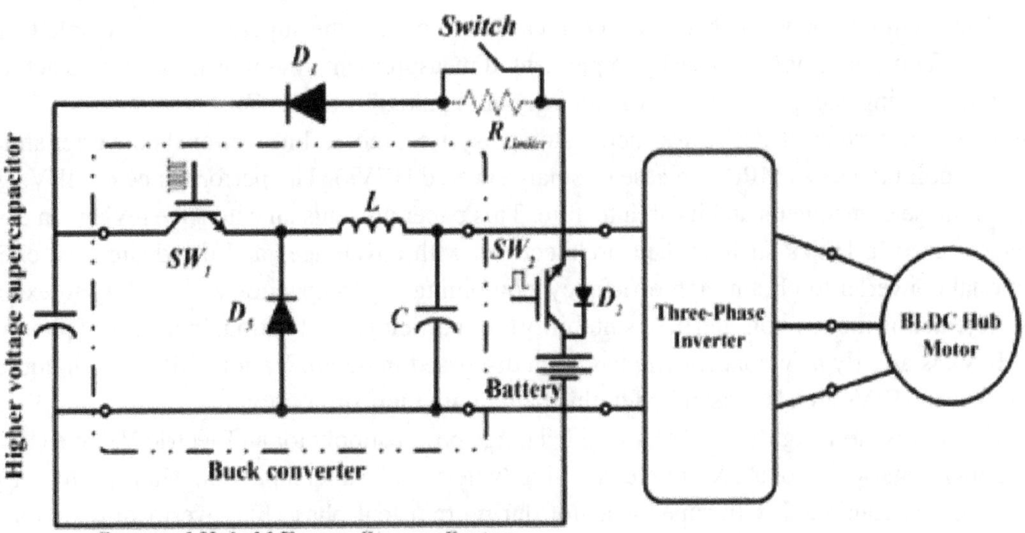

One of the most important features of electric vehicles is that the kinetic energy of vehicle mass in the braking process can be converted into other forms of energy and stored in the storage devices. Those regenerative braking energy can be converted to the kinetic energy of vehicles by controllers when starting or accelerating again. The energy regeneration system can be classified into three categories: flywheel energy-storage system, hydraulic energy-storage system and electrochemical energy storage system. Electrochemical energy-storage system was proved to be a promising technical means to realize the energy regeneration in vehicles. If an electric vehicle runs at a high-speed mode, the transient current due to braking feedback in the motor bus will increase up to 200 A or more, and this current will cause enormous damage to traditional batteries such as lead acid and lithium batteries. In contrast to the traditional batteries, the supercapacitors have higher power density, and it is more reasonable for the large amount of braking energy to be quickly charged into supercapacitors by proper transformation from kinetic energy to electrical energy. Therefore, the supercapacitors can greatly enhance energy savings and consequently extend the driving range. On the other hand, supercapacitors could output huge current instantaneously, and then reduce the power output of the batteries. Moreover, the accelerating capability of electric vehicles and battery life will also be improved accordingly (Ashar, 2016; Bottu et al., 2013; Michalczuk et al., 2012; Parker, 2015; Plett, 2004).

Supercapacitor vehicles can be divided into two types: pure supercapacitor electric vehicles and hybrid supercapacitor vehicles. A hybrid supercapacitor vehicle is composed of supercapacitors and batteries or fuel cells. It is obvious that hybrid electric vehicles with supercapacitors have a complex energy management system. A pure electric vehicle with supercapacitors as the only energy source will have a simpler energy management system, faster recharging speed and lower cost compared with hybrid electric vehicles. Therefore, pure supercapacitor electric vehicles have more advantages in the condition of short distance and frequent start–stop. Energy regeneration technology has been widely used in electric vehicles or electric motorcycles; however, the regenerative braking energy is not well utilized because the power density of batteries like lead–acid, Ni-MH and Lithium is lower than that of supercapacitors so that it is difficult to increase the overall energy efficiency of vehicles. Furthermore, the optimal design of control strategy over regenerative braking is also required. Electric vehicles powered only by supercapacitors have become a recent focus of research due to their potentially high performance. The issue is that there are no experimental evaluations for regenerative energy in supercapacitor vehicles at present. The accurate evaluation becomes a critical work to push the supercapacitor vehicle technologies forward. Thus, developing an easily implemented measurement system is necessary to achieve the regenerative-braking energy evaluation (Schneuwly et al., n.d.; Tuite, 2007).

The main components of HEVs are energy storage system, motor, bidirectional converter and maximum power point trackers (MPPT, in case of solar-powered HEVs). The performance of HEVs greatly depends on these components and its architecture. This paper presents an extensive review on essential components used in HEVs such as their architectures with advantages and disadvantages, choice of bidirectional converter to obtain high efficiency, combining supercapacitor with battery to extend the battery life, traction motors' role and their suitability for a particular application. Inclusion of photovoltaic cell in HEVs is a fairly new concept and has been discussed in detail. Various MPPT techniques used for solar-driven HEVs are also discussed in this paper with their suitability.

This paper presents a highly reliable, extended range power supply for an Electric Vehicle (EV). The power supply consists of solar PV source, a Li-ion battery and Supercapacitor. Main source of power is battery, and is connected with supercapacitor during transient phase like overloading and starting. Photovoltaic cell works for the steady condition. The total effect of such arrangement is to improve travel range, reduced size of battery, enhanced excellent response while the overloading condition and li-ion battery life . Improved performance which gives optimal use of energy, smooth ride and minimum size of sources of energy. Many stand-alone photovoltaic systems need storage device for providing steady state energy to the load when photovoltaic irradiation are not sufficient. Generally, Batteries are used for such application. Thus, providing a large peak current, like starting of motor, degrades plates of battery, results in devastation of battery. An alternate solution of providing heavy current have to connect battery with supercapacitors forming an hybrid energy storage system, for which battery could provide steady state energy also the supercapacitor can provide the peak power to connected load. A secondary source, solar panel module is available to charge battery and supercapacitor (Dixon & Ortuzar, 2002; Pay & Baghzouz, 2003).

3. SUPERCAPACITOR/BATTERY BASED HYBRID POWERED ELECTRIC BICYCLE

This paper presents a smart power converter to enable an electric bicycle to be powered by a battery/supercapacitor hybrid combination. A rear hub motor was retrofitted onto a normal geared bike powered by a lead acid battery pack. A supercapacitor module was connected in parallel to the battery pack via a custom made Field Programmable Gate Array (FPGA) controller-based power converter which arbitrates power between the battery and supercapacitor. The control method for the power converter was developed using a practical approach by using various inputs (battery/supercapacitor current and voltage) and comparing the robustness of conventional control scheme. Also energy efficient components were used in designing the power converter to ensure maximum power transfer efficiency (Farzanehfard et al., 2008; George & Choi, 2003; Paymen & Oierfederici, 2008).

Bicycles have been a transportation mainstay because the work place and housing areas in most of these densely populated cities are within walking or cycling distance. This reliable yet overlooked form of transportation has evolved over the years from simple utility bicycles to powerful geared mountain bikes and now electric assisted bicycles. Environmental concerns in terms of emissions and depleting fuel reserves has revived the electric vehicle industry and research community. Electric assisted bicycles still retain the characteristics of a conventional bicycle with an added advantage of extra power, say when riding up a hill. This enables the elderly or not so physically fit people to still enjoy riding a bicycle up a slope.

Batteries are the weak leak at the moment for any electrically propelled vehicle including the bicycle. The lack of a single reasonably priced energy storage device that can simultaneously provide high power density and high energy density has been the main stumbling block to the acceptance of electric propulsion as the main form of private and public transportation. Presently the only viable solution to this problem is to combine a high energy storage device such as an electrochemical battery or fuel cell with a high power device such as an Electric Double Layer Capacitor (EDLC) or supercapacitor or more often called a supercapacitor. Usually, some form of power converter executing an energy management control technique is used to interface the battery bank and supercapacitor array to the load bus. It is the aim of this research work to design a smart power converter with a heuristic based energy management technique which will optimize the power flow from the battery pack to the load. As the name implies, a supercapacitor is a capacitor with capacitance greater than any other, usually in excess of up to 3400 Farad. Supercapacitors do not have a traditional dielectric material like ceramic, polymer films or aluminum oxide to separate the electrodes instead a physical barrier made of activated carbon. A double electric field which is generated when charged, acts a dielectric. The surface area of the activated carbon is large thus allowing for the absorption of large amount of ions.

Advantages of Supercapacitors

- Cell voltage determined by the circuit application not limited by cell chemistry
- Very high cell voltages possible
- High power density
- Can withstand extreme temperatures
- Simple charging methods
- Very fast charge and discharge
- Overcharging not possible
- Long life cycle
- Low impedance

Disadvantages/Shortcomings

- Linear discharge voltage characteristic prevents use in some applications
- Power only available for very short duration (short bursts of power)
- Low capacity
- Low energy density
- Voltage balancing required when banking
- High self-discharge rate

4. BI-DIRECTIONAL DC-DC CONVERTER DESIGN

In order to size the power converter appropriately, the electric bicycle was powered with the 36V 12Ah Lead acid batteries. The track used was a mixture of flat terrain, up hills and down hills. This is shown in the table below.

From the table above, the average voltage is 39.0V. This voltage is actually equivalent to the total batteries voltage at full charge. Even though the datasheet mentions that each battery is only supplied 36V, however there is some tolerance in battery which causes each battery to go up to 39 V. When the motor is running at maximum load (i.e. uphill), the maximum current drawn from the battery is 13.2A whereas when the motor is running at constant load (usually on a flat terrain) the average current is 9.2A. Thus, the supercapacitor is required to supply at least 16.18A to ensure that batteries only supplied average current. It can be operated in two modes which are continuous current mode (CCM) and discontinuous current mode (DCM. This converter functions based on pulse width modulation (PWM); PWM is sent from a controller/driver in order to control the MOSFET to switch on or switch off as shown in Figure 5.

The simple control strategy adopted for calculating the duty cycle of the PWM pulse which is required to turn on the MOSFET for the power converter. Inputs from a Hall Effect speed sensor and current sensors are required to turn on the converter. This ensures that the E-bicycle is on and ready for

Figure 5. Schematic diagram hybrid energy management system.

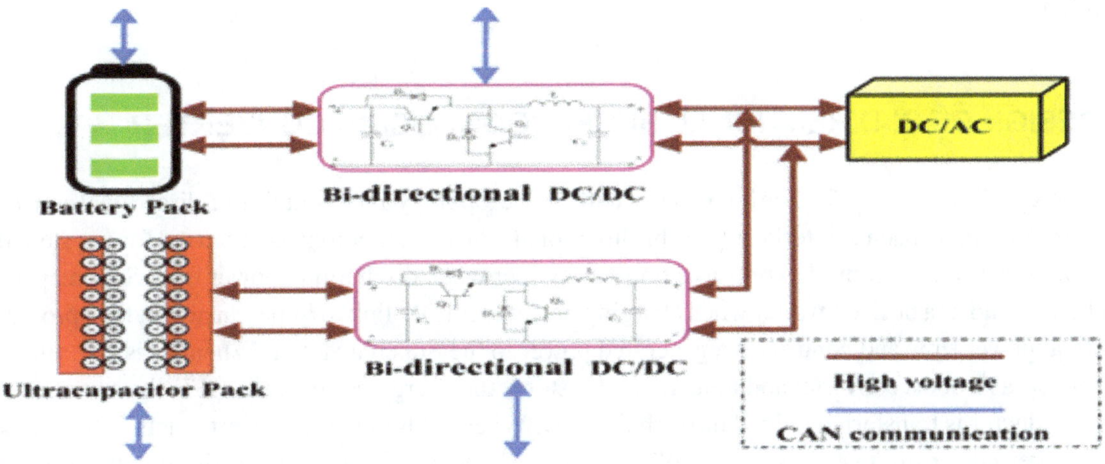

an impending acceleration and also prevents unnecessary usage of the supercapacitor's limited energy. Voltage divider circuits at the input side (supercapacitor) and output side (battery/converter out) feed

Table 1. Specification of Power Converter

Parameters	Values
Average Voltage	39.0 V
Average Current	8.2 A
Average Power	245.0 W

Figure 6. Bi-directional buck–boost DC/DC converter

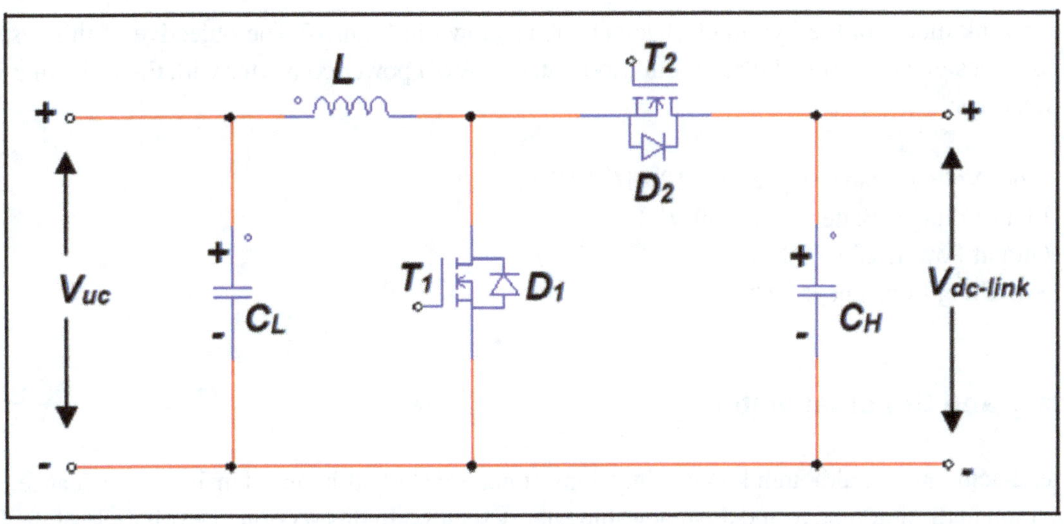

input signals to the controller. By comparing these two signals, the duty cycle can be appropriately adjusted as shown in Figure 6.

5. DESIGN OF BI-DIRECTIONAL BUCK–BOOST DC/DC CONVERTER

The survey of literature prefers the buck–boost half bridge topology as the suitable bi-directional converter for battery/Supercapacitor interface. This bi-directional converter topology is preferred for a system that has size and cost constraint. It works as a boost converter when discharging power from Supercapacitor to DC-link and as a buck converter when charging power from DC-link to Supercapacitor, thus providing bilateral power flow and controlled regenerated power to the Supercapacitor. When T1 is ON, the converter acts as a boost converter and inductor and L stores the energy. When T1 is OFF, the stored energy of the inductor is transferred to load through diode D. When T2 is ON, the converter acts in buck mode, and it takes power from DC link to UC. When T2 is OFF, the current freewheels through the diode D2. The high-voltage side (DC-link) and low-voltage (LV) side (Supercapacitor) has voltage relationships.

To maximize the benefits of the battery/Supercapacitor Hybrid Energy Storage System, proper management of the power flow from and into the system is essential. The significance of the power flow management and various energy management schemes has been described in literature. As discerned from the literature, the filtering method and rule-based strategy have been mostly preferred for real time applications. The filtering method does not use heuristic knowledge to consider the energy capacity of Hybrid Energy Storage System, therefore results in improper sizing of Energy Storage System with the separation of high- and low-frequency component. But the rule based algorithm uses the set of rules based on prior knowledge for the management of Energy Storage System. Therefore, an effective power-split strategy incorporating filtering technique and deterministic rules has been proposed for effective control of Hybrid Energy Storage System. This proposed strategy combines the advantages of rule-based algorithm and the frequency separation strategy, thus providing smoother battery current.

6. SIMULATION RESULTS

The simulink model of the hybrid electric vehicle is shown in Figure 7.The objective of this research work is to design and build a Pulse Width Modulated (PWM) power converter with the following ideal specifications:

- Input Voltage (Supercapacitor): 12 - 16.2 VDC
- Input Voltage (Battery): 36 - 39 VDC
- Output Power: 200 – 300 W
- Switching Frequency: 50 kHz

Battery and Supercapacitor

The lead-acid battery, although known since long time, are studied in an intensive way because of its application in the automotive and the renewable energy sectors. In this section, the principle of the lead-

Figure 7. Simulink model of hybrid Powered Electric Bicycle

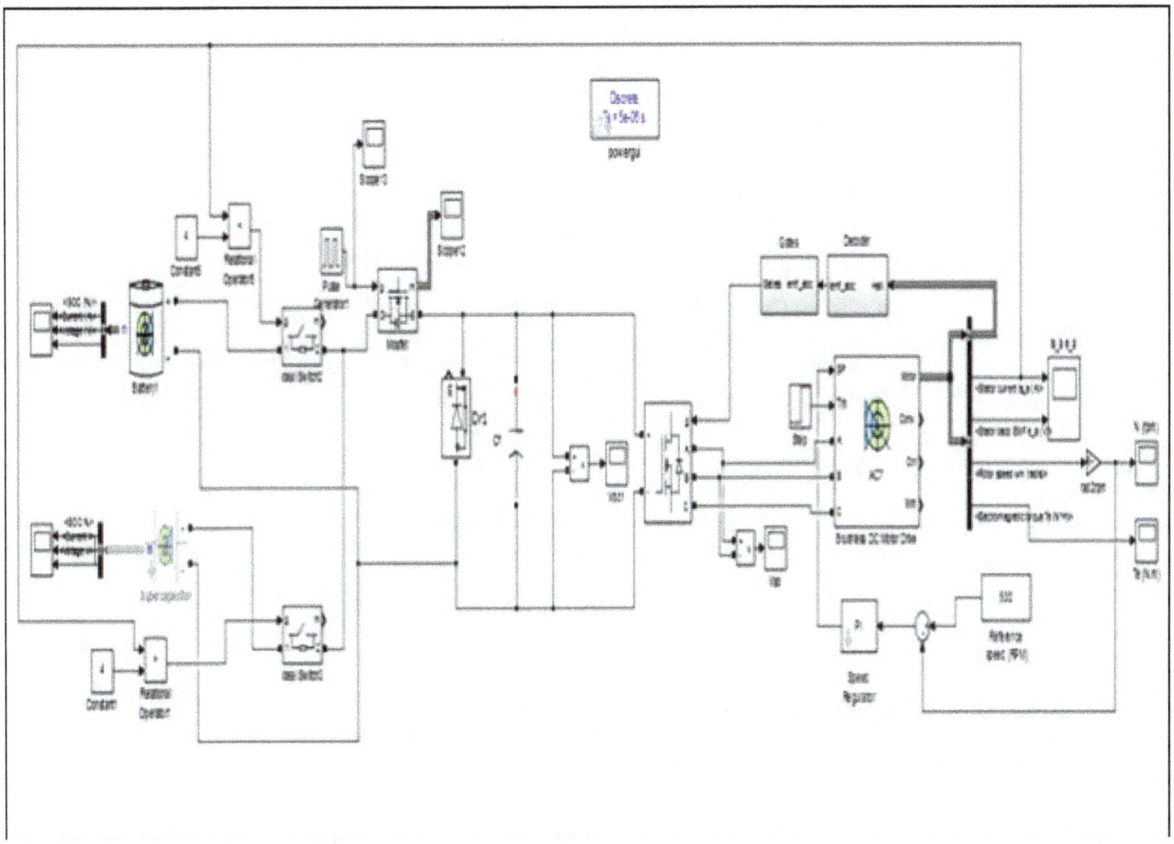

acid battery is presented. A simple, fast, and effective equivalent circuit model structure for lead-acid battery is implemented. The identification of the parameters of the proposed lead-acid battery model is treated. This battery model is validated by simulation using the Matlab/Simulink.

The main power source consists of a 36V~40V, 12Ah lead acid battery pack with cell balancing circuitry which weighs 7 kg. This can comfortably provide a continuous discharge current of 12A. The supercapacitor consists of a 500g 16V, 350F module to be connected in parallel with the battery pack via a power converter which is designed to harvest the maximum energy from it. The propulsion device consists of a state of the art rear hub motor rated at 250W, eliminating the need for transmission and the losses associated with it.

Figure 8 & 9 shows that charging voltage of battery is about 42.5 V. The charging current of battery is about 7.5 A. State of charge of battery is about 80%.It is the present capacity of the battery. It is the amount of capacity that remains after discharge from a top-of-charge condition. Depth of charge is the percentage of battery capacity to which a battery is discharged. The battery maximum temperature is about 28 C.

Figure 10 & 11 shows that charging voltage of supercapacitor is about 16.0 V. The charging current of supercapacitor is about 23.5 A.

The simulated results (rotor speed, electromagnetic torque, stator current and back EMF) of BLDC motor are shown in Figures [12-15] respectively.

Figure 8. Battery voltage characteristics

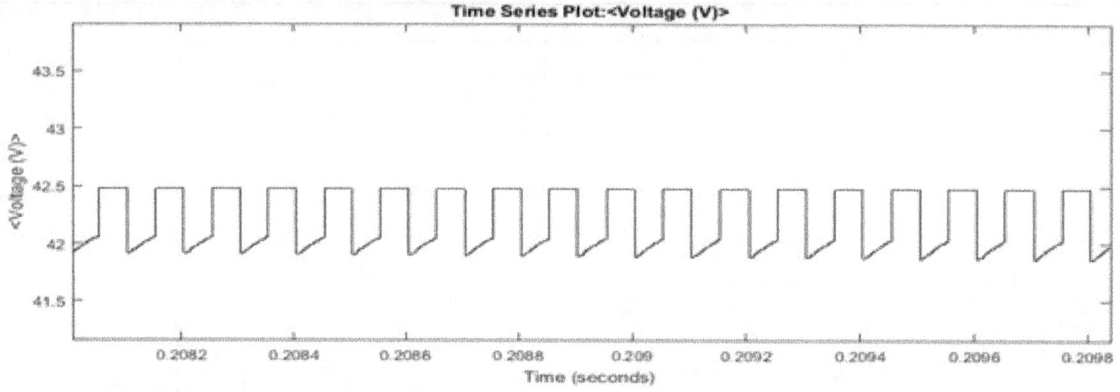

Figure 9. Battery current characteristics

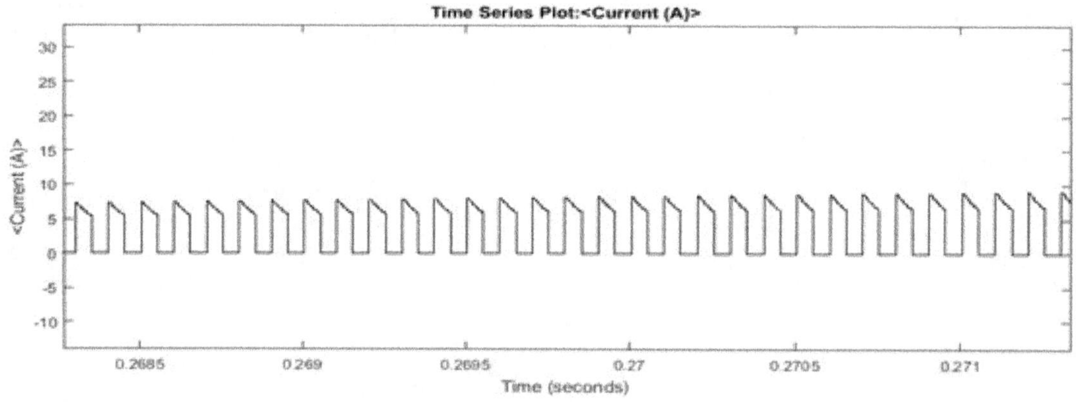

Figures 12 & 13 show that the BLDC motor speed is settled 350 rpm and torque is about 4 Nm. Figures 14 & 15 show that the BLDC motor stator current and back emf voltage.

Figure 10. Supercapacitor voltage characteristics

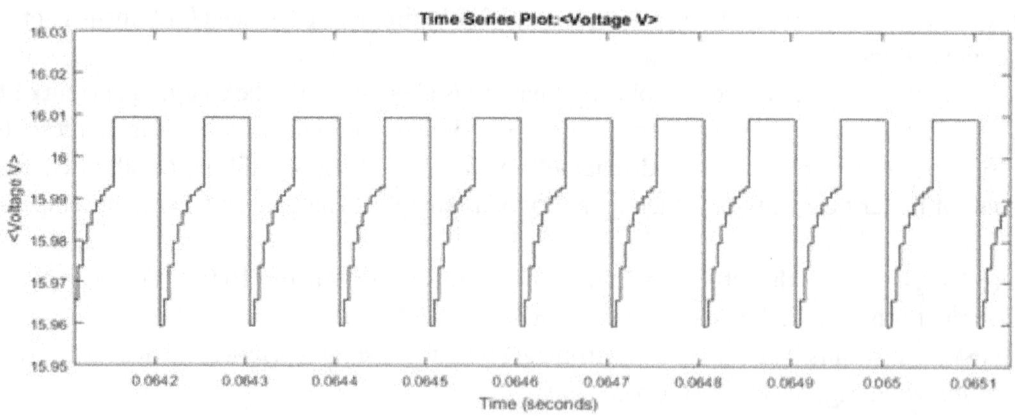

Figure 11. Supercapacitor current characteristics

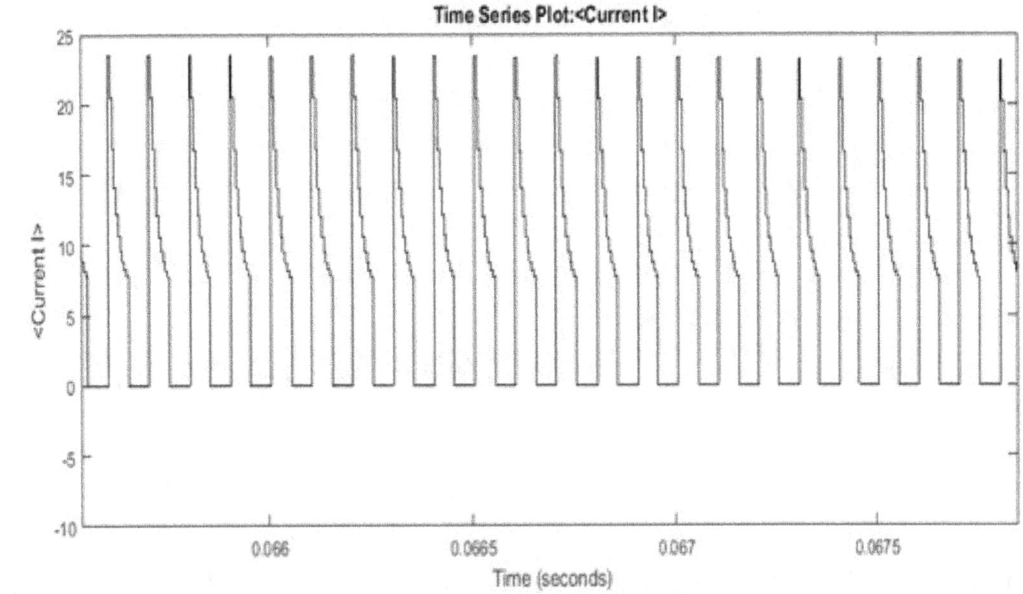

7. HARDWARE IMPLEMENTATION OF E-BICYCLE:

The hardware prototype of the proposed bi-directional dc-dc converter is developed using MOSFETs as the power device, along with driver circuit. The gating pulses were obtained from a FPGA controller. The hardware set-up for bi-directional dc-dc converter with FPGA board is shown in Figures 16.Figure 17 shows that battery and supercapacitor charging setup with bi directional converter.

Figure 12. BLDC motor speed characteristics

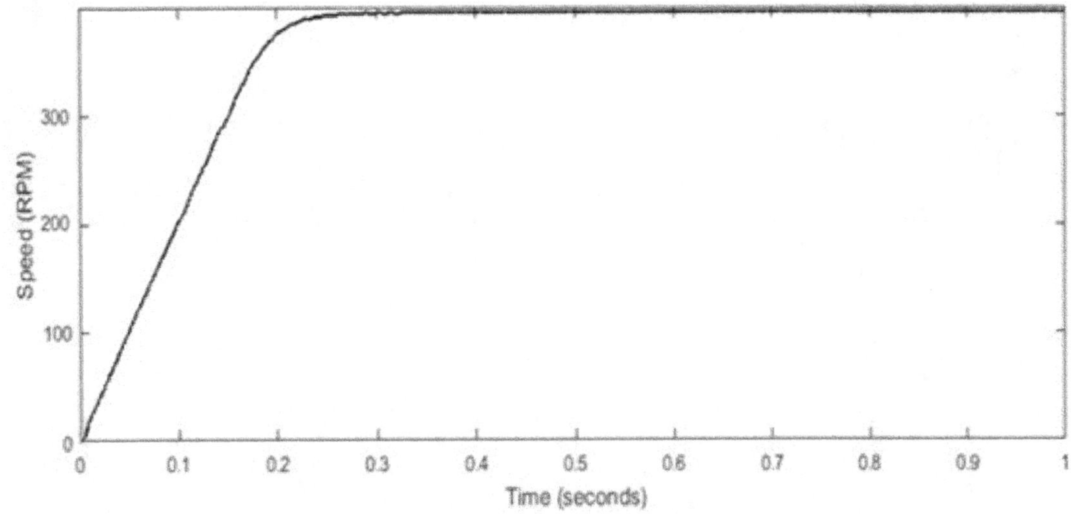

Figure 13. BLDC motor torque characteristics

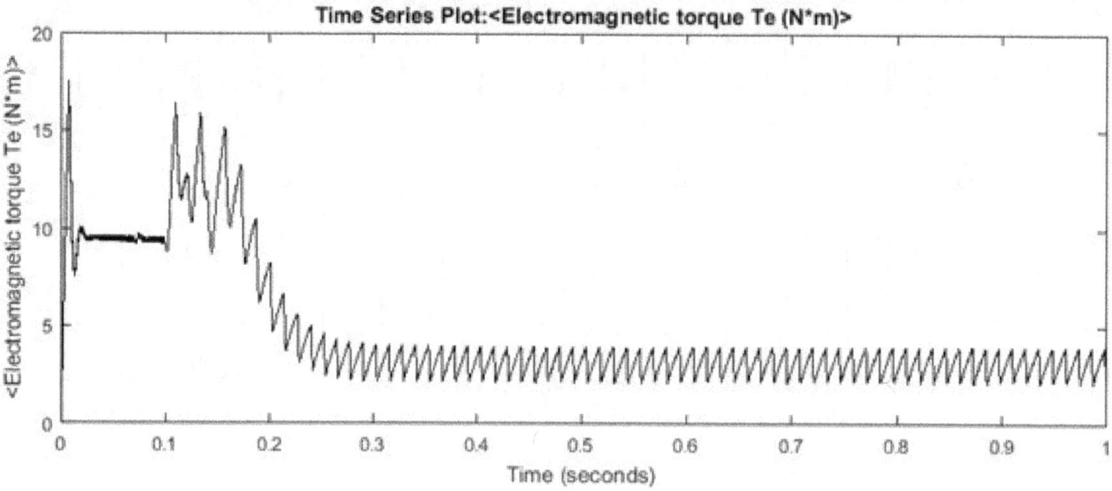

The dynamic charging and discharging characteristics of battery is measured using scope corder and it is shown in Figures.18 & 19

The dynamic charging and discharging characteristics of supercapacitor is measured and it is shown in Figures.20 & 21.

Figures 22-23 show the phase voltage and stator current of hub BLDC drive with the following testing conditions: starting voltage: 30 V, minimum speed: 50 rpm, maximum voltage: 36 V,300 rpm, high frequency - 10 KHz and low frequency-50 Hz. Figure 24 shows that with the e bicycle developed motor shaft speed measured was 354 rpm which is verified experimentally. Figure 25 shows that Battery with Supercapacitor set up and Final working model E-bicycle.

Figure 14. BLDC motor stator current characteristics

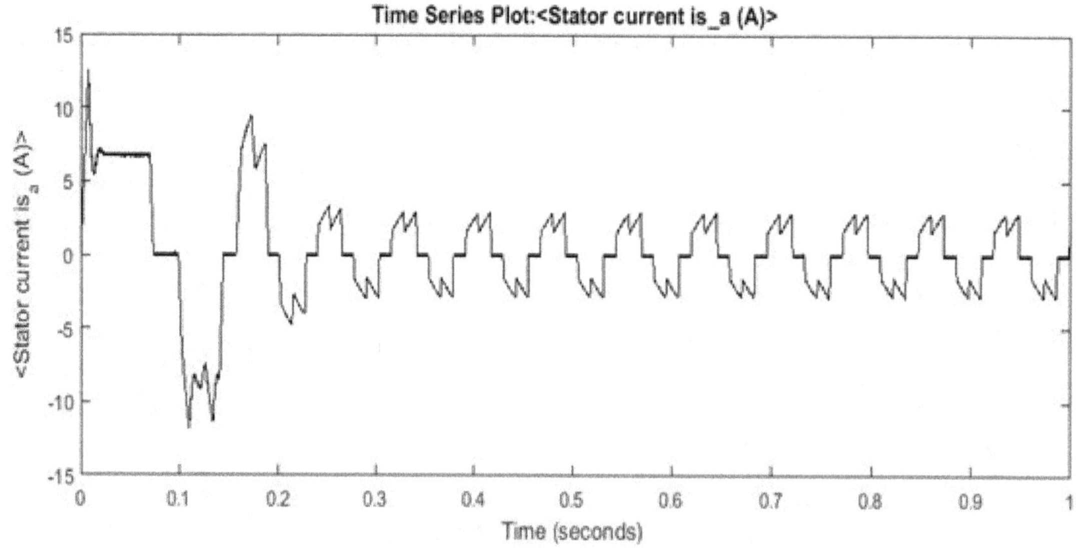

Figure 15. BLDC motor back EMF characteristics

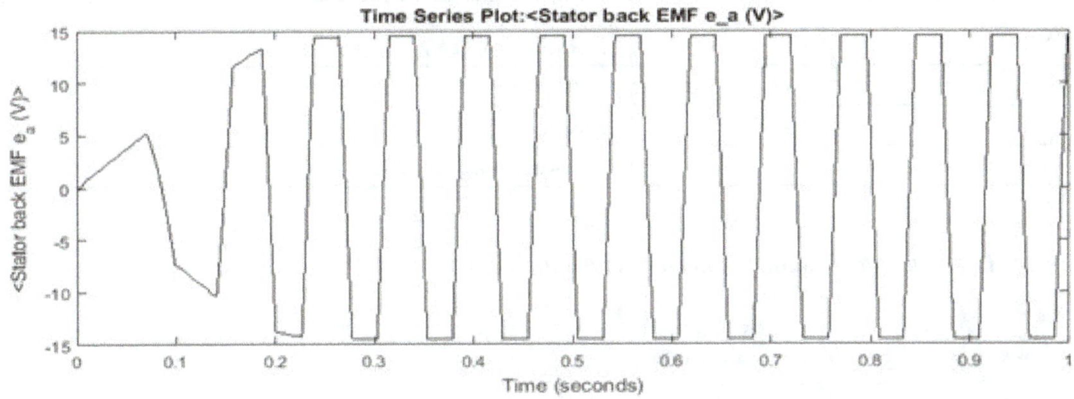

Figure 16. Hardware set-up Battery/Supercapacitor

Figure 17. Battery and Supercapacitor set up with bi-directional converter

Table 2. Specification of Battery set up

Voltage	36 V (no of series = 3, 12 V)
Current	12 Ah
Type	Lead acid
No of cells	6 cells with a voltage of 12 V

Figure 18. Experimental results for Battery charging characteristics

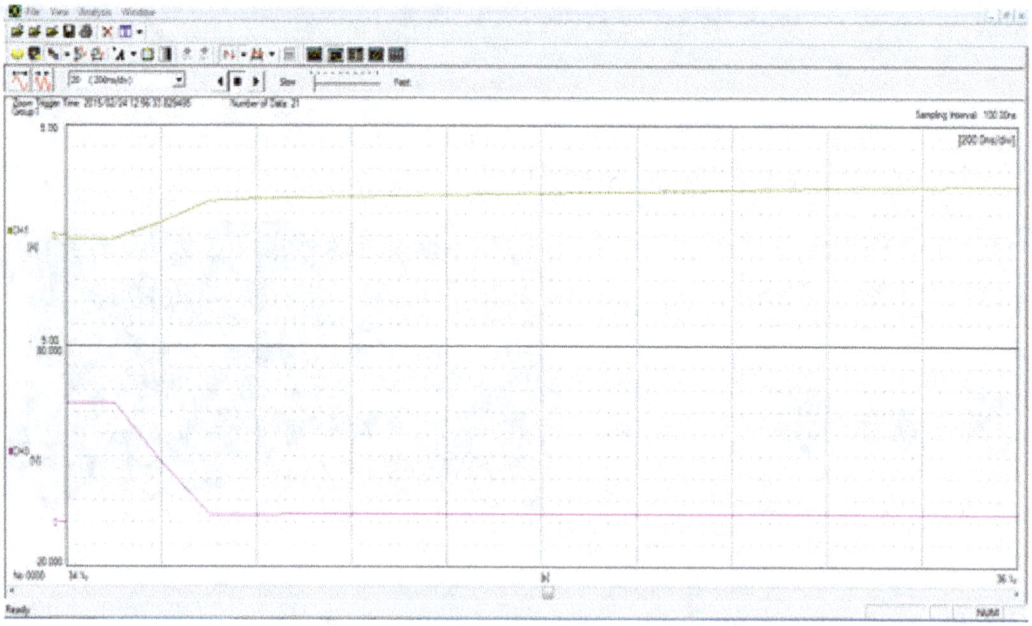

Figure 19. Experimental results for Battery discharging characteristics

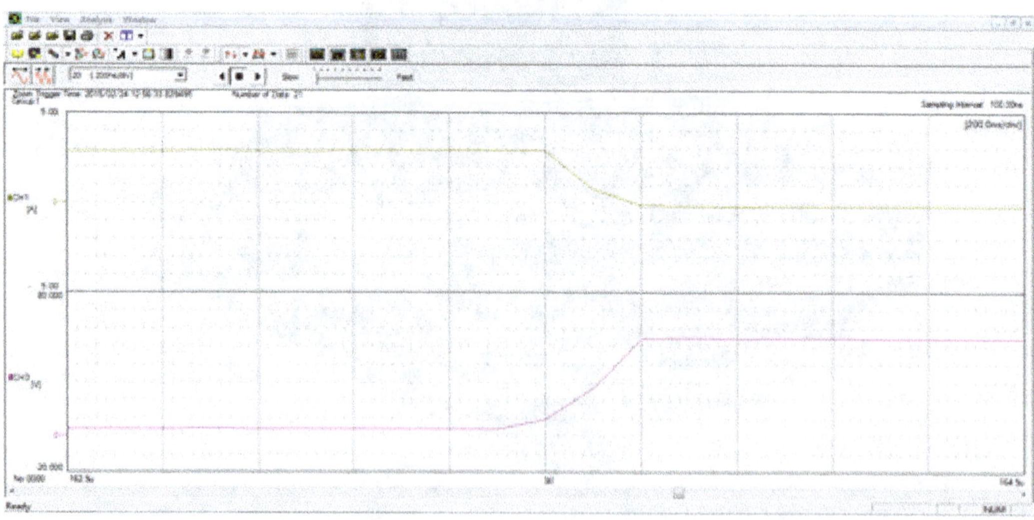

Figure 20. Experimental results for Supercapacitor charging characteristics

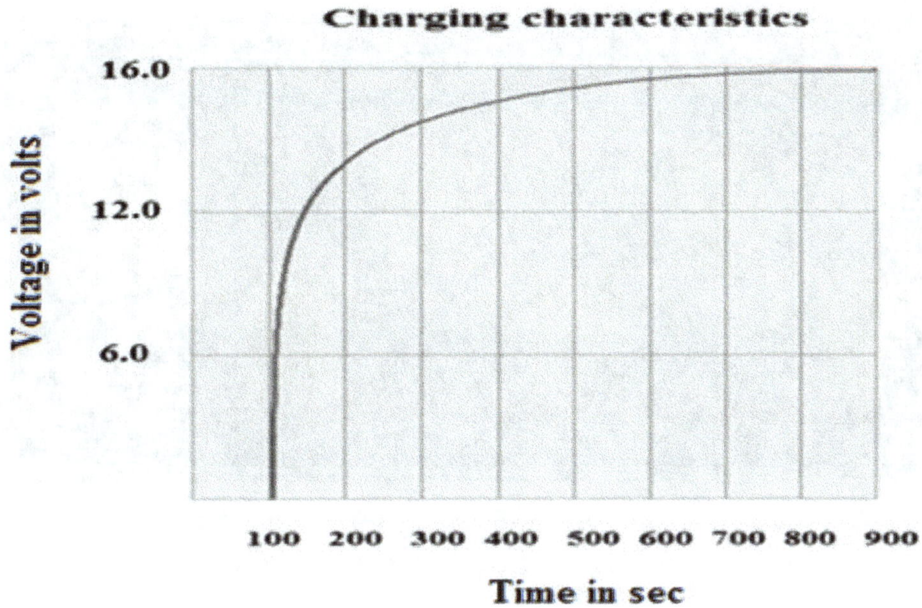

A Simulation and experimental results discussed that the battery voltage, battery current, supercapacitor voltage and supercapacitor current. This was used to evaluate the performance of the electric assisted bicycle in terms of motor voltage, stator current and motor speed. The parallel hybrid power source has no significant effect on the maximum speed achievable over a drive cycle except that the rider felt an improvement in uphill acceleration as compared to battery alone.

Figure 21. Experimental results for Supercapacitor discharging characteristics

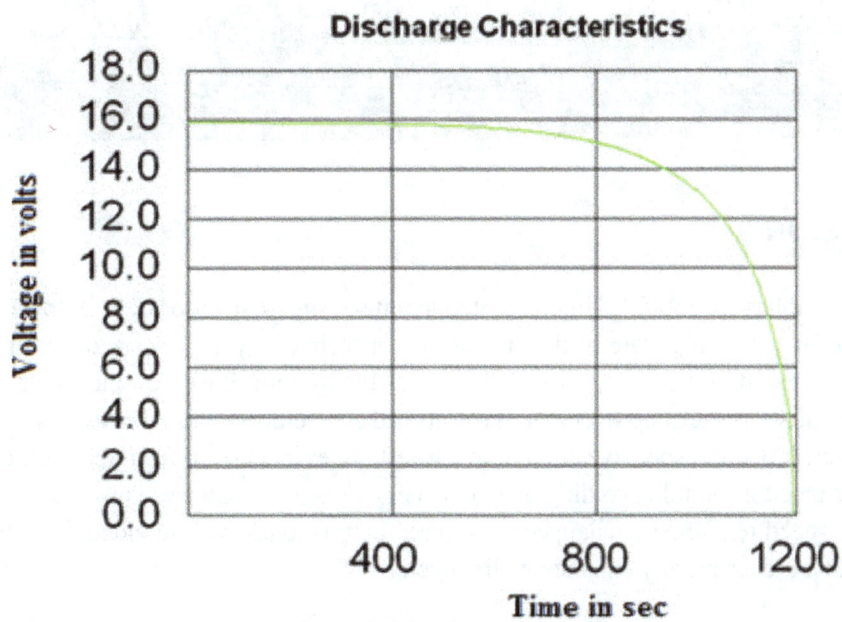

Figure 22. Output voltage of BLDC drive

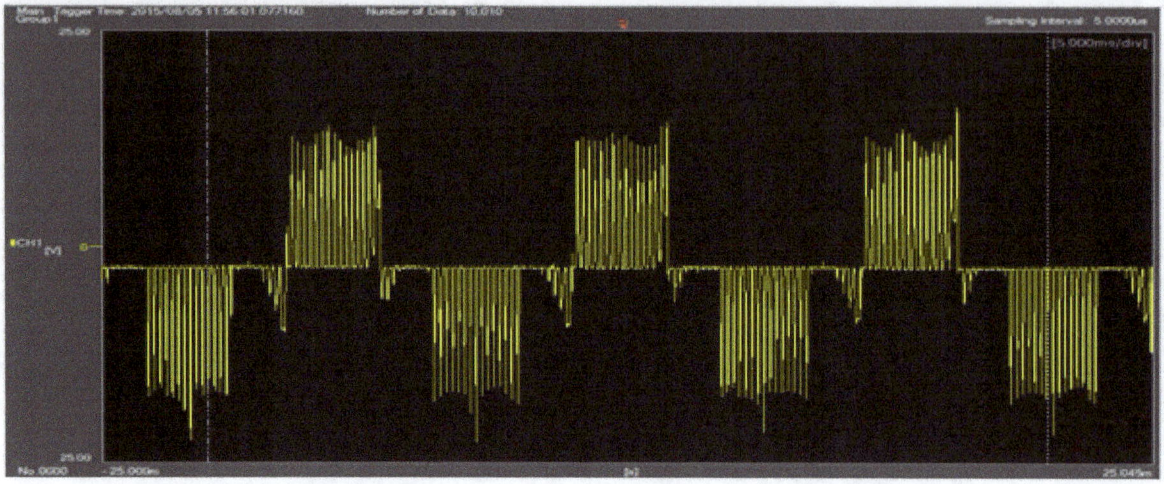

Figure 23. Stator current of BLDC drive

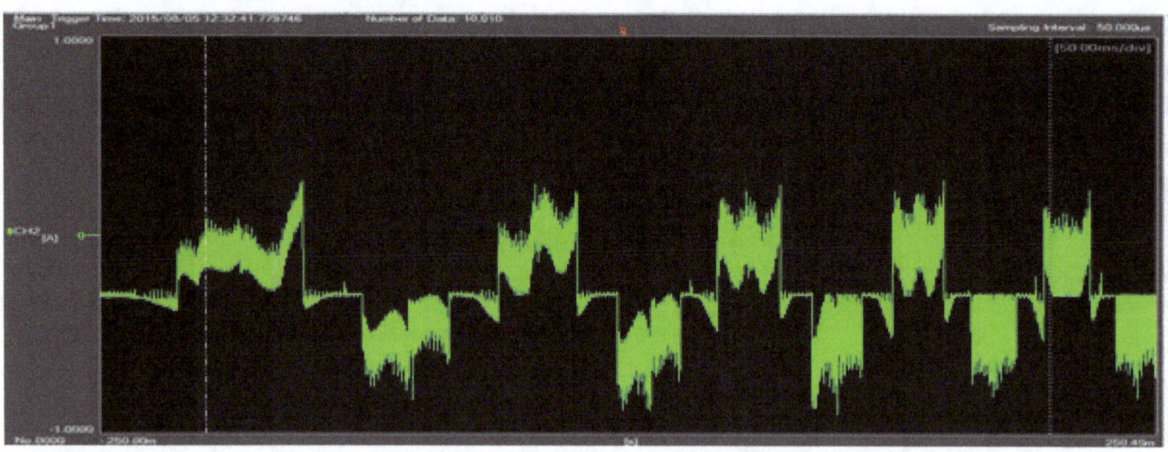

8. CONCLUSION

This research work has successfully implemented a battery/supercapacitor hybrid power source for an electric assisted bicycle using state of the art hub motor technology. A power converter was designed and implemented based on the energy requirements of the system. Based on the implemented system experimental results show an improvement in the up-hill acceleration of the bicycle as a direct result of the power converter being responsive enough to harvest the extra current from the high power complementary supercapacitor module avoiding deep discharges from the battery. This enhanced battery life. The maximum speed remained unchanged. The main battery pack was shielded from high discharge currents which would eventually enhance its life cycle.

Figure 24. Motor speed in BLDC hub motor

ACKNOWLEDGMENT

The authors wish to thank the management of SSN College of Engineering for providing the financial support to carry out this research work.

Figure 25. Final working model E-bicycle

REFERENCES

Amari, M., Bacha, F., & Ghouili, J. (2015). Average model of dual active bridge interfacing supercapacitor used in electrical vehicle. *International Journal of Energy Optimization and Engineering, 4*(1), 36–54. doi:10.4018/ijeoe.2015010103

Ashar, M. (2016). Integration of supercapacitor with battery using DC-DC bidirectional buck boost converter in an electric vehicle. *International Research Journal of Engineering and Technology, 3*, 1513–1517.

Ates, Y., Erdinc, O., Uzunoglu, M., & Vural, B. (2010). Energy management of an FC/UC hybrid vehicular power system using a combined neural network-wavelet transform based strategy. *International Journal of Hydrogen Energy, 35*(2), 774–783. doi:10.1016/j.ijhydene.2009.11.021

Ban, S., Zhang, J., Zhang, L., Tsay, K., Song, D., & Zou, X. (2013). Charging and discharging electrochemical supercapacitors in the presence of both parallel leakage process and electro chemical decomposition of solvent. *Electrochimica Acta, 90*, 542–549. doi:10.1016/j.electacta.2012.12.056

Bottu, M., Crow, M. L., Elmore, A. C., & Atcitty, S. (2013). A Power electronic conditioner using supercapacitors to improve wind turbine power quality. *Smart Grid and Renewable Energy, 4*(01), 69–75. doi:10.4236gre.2013.41010

Cultura, A. B., & Salameh, Z. M. (2015). Modeling, evaluation and simulation of a supercapacitor module for energy storage application. *Proceedings of the International Conference on Computer Information Systems and Industrial Application*, 876–882. 10.2991/cisia-15.2015.235

Dixon, J. W., & Ortuzar, M. E. (2002). Supercapacitors + DC-DC converters in regenerative braking system. *Aerospace and Electronic Systems Magazine, IEEE, 17*(8), 16–21. doi:10.1109/MAES.2002.1028079

Farzanehfard, H., Beyragh, D. S., & Adib, E. (2008). A Bidirectional soft switched supercapacitor interface circuit for hybrid electric vehicles ". *Energy Conversion and Management, 49*(12), 3578–3584. doi:10.1016/j.enconman.2008.07.004

Fletcher, S., Kirkpatrick, L., Drink, R., Puttock, R., Thring, R., & Howroyd, S. (2017). The modelling of carbon-based supercapacitors: Distributions of time constants and pascal equivalent circuits. *Journal of Power Systems, 345*, 247–253. doi:10.1016/j.jpowsour.2017.02.012

Fuyuan, Y., Languang, L., Yuping, Y., & He, Y. (2010). Characterization, analysis and modelling of an supercapacitor. *World Electric Vehicle Journal, 4*(2), 358–369. doi:10.3390/wevj4020358

George, M., & Choi, S. (2003). Contrasting Hybrid Electric Bicycles and Electric Bicycle. *Inter-Professional Project, 315*(Spring), 20–25.

Hala, S., Khalil Faris, N., & Khaled Faris, N. (2015). Supercapacitor: Modelling and characterization for electric vehicle application. *International Journal of Engine Research, 5*, 681–686.

Jin, L.-Q., Zheng, Y., Li, J.-H., & Liu, Y.-L. (2015). A study of novel regenerative braking system based on supercapacitor for electric vehicle driven by in-wheel motors. *Advances in Mechanical Engineering, 7*(3), 1–12. doi:10.1177/1687814015573762

Marzougui, H., Amari, M., Kadri, A., Bacha, F., & Ghouili, J. (2016). Energy management of fuel cell/ battery/supercapacitor in electrical hybrid vehicle. *International Journal of Hydrogen Energy*, *42*(13), 8857–8869. doi:10.1016/j.ijhydene.2016.09.190

Michalczuk, GrzesiakLech, & Ufnalski. (2012). Alithium battery and supercapacitor hybrid energy source for an urban electric vehicle. *Politechnika Warszawska*, *4*, 158–162.

Parker. (2015). *Electric Power-Assisted Bicycles Reduce Oil Dependence and Enhance the Mobility of the Elderly electric power assisted bicycle*. Presented at 29th Australian Transport Research Forum.

Pay, S., & Baghzouz, Y. (2003). Effectiveness of battery-supercapacitor combination in electric vehicles. *Power Tech Conference Proceedings, 2003 IEEE Bologna, 3*, 23-26. 10.1109/PTC.2003.1304472

Payman, A., Pierfederici, S., & Meibody-Tabar, F. (2008). Energy control of supercapacitor/fuel cell hybrid power source. *Energy Conversion and Management*, *49*(6), 1637–1644. doi:10.1016/j.enconman.2007.11.012

Paymen, A., & Oierfederici, S. (2008). Energy control of supercapacitor/fuel cell hybrid power source. *Energy Conversion and Management*, *49*(6), 1637–1644. doi:10.1016/j.enconman.2007.11.012

Plett, G. L. (2004). Extended Kalman filtering for battery management systems of Li PB-based HEV battery packs—Part 1. Background. *Journal of Power Sources*, *134*(2), 252–261. doi:10.1016/j.jpowsour.2004.02.031

Schneuwly, Maher, & Auer. (n.d.). *Supercapacitors, the New Thinking in the Automotive World*. Maxwell Technologies Inc.

Shi, L., & Crow, M. L. (2008). Comparison of supercapacitor electric circuit models. *Proceedings of the IEEE Power and Energy Society 2008 General Meeting: Conversion and Delivery of Electrical Energy in the 21st Century*.

Stefan, S. (2009). *Power Sources for Hybrid Electric Vehicles* [M. thesis]. Lund University, Faculty of Engineering.

Tuite, D. (2007). *Get the Lowdown on Supercapacitors*. Technology report electronic design online. http://electronicdesign.com/Articles/ArticleID/17465/17465.html

Zhang, L., Wang, Z., Hu, X., Sun, F., & Dorrell, D. G. (2015). Acomparative study of equivalent circuit models of supercapacitors for electric vehicles. *Journal of Power Sources*, *274*, 899–906. doi:10.1016/j.jpowsour.2014.10.170

Zhang, L., Wang, Z., Sun, F., & Dorrell, D. G. (2014). Online parameter identification of supercapacitor models using the extended Kalman filter. *Energies*, *7*(5), 3204–3217. doi:10.3390/en7053204

Zhang, L., Zhenpo, W., Sun, F., & Dorrell, D. (2014). Supercapacitor modelling and parameter identification using the extended Kalman filter. *ITEC Asia Pacific*, *7*, 3204–3217.

Chapter 3
Error-State Extended Kalman Filter-Based Sensor Fusion for Optimized Drive Train Regulation of an Autonomous PHEV

Parag Jose Chacko
CHRIST University, India

Haneesh K. M.
CHRIST University, India

Joseph X. Rodrigues
CHRIST University, India

ABSTRACT

An efficient state estimator is critical for the development of an autonomous plug-in hybrid electric vehicle (PHEV). To achieve effective autonomous regulation of the powertrain, the latency period and estimation error should be minimum. In this work, a novel error state extended kalman filter (ES-EKF)-based state estimator is developed to perform sensor fusion of data from light detection and ranging sensor (LIDAR), the inertial measurement unit sensor (IMU), and the global positioning system (GPS) sensors, and the estimation error is minimized to reduce latency. The estimator will provide information to an intelligent energy management system (IEMS) to regulate the powertrain for effective load sharing in the PHEV. The integration of the sensor fusion data with the vehicle model is simulated in MATLAB environment. The PHEV model is fed with the proposed state estimator output, and the response parameters of the PHEV are monitored.

DOI: 10.4018/978-1-7998-7626-7.ch003

INTRODUCTION

The widespread utilization of alternative hybrid powertrains will be inevitable and many sustainable opportunities develop. The concerns about the environmental impact of fossil fuel based vehicles, economy's foreign oil dependence, and the climate change has made countries to concentrate upon the improved propulsion technologies. Plug-in hybrid electric vehicles (PHEVs) has high potential to curtail the petroleum consumption and reduce the greenhouse gas (GHG) emissions. The scope of Plug-in Hybrid Electric Vehicles (PHEVs'), is huge. In India, where majority of vehicles are IC Engine based, the transition to Electric Vehicles is not easy even though the Government of India is promoting Electric Mobility (Niti Aayog,2018). The PHEV powertrain has an optimized combination of the conventional IC engine and an alternative traction motor. This extra degree of freedom provided by the traction motor, enables the operation of the IC engine operated in the best specific fuel consumption (sfc) range. . The regulation of the power train requires an efficient and Intelligent Energy Management System (IEMS). This controller performs the optimized power-split between the two power trains. The PHEVs in the market today, utilizes heuristic or ad-hoc power management strategies that may not be optimal under a real driving scenario. These approaches has the sparing utilization of the traction motor to prevent depletion of the traction battery. The efficient powertrain optimization requires the determination of the best power-split ratio in real time scenarios. To perform this it is critical to anticipate the vehicle speed trajectory a priori. This would not be practical without real-time data collection about the surroundings of the ego PHEV. Therefore, employing sensors to capture the proximity information would enable the optimal controller operation by anticipating the upcoming vehicle trajectories accurately and then select the appropriate power source intelligently. For example, if the ego PHEV is approaching an intersection and the sensors informs the ego PHEV that a vehicle is going to cross the road and therefore the ego PHEV has to stop, then the controller can choose to use more electrical power to drive the vehicle approaching the intersection. By doing this the engine operation would be reduced to the best sfc region, thereby less fuel would be consumed, while the traction battery can be recharged through regenerative braking for the duration the ego PHEV decelerates to a stop.

This work introduces the modelling and development of a split parallel PHEV. Section II describes the selection of the sensors and its calibration. Section III has the implementation process explained with focus on the Error State Extended Kalman Filter (ES-EKF) State Estimator modelling and implementation. The Intelligent Energy Management System and its role in regulation of the developed PHEV model is discussed in Section IV. Section V has the Results & discussions also considering the effects of sensor calibration error on the vehicle state estimation.

Background

The interest towards autonomous and self-driving cars are also on the rise.. The self-driving capability would require efficient interaction between the IEMS and the State Estimator (Xia,2016). The adoption of PHEVs' by conversion of existing IC Engine vehicles is an effective approach. But PHEVs' requires complex regulation of the powertrain (Yang,2016). This complex regulation involves optimizing the load sharing between the mechanical and electrical powertrain. (Son,2016). The PHEV will consider the road terrain and traffic conditions to predetermine the regulation to be implemented (Jose,2020). State estimation of the Vehicles is critical in making it autonomous. This requires the usage of different sensors. The process of sensor fusion involves fusing information from different sensors to regulate the self-driving capabilities of an Autonomous vehicle (Gravina,2017). A typical autonomous vehicle has to utilize a combination of different sensors including Inertial Measurement Unit (IMU), Global Positioning System (GPS) and Light Detection and Ranging (LIDAR) to effectively determine its position and also utilize cameras for safe driving regulations (Li,2014). The IMU makes use of a combination of gyroscopes and accelerometer to determine the Degrees of Freedom when the vehicle moves. The GPS unit is important for the estimator to correctly locate the position of the ego vehicle. The GPS unit requires it to be modelled considering pseudo ranging and trilateration. The camera could be used to detect the lane and also to detect the road side signs to regulate the powertrain of the PHEV (Wang,2018). The IMU would supplement the GPS unit in position estimation, while the GPS unit data could mitigate the errors caused due to IMU calculations(Yuan,2019). LIDAR would be providing very accurate position estimates using an existing map and in instances where the GPS connection is lost thereby will provide an effective localization (Wang,2019). These are technologies following the vision based warning system proposed, called as AURORA (Chen,1995). The primal reason for safe driving is the driver's alertness. Two main reasons for majority of accidents are driver fatigue (Khan,2019) and road side distractions(Young,2019) which has to be considered for optimized driving. Studies done by the World Health Organization (WHO), (WHO,2013),(WHO,2014) (WHO,2009), showcases that on an average, 1–1.24 million people are killed in accidents around 20–50 million people are subjected to injuries. Therefore, it is critical that the driving assistance and the vehicle monitoring systems should work in tandem. The main difference in operation is that the driving monitoring system understands the driving situation, while the role of the driver assistance system is to assist the drivers to properly respond to the situation. It could be seen that the monitoring systems (Li,2015) are more oriented towards the safety whereas the assistance systems prioritize the drivers' comfort (Mukhtar,2015).

Integration of these sensors however is complex and would require a powerful estimator. Improper fusion of the signals can lead to wrong decisions leading to unpredictable responses from the ego vehicle. State estimation using Kalman Filter has proven to be one of the most effective tool (Wischenewski,2019). Real life situations, offer non-linear conditions which would not be effectively optimized by a linear Kalman filter (Barrau,2015). The Extended Kalman Filter enables to adapt the linear Kalman filter to non-linear conditions (Akram,2019). This filter model works on linearizing a non-linear condition. But, this local linearization approach leads to linearization error. In conditions where the non-linear situations are almost flattened the linearization error will be minimum and the regulation could be performed by Extended Kalman Filter. But in autonomous driving conditions, the operation point would never be near to a flattened, non-linear condition and therefore would lead to large linearization error. A powerful estimator is required in such a scenario The sensor sampling rates are slow and therefore the estimator should be able to foresee the future conditions based on effective estimations. In the Error State-Extended Kalman Filter, the Estimator divides the True Driving state into a large nominal state and an error state (Farrell,2017). The nominal state is the non-linear driving model created for the vehicle to drive through linearization with linearization error present. This is overcome by the error state component. The error state is effectively used as a correction factor by the estimator to regulate the non-linear driving of the vehicle.

Figure 1. Process flow diagram

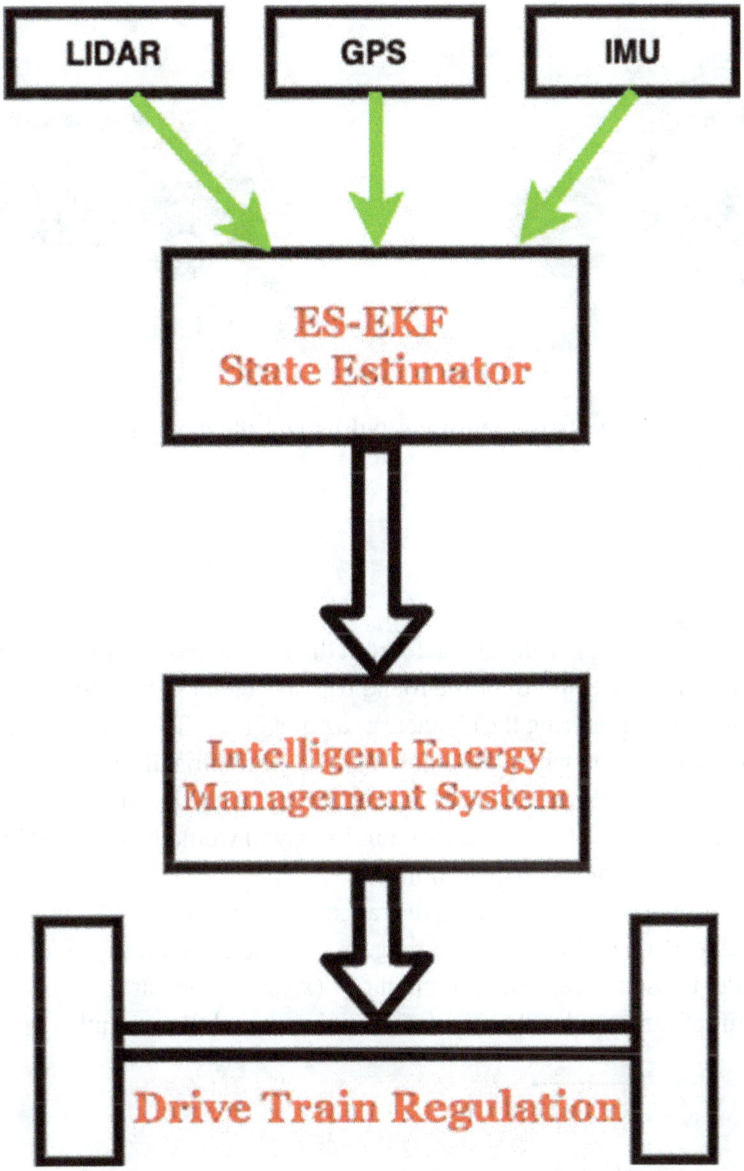

SENSOR FUSION AND CALIBERATION

Fig. 1 shows the process flow diagram of the entire model. The LIDAR, the IMU GPS are the three sensors considered to provide autonomous regulation to the PHEV drivetrain through the IEMS.

LIDAR: The LIDAR performs continuous 2D & 3D scans of the ambient environment around the vehicle. The LIDAR unit encases a laser, a photodetector, and a stopwatch.

The laser unit initially emits a short pulse of near infrared frequency light in a particular direction with the stopwatch recording time. The laser pulse upon hitting an obstacle will scatter with some reflected light travelling back to the photodetector.

Figure 2. Point Cloud Representation for LIDAR processing

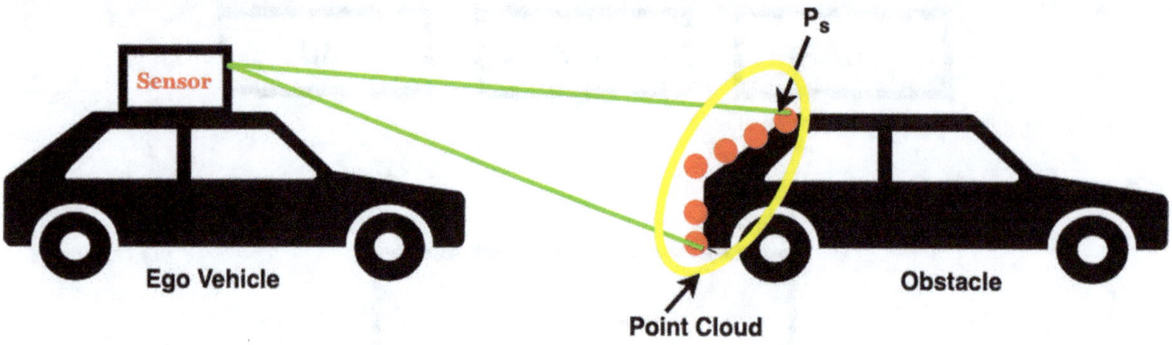

The photodetector then collects the returning pulses and the stopwatch shows the distance of the obstacle from the sensor unit.

This is provided by,

$$r = 1/2\ c*t \quad (1)$$

where r is the distance (range) of the obstacle from the PHEV, c is the speed of light emitted by the photo detector and t is the time elapsed for the round trip, as noted by the stop watch. The evaluation is performed with the sensor considering the obstacle as a point cloud. This implies the obstacle is divided into large number of points upon which the light is expected to hit and reflect. Fig.2, shows the point cloud representation of an obstacle in front of the PHEV ego vehicle. The obstacle vehicle would be divided into smaller points. Out of these points point Ps (x,y,z) would be detected by the sensor and its range, elevation and azimuth angles are determined.

For 3D LIDAR sensing the sensor captures the range r, the elevation of the point, ε, and the azimuth angle, α. For 2-D LIDAR operations only the range r, and the azimuth angle, α, are considered. The forward sensor model is used to represent the point P_s (x,y,z) in the earth frame to be brought to the sensor frame in terms of range r, the elevation of the point, ε, and the azimuth angle, α.

$$\begin{bmatrix} r \\ \alpha \\ \varepsilon \end{bmatrix} = h\,(x,y,z) = \begin{bmatrix} \sqrt{x^2+y^2+z^2} \\ tan^{-1}\left(\dfrac{y}{x}\right) \\ sin^{-1}\left(\dfrac{z}{\sqrt{x^2+y^2+z^2}}\right) \end{bmatrix} + V \quad (2)$$

Equation (2) showcases the transformation. However, there are chances for noises to impact the signal reflected back from the point object. This can be due to various factors like, uncertainty in determining the exact time of arrival of the reflected signal, the uncertainty in determining the exact orientation of the sensor, the surface absorption of the reflected signal that can occur at the obstacle and the variation in the propagation speed through various mediums. Therefore, a noise factor. V is also considered for accurate determination.

Upon determination of the point P_s in the point cloud, it has to be subjected to three types of analysis for accurate determination of the point. The operations include translational operation, rotational operation and scaling. The point coordinates in the transformed frame will be given as

$$p_{s'}^{(j)} = S_{s's} C_{s's} \left(p_s^{(j)} - r_s^{s's} \right) \tag{3}$$

where $S_{s's}$ is the scaling matrix, $C_{s's}$ is the rotational matrix and $r_s^{s's}$ is the translational matrix. For the instantaneous pose estimation of the point, p_s, the Iterative Closest Point (ICP) approach is used. This involves the following steps which are,

- Obtain an initial guess of { $S_{s's}, C_{s's}, r_s^{s's}$ }
- Associate each point $p_{s'}$ with the nearest point p_s.
- Solve for the optimal transformation matrix { $S_{s's}, C_{s's}, r_s^{s's}$ }
- Repeat the steps until convergence.

GPS: Global Positioning System is a satellite based system that can be utilized for pinpointing a PHEV position with respect to a destination. The GPS is composed of 24 to 32 satellites which operates in 6 orbital planes at an altitude of around 20,200 km. The GPS has an orbital period of around 12 hours and broadcasts the signals at two frequencies which are 1575.42 MHz and 1227.6 MHz. Around 4 satellites would be required for obtaining a 3D position for autonomous driving applications. To determine the GPS position with respect to the Earth frame, the receiver uses the speed of light to compute the distances to each satellite based on the time of signal arrival. The GPS unit performs trilateration to determine the pseudo range; which is the range to the destination, accurately. The GPS units are of three types which are, basic GPS which provides an accuracy up to 10m, Differential GPS; where the receiver is supported by a base station to quantify for the errors due to atmospheric effects providing an accuracy of up to 10cm. The third type of GPS units used is the Real Time Kinematic GPS where analysis of the phase of the carrier signal is performed to have an improved accuracy up to 2 cm.

IMU: The Inertial Measurement Unit has an inbuilt gyroscope to measure the angular rotation rates of the vehicle in 3 D and accelerometers to measure the accelerations, also in 3 D. In the measurement model the gyroscope provides information on the angular velocity of the sensor with respect to the sensor frame,

$$\omega(t) = \omega_s(t) + b_{gyro}(t) + n_{gyro}(t) \tag{4}$$

while the accelerometer provides information on the orientation of the sensor with respect to the sensor frame. This is determined by integration of the rotational rates obtained from the gyroscopes

$$a(t) = C_{sn}(t)\left(r_n(t) - g_n\right) + b_{accel}(t) + n_{accel}(t) \tag{5}$$

The IMU however could only be used as a support unit to the GPS in motion modelling. This is because for localization purposes when IMU is used, inaccurate determination of sensor orientation

Figure 3. ES-EKF state estimator integration with IEMS

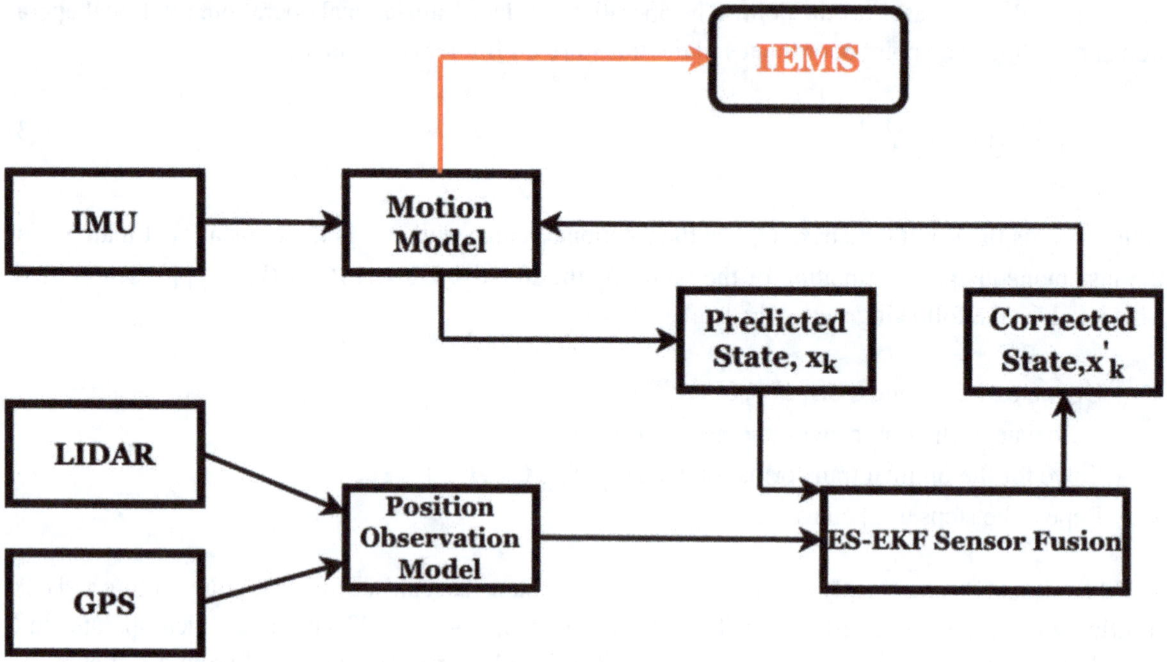

can occur which can lead to wrong position estimation. Moreover, IMU does not consider the effects of Earth's rotation which further would affect the accuracy level.

ES-EKF and IEMS Integration

The Error State Extended Kalman filter as a State Estimator and its integration with the Intelligent Energy Management system is shown in Fig.3. The GPS unit is initially used to provide information about the surroundings of the Ego PHEV. This would provide a detailed description on the road terrain, traffic and profile. Based on this base model the sensor data are utilized.

The sensors are used to perform two discrete operations. The IMU data is used to develop the motion model. The motion model developed based on the IMU sensor data produces a predicted state of operation. The GPS and the LIDAR units provide the correction parameters. These units acts as the Position Observation model to compare the predicted state of motion with the required state of motion. The error thus determined for the considered state is used to correct the predicted state. Therefore, this corrected information is then provided to the Motion model to update its operational state. This corrected position is then transferred to the IEMS for the powertrain regulation. When the three sensors; IMU, LIDAR and GPS is considered for sensor fusion, the error dynamics of each unit is completely different and uncorrelated from each other. The GPS provides the absolute positioning in the region where the PHEV is present. The identification of the features and terrains in the region through GPS is then supported by the IMU. The IMU provides a smoothing effect to the GPS data which may be lost or may not be pin point accurate due to signal loss or during maneuvering.

Fig.2 shows the model in which the Sensors are placed and State estimator is regulated. While the GPS-IMU units provide information regarding the motion model, the LIDAR provides information regarding the position of the PHEV with respect to the environment.

In the Error State-Extended Kalman Filter, the Estimator divides the True Driving state into a large nominal state and an error state. The nominal state is the non-linear driving model created for the vehicle to drive through linearization with linearization error present. This is overcome by the error state component. The error state is effectively used as a correction factor by the estimator to regulate the non-linear driving of the vehicle.

The motion model is the linearized considering the rotational error and the measurement noise in terms of the Error state and Error Dynamics. Error state representation is shown in equation (6) as,

$$\delta x_k = \begin{bmatrix} \delta p_k \\ \delta v_k \\ \delta q_k \end{bmatrix} \tag{6}$$

where δqk represents the rotational error and the error dynamic equation is represented as shown in equation (7), the parameters represented in the abbreviations.

$$\delta x_k = F_{k-1}\delta x_{k-1} + L_{k-1}n_{k-1} \tag{7}$$

where n_{k-1} represents the measurement noise.

The Jacobian matrices L and F are used for the error determination. The Jacobian matrix L_{k-1}, is constant. The Jacobian matrix F_{k-1} is designed considering a global orientation representation (Barfoot,2019) . This error determination is critical for ES-EKF approach as the state uncertainty has to be propagated forward for updating the predicted state to the actual state as will be discussed in the following paragraphs.

Fig.4 explains the process flow in determination of the corrected state from the initial predicted motion model.

The steps involved are

- Operation of the LIDAR, IMU and GPS units to create the motion model and the position observation model.
- The Position observation model would be based on the processing of the point clouds generated around the obstacle in the path of the ego PHEV.
- The IMU is used to develop the predicted model,

The vehicle state is represented using a unit quaternion and has the information that includes the position, velocity and orientation data represented as,

$$x_k = \begin{bmatrix} p_k \\ v_k \\ q_k \end{bmatrix} \tag{8}$$

Figure 4. Corrected state estimation process

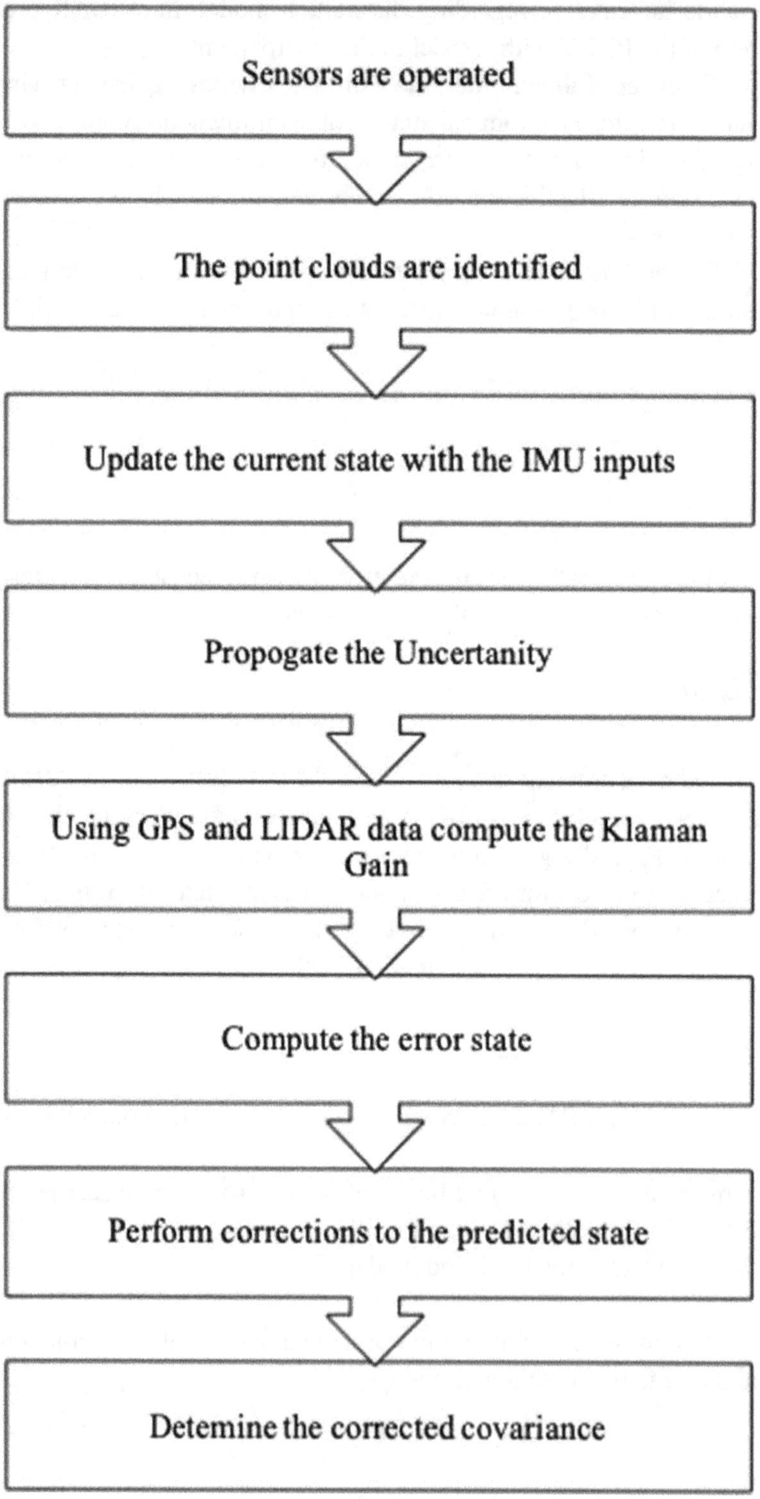

where,

$$p_k = p_{k-1} + \Delta t v_{k-1} + \frac{\Delta t^2}{2}\left(C_{ns} f_{k-1} - g\right)$$

$$v_k = v_{k-1} + \Delta t\left(C_{ns} f_{k-1} - g\right)$$

$$q_k = \Omega\left(q\left(\omega_{k-1}\Delta t\right)\right) q_{k-1})$$

and has

$$C_{ns} = C_{ns}\left(q_{k-1}\right)$$

- The predicted model has errors which is termed as the uncertainty which is propagated as the predicted state,

$$P"_k = F_{k-1} P_{k-1} F_{k-1}^T + L_{k-1} Q_{k-1} L_{k-1}^T \tag{9}$$

- The Position observation model is used to compute the Kalman gain,

$$K_k = P"_k H_k^T \left(H_k P"_k H_k^T + R\right)^{-1} \tag{10}$$

Here the measurement Jacobian matrix, H_k is constant. (Barfoot,2019).

- Upon determination of the Kalman gain the error state is also determined based on the information from the GPS and LIDAR units to rectify the predicted motion state.

$$\delta x_k = K_k\left(y_k - p_k\right) \tag{11}$$

- The Corrected state is therefore determined considering the computed error state as,

$$p'_k = p_k + \delta p_k \tag{12}$$

$$v'_k = v_k + \delta v_k$$

Figure 5. PHEV powertrain modelling

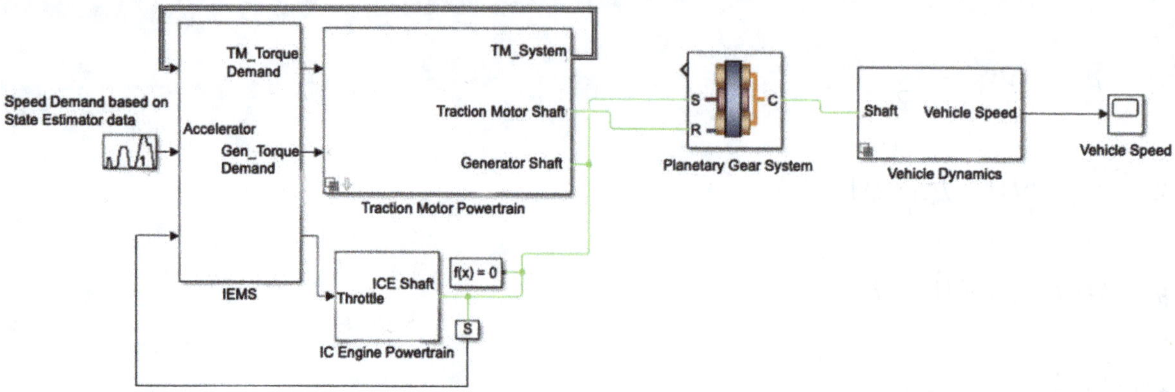

$$q'_k = \Omega\left(q\left(\delta q_k\right)\right)q_{k-1})$$

The orientation state updation is done using quaternion multiplication as shown above.

- The correction in the covariance is determined which is then fed to the Motion model to have the path rectified. This rectified motion model is then given as input to the IEMS for regulation of the powertrain.

$$P_k = \left(1 - K_k H_k\right)P''_k \tag{13}$$

IEMS Operation and PHEV Powertrain Regulation

Using the information gathered from the sensors, the vehicle speed estimated over the distance of 200 m covered is given as input to the IEMS to regulate the power train. This estimated speed and the variation in the speed during the coverage of the charted path is important for the exact performance of the PHEV. The IEMS thereby has to determine the load sharing that is to be performed between the IC Engine and the Traction motor. The parallel PHEV arrangement utilizes a Planetary gear system as the mechanical coupler as shown in Fig.5.

This gear system acts as a speed coupler thereby coupling the different speeds produced by the IC engine and the Traction Motor. The IC Engine is coupled with the Sun gear of the planetary gear system while the Traction Motor is coupled to the ring gear. The planets are in turn connected with the connecting rods which are coupled with the differential and transmission of the vehicle, depicted as the vehicle dynamics in Fig.5. The PHEV has a provision to operate in two modes which are, Hybrid mode; where the Traction motor and the IC Engine both share the load requirement; and the Engine traction & charging mode. In the engine traction & charging mode, the IC Engine is used to operate a generator which would be coupled to the sun gear through a lock mechanism. During this mode the engine would be providing power to the generator to charge the battery pack while at the same time also providing power for the motion of the vehicle. To have this mechanism implemented the IEMS follows the Maximum

Figure 6. IEMS control logic & operation mode

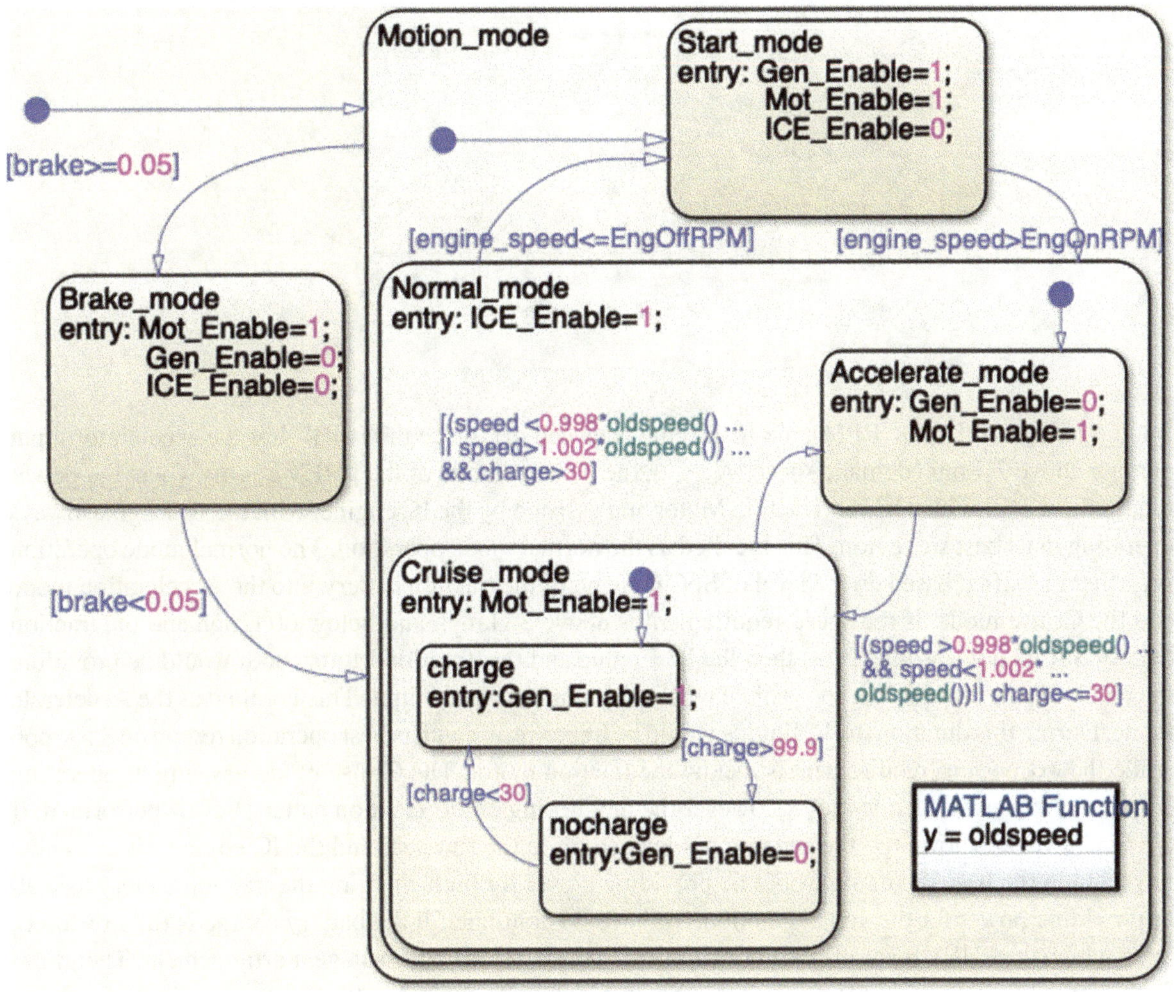

SoC of PPS approach. In this approach the system tries to charge the traction battery pack when the load requirement is less than the best operating point of the IC engine.

The operation mode is as shown in Fig.6. Based on the input from the state estimator the PHEV accelerator and brake is controlled. Initially at start the motor and the generator are enabled for operation while the IC engine would be off. This is because the PHEV would be requiring high torque initially which would be ideal for the traction motor to operate. It would be detrimental for the IC Engine to operate since its best sfc would be at a later speed level.

The PHEV motion mode starts with the initial operation would be performed by using the traction motor. The speed requirement as obtained as the input in terms of the accelerator position would be the deciding factor for regulation of the Traction motor and IC Engine throttle control. The IC Engine is found to be best operational from a speed of 8.5 kmph to 50kmph. Therefore the Engine on speed (EngOnRPM) would be when the accelerator demands a speed above 27 kmph. The IC Engine would turn on at that point and increment its speed to the best operating speed range of 8.5 kmph to 50 kmph. The moment the accelerator position shows the demand to be less than 24kmph, which is called as the

Figure 7. Actual travel path and state estimator based estimated travel path

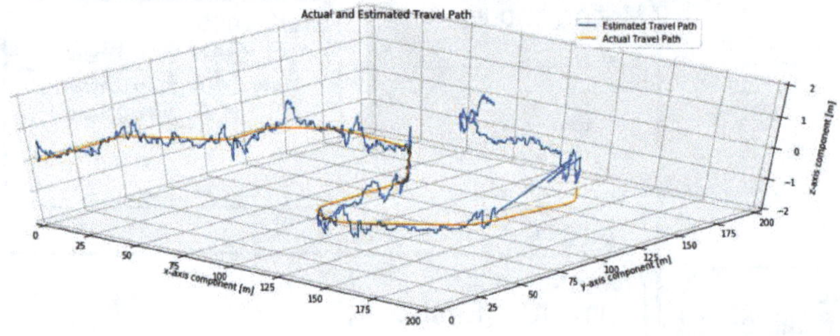

Engine off speed (EngOffRPM), the IC Engine is turned off. Once the PHEV has the accelerator input coming above 27kmph demand speed (N_{demand}), the Normal mode of the PHEV begins where the power requirement is provided by the Traction Motor and assisted by the IC engine, with the IC Engine always operating in its best sfc region. This is called as the normal mode operation. The normal mode operation is further classified based on the level of SoC demand of the traction battery into the Acceleration mode and the Cruise mode. If the speed requirement is above 27 kmph and below 60 kmph and the traction battery SoC is not less than 30%, then the IC Engine and the Traction Motor; both would be providing the required power to operate the vehicle within the required speed limit. This is called as the Accelerate mode. During this duration the IC Engine would be lingering around its best operation region on 8.5 kmph while the extra speed requirement is met by the traction motor. The Cruise mode has similar operating conditions as the Accelerate mode, however the monitoring of the Traction battery SoC is performed. If the SoC level is below 30% the generator is latched on to the sun gear and the IC engine which is now lingering in the best sfc region would be providing power for both charging the traction battery as well as providing power for the vehicle motion. This would continue till the battery charge is fully restored; upon which the PHEV turns off the generator and locks it out from the sun gear arrangement. Therefore, from then on the power requirement would be provided by both the traction motor and the IC engine.

RESULTS AND DISCUSSIONS

Sensor Fusion Output

The sensor fusion operation using the LIDAR, IMU and the GPS unit has yielded the output as shown in Fig.7. The actual travel path is shown using the orange path while the estimated path is shown in blue.

The fluctuations in the determined path can be accounted to the error that occurs during the estimation process. The predicted path is undergoing variation as per the updation occurring from the position observation model. The evaluation is done for a distance of 200 meters from the point or origin. The chartered path as per the GPS value is initially evaluated using the signal from the IMU. The IMU data however could not be dependent upon fully without the support of the position observation sensors which are the LIDAR and the GPS units.

Figure 8. Prototype PHEV

The IEMS receives the updated state values which then utilizes the data to regulate the powertrain. The powertrain is designed considering a Parallel PHEV with a 150cc single cylinder engine as the Mechanical powertrain and a 3kW BLDC motor as the Electrical powertrain component. The 3kW BLDC motor is powered from a 48V,100Ah Li ion battery pack. The PHEV is modelled in the form of an ATV as shown in Fig.8. The simulation studies are done using MATLAB environment and the analysis performed on the various PHEV aspects is shown in Fig.9 and Table I. Over the course of the journey of 200 m, the factors analyzed include the weight distribution at the front end and rear end as shown in Fig.9(a) and Fig.9(b). A maximum of 16kN has been observed at a sharp turn with the vehicle travelling at 400 rad/sec. The load sharing is based on the torque required by the vehicle which depends on the force exerted on the accelerator. From Fig.9 (c) it can be seen that the initial motion is provided by the traction motor coupled to the ring gear. The IC Engine is operational until its EngOnRPM, the IC Engine is not coupled to the planetary gear system. Once the accelerator demand goes above 27kmph the IC Engine turns on and provides the extra power required. During the IC engine operation, it ensures to be in the best sfc region, thereby obtaining maximum mileage from the PHEV.

Table I shows the various parameters observed based on the simulation study done on the PHEV model in Matlab. The evaluation of the forces acting on the various parts of the PHEV chassis is important to evaluate the ability of the PHEV to withstand the drive during the autonomous driving condition. The force acting on all the four wheels are analyzed and the impact of braking is also studied. The presence of centrifugal force during sharp turns and the lateral forces experienced during these turns are also captured in Table. I.

Figure 9. PHEV vehicle dynamic responses (a) Weight distribution-front end (b) Weight distribution-rear end (c)Planetary gear set operational responses

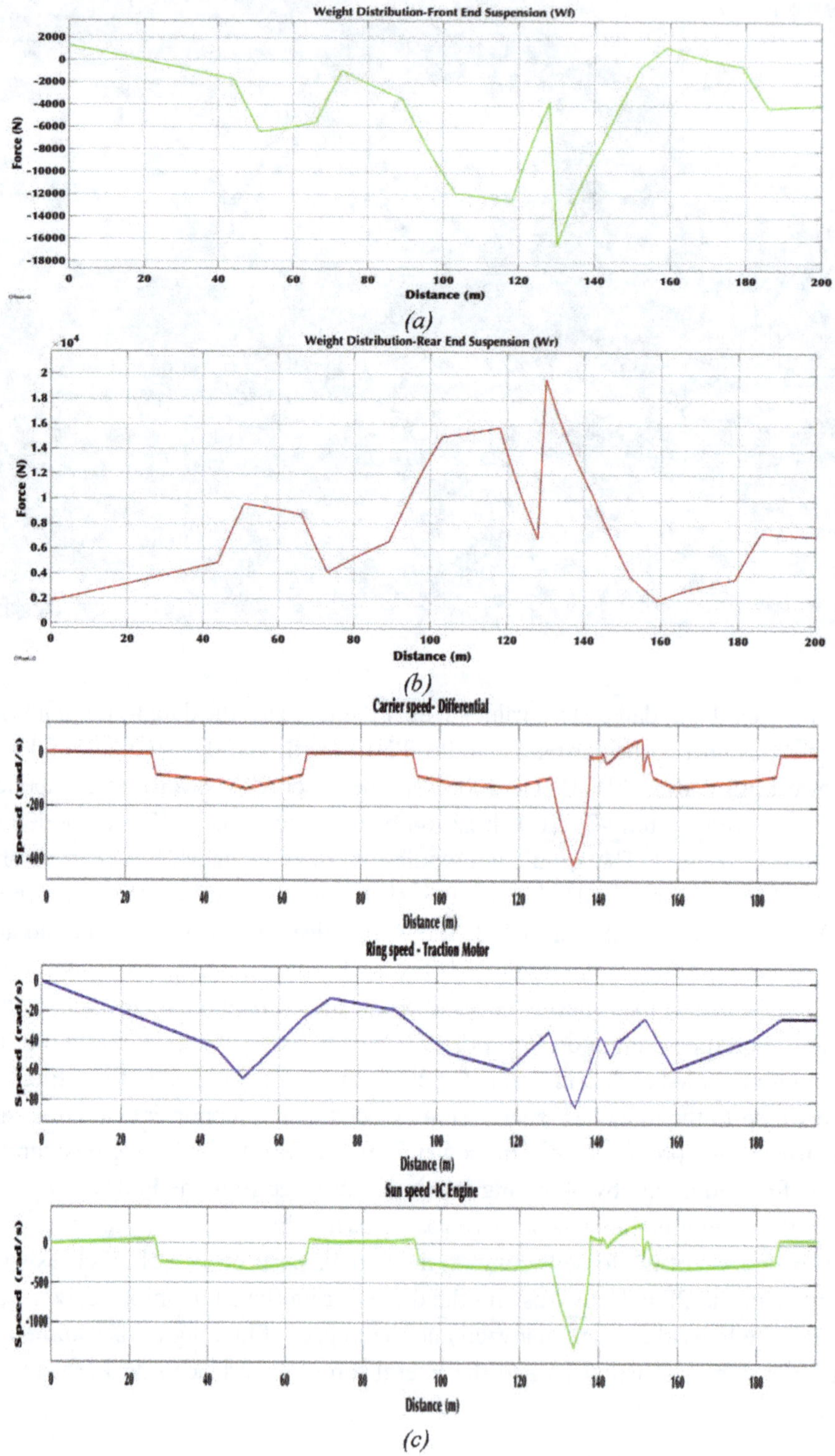

Table 1. PHEV force distribution study for the IEMS drive cycle

Vehicle Parameter	Value
Maximum deceleration observed	0.915g
Static axle load distribution	0.6
Normal Force at maximum condition front	1595 N
Normal Force at maximum condition rear	1544 N
Normal force of inner wheel front at maximum deviation	517.6 N
Normal force of outer wheel front at maximum deviation	1080 N
Normal force of inner wheel rear at maximum deviation	605.7 N
Normal force of outer wheel rear at maximum deviation	935.8 N
Maximum lateral force of outer front wheel	593.9 N
Maximum lateral force of inner front wheel	284.6 N
Maximum lateral force of outer rear wheel	800 N
Maximum lateral force of inner rear wheel	517.8 N
Rear wheels tire road friction coefficient to prevent wheel lock up	0.4779
Front wheels tire road friction coefficient. to prevent wheel lock up	0.6929

FUTURE RESEARCH DIRECTIONS

The future is for Autonomous and connected vehicles, where focus would be to save energy through regulation of vehicle dynamics and powertrain optimization. The vehicles would be driven more proactively anticipating future driving conditions and the powertrain would be operated in the most efficient region. The chapter demonstrates the potential benefits from PHEV and powertrain co-optimization by investigating an autonomous power train regulation. Further research in this area can bring about more mature solutions to implement such optimization in a real production vehicle. The sensor fusion operation through inclusion of stereophonic cameras to provide steering regulation would provide full autonomy to the PHEV. Inclusion of powerful Deep learning based approach would enhance the performance of the powertrain and steering regulation.

CONCLUSION

The regulation of an autonomous PHEV using an improved Error State Extended Kalman Filter as the State Estimator to perform sensor fusion is developed. The IMU sensor is used to create a predicted motion model which is updated with GPS and LIDAR data thereby reducing the calibration and estimation errors. The ES-EKF estimated the travel path as per the actual path proposed for the model. The estimated motion model is then provided to a PHEV that was subjected to evaluation of vehicle parameters based on the State Estimator model. The PHEV load sharing was performed by an Intelligent Energy Management system which monitors the load sharing with respect to the load demand in terms of the accelerator position. The PHEV is designed for an All-terrain vehicle equipped with a 150cc single cylinder IC Engine and a 3kW BLDC motor as its mechanical and electrical powertrains respectively.

ACKNOWLEDGMENT

We thank E-Mobility Research & Development Centre at School of Engineering & Technology, CHRIST (Deemed to be University), Bangalore, for providing facilities to develop this project.

This research was supported by the CHRIST (Deemed to be University), Bangalore, India [MNGDFE1811].

REFERENCES

NITI Aayog & World Energy Council. (2018). *Zero Emission Vehicles (ZEVs): Towards a Policy Framework*. Author.

Akram, M.A., Liu, P., Tahir, M.O., Ali, W., & Wang, Y. (2019). Article A State Optimization Model Based on Kalman Filtering and Robust Estimation Theory for Fusion of Multi-Source Information in Highly Non-linear Systems. *Sensors, 19*(1687), 1-22.

Barfoot, T. D. (2019). *State Estimation for Robotics*. Cambridge University Press.

Barrau, A. (2015). *Non-linear state error based extended Kalman filters with applications to navigation*. Automatic.

Chen, M., Jochem, T., & Pomerleau, D. (1995). AURORA: A vision-based roadway departure warning system. *Proc. IEEE/RSJ IROS*, *1*, 243–248.

Farrell, J. A., & Roysdon, P. F. (2017). Advanced Vehicle State Estimation: A Tutorial and Comparative Study. *IFAC-Papers Online*, *50*(1), 15971–15976.

Gravina, R., Alinia, P., Ghasemzadeh, H., & Fortino, G. (2017). Multi-sensor fusion in body sensor networks: State-of-the-art and research challenges. *Information Fusion*, *35*, 68–80. doi:10.1016/j.inffus.2016.09.005

Jose, C. P., & Meikandasivam, S. (2020). Optimization & Validation of Intelligent Energy Management System for Pseudo Dynamic predictive regulation of Plug-in Hybrid Electric Vehicle as Donor Clients. *Transportation*, *3*(1), 1–16.

Khan, M. Q., & Lee, S. (2019). A Comprehensive Survey of Driving Monitoring and Assistance Systems. *Sensors (Basel)*, *19*(2), 1–32. doi:10.339019112574 PMID:31174275

Li, L., Wen, D., Zheng, N., & Shen, L. (2015). Cognitive Cars: A New Frontier for ADAS Research. *IEEE Transactions on Intelligent Transportation Systems*, *13*(1), 395–407. doi:10.1109/TITS.2011.2159493

Li, Q., Chen, L., Li, M., Shaw, S., & Nuchter, A. (2014). A Sensor-Fusion Drivable-Region and Lane-Detection System for Autonomous Vehicle Navigation in Challenging Road Scenarios. *IEEE Transactions on Vehicular Technology*, *63*(2), 540–555. doi:10.1109/TVT.2013.2281199

Mukhtar, A., Xia, L., & Tang, T. B. (2015). Vehicle detection techniques for collision avoidance systems: A review. *IEEE Transactions on Intelligent Transportation Systems*, *16*(5), 2318–2338. doi:10.1109/TITS.2015.2409109

Son, H., & Kim, H. (2016). Development of Near Optimal Rule-Based Control for Plug-In Hybrid Electric Vehicles Taking into Account Drivetrain Component Losses. *Energies*, *9*(420), 1-18.

Wang, H., Lou, X., Cai, Y., Li, Y., & Chen, L. (2019). Real-Time Vehicle Detection Algorithm Based on Vision and Lidar Point Cloud Fusion. *Journal of Sensors*, *1*(7), 1–9. doi:10.1155/2019/8473980

Wang, W., Zhao, D., Xi, J., & Han, W. (2018). A learning based approach for lane departure warning systems with a personalised driver model. *IEEE Transactions on Vehicular Technology*, *1*(3), 1–13.

Wischenewski, A., Stahl, T., Betz, J., & Lohmann, B. (2019). Vehicle Dynamics State Estimation and Localization for High Performance Race Cars. *IFAC-PapersOnLine*, *52*(8), 154–161. doi:10.1016/j.ifacol.2019.08.064

World Health Organization. (2009). *Global Status Report on Road Safety: Time for action*. WHO.

Xia, Q., Chen, L., Xu, X., Cai, Y., Jiang, H., Chen, T., & Pan, G. (2016). Running States Estimation of Autonomous Four-Wheel Independent Drive Electric Vehicle by Virtual Longitudinal Force Sensors. *Mathematical Problems in Engineering*, *1*(1), 1-17.

Yang, Y., Ali, K. A., Roeleveld, J., & Emadi, A. (2016). State-of-the-art electrified powertrains - hybrid, plug-in, and electric vehicles. *International Journal of Powertrains*, *5*(1), 1. doi:10.1504/IJPT.2016.075181

Young, M. A. R., John, D. L., Kristie, Y., & John, D. L. (2019). Driver Distraction: Theory, Effects, and Mitigation. CRC Press.

Yuan, Y., Lu, Y., & Wang, Q. (2019). Adaptive Forward Vehicle Collision Warning Based on Driving Behavior. *Neurocomputing*, *408*, 64–71. doi:10.1016/j.neucom.2019.11.024

Abbreviations

Symbol	Representation
c	Speed of light (m/s)
t	Time elapsed for round trip of light rays (s)
V	Noise factor
$p_{s'}^{(j)}$	Point in the sensor frame (LIDAR)
$S_{s's}$	Scaling matrix wrt LIDAR frame
$C_{s's}$	Rotational matrix wrt LIDAR frame
$r_s^{s's}$	Translational matrix wrt LIDAR frame
$\omega s_(t)$	Angluar veocity wrt IMU sensor frame
$b_{gyro}(t)$	Bias factor for gyroscope wrt IMU sensor frame
$n_{gyro}(t)$	Noise factor for gyroscope wrt IMU sensor frame
$C_{sn}(t)$	Orientation of sensor wrt IMU sensor frame
$(r_n(t))$	Range wrt IMU sensor frame
$b_{accel}(t)$	Bias factor for accelrometer wrt IMU sensor frame
$n_{accel}(t)$	Noise factor for accelrometer wrt IMU sensor frame
p_k	Current position state
v_k	Current velocity state
q_k	Current orientation state
C_{ns}	Direction cosine matric
P_k''	State Uncertinity propogation matrix
F_k	Jacobian Matrix considering a global orientation error representation
K_k	Kalman Gain matrix
H_k	Measurement Jacobian matrix
P_k	Corrected nominal state covariance matrix
y_k	Measured vehicle position
L_{k-1}	Jacobian Matrix

APPENDIX

Chapter 4
Integration Strategies, Challenges, and Merits of Renewable Resources in Electric Vehicles

Sudhakar Hallur
KLS Gogte Institute of Technology, Belagavi, India

Roopa Kulkarni
Dayanand Sagar Academy of Technology and Management, Bengaluru, India

Prashant P. Patavardhan
RV Institute of Technology and Management, Bengaluru, India

Vishweshkumar Aithal
KLS Gogte Institute of Technology, Belagavi, India

ABSTRACT

A non-sustainable wellspring of energy is for the most part non-plentiful in nature and can be utilized until they exist in plenitude. As and when the acknowledgment of the decrease of the bountiful assets and natural risks hit, an auxiliary idea of use of non-pollutable-sustainable power sources shows up into place. Different vulnerabilities as for the plan and supplies of a vehicle emerge with less or weakened association of the sustainable assets. A few vulnerabilities will emerge from the capital venture part and extents from the general population and private segments. Another challenge is the contribution of the innovation to travel the flow ordinary vehicles to electrical vehicles (EV) and to coordinate the renewable energy sources (RES), for example, wind-vitality, sunlight-based photovoltaics, and differing assortments of bio-energies to constantly work according to the prerequisite to reduce greenhouse gas (GHG). This chapter deals with the integration and charging challenges and strategies, coordination of vehicles, demand integration, and solutions to them.

DOI: 10.4018/978-1-7998-7626-7.ch004

INTRODUCTION

Great potential embeds great energy usage. A non-renewable source of energy is generally non-abundant in nature and can be made use of them only until they exist in abundance. As and when, the realization of the decline of the abundant resources and environmental hazards hit, a secondary thought of utilization of non pollutable-renewable energy sources arrives into place.

Advancing the jolt in the vehicle division just as a power creation dependent on RES would empower the accomplishment of the objectives of utilization of renewable resources. Various challenges involve in integration of renewable energy sources and transiting a vehicle from adjusting and making complete utilization of it to perform efficiently. Some of the challenges include the intermitted power creation of RES, the comparing vitality frameworks ought to remember noteworthy stores for a type of regular vitality sources or critical limits of vitality stockpiles. Expansion in the costs of power and the general vitality framework and limits a more extensive conveyance of RES. The expanded utilization of electric vehicles (EVs) gives a huge limit of the conveyed battery chargers, being associated with the network, for making additional opportunities to incorporate RES in the force frameworks.

Inclusion and utilization of the renewable resources of energy into the Electric Vehicles (EV) may also initiate a reasonable discount in the cost, security and durability of the vehicles and the sources. Also, the current involvement of the non-renewable energy resources may be restricted only when needed saving natural resources and environment.

The chapter also deals with the technicality of various methods involved in capturing, conversion and storing the renewable and abundant energy in the batteries of the Electric Vehicle (EV) and also discusses various strategies for the minimalist utilization of the stored energy. The cost and the air pollution aspect during integration of the energy sources are addressed by inclusion and describing various strategies. The chapter also discusses various solutions with respect to the challenges and issues.

Background

The increase in the pollution has led to the greenhouse effect with the harmful emissions of various gases such as carbon monoxide, carbon dioxide and various oxides of sulphur and this has led to the increase of the global temperature by about 2oC. An efficient approach is the integration of the power generated from the renewable energy resources locally and provide it to the charging of the Electric Vehicles. (Honarmand et al., 2014a; Rahbari et al., 2017; Schuller et al., 2015; Van der Kam & van Sark, 2015) Electrification of the transportation making use of Renewable Energy Sources (RES) would lead to an efficient and pollution-free system. (Mwasilu et al., 2014) The investment in the charging of the Electrical vehicles are much inefficient than investment inclusion of the Renewable Energy resources. (Figueiredo et al., 2017; Honarmand et al., 2014b) The renewable energy resources prove to be a cost-effective solution offering quick charge while in the absence of the electricity. (Introduction to System Integration of Renewables – Analysis, 2020) Also, reverse is also possible such that extracting the charge from the batteries of the Electric Vehicles (EV) to utilize for domestic purpose. This model is known as Vehicle-to-Grid (V2G) model. (Igbinovia & Krupka, 2018) Harvesting the solar and wind energy and storing it as a reserve in the batteries of the electric vehicles serves the purpose. (Güneralp et al., 2017) Various optimizing strategies and algorithmic approaches are used in combination in the Electric vehicles. (Harnessing Renewable Energy for Sustainable Development, 2018; Karunathilake et al., 2018; Simaremare et al., 2018) Integration of Electric Vehicles with micro grids(Van der Kam & van Sark,

2015) with the utilization of RES considering the battery conditions and charging strategies along with the duration of charging. (Li et al., 2019; Li et al., 2017). Managing the charging and discharging process at the parking lot corresponds to the demand integration with the parking lot owner's profitability and losses. (Cetinkaya et al., 2020; Rahbar et al., 2018) Various uncertainties and challenges that occur in the integration of various components of which some of them are major ones. Solutions to the same improve the efficiency of the Electric Vehicles (EV).

GREENHOUSE EFFECT, INTEGRATION STRATEGIES AND CHALLENGES

Greenhouse Effect

Despite the fact that EVs are a few times more proficient than regular fuel-controlled vehicles and have no discharges at the tailpipe, the decrease in Green House Gas (GHG) outflows and generally speaking air quality advantages they bring rely upon the blend of age sources on the framework used to charge them. Various strategies on how the EVs could bring about considerable GHG emanations or even assistance broaden the life of petroleum derivatives, whenever accused fundamentally of non-renewable energy source-based generators are discussed and studied. (Puscasiu et al., 2019)

The shift from the fossil fuel-based combustion engine to the electric vehicles have significantly reduced the emissions of CO2 generated by the industries and small-duty vehicles. (Puscasiu et al., 2019) Usage of electricity increases the efficiency of the engine by about three times the efficiency of the internal combustion engine. The limitation in the amount of the CO2 emissions outweighs the amount of electricity generated. The increase in the number of Electric Vehicles (EV) increases the amount of electricity that needs to be generated for the supply. (Puscasiu et al., 2019) (Yan et al., 2019) Table 1 below shows the distribution of Green-House Gas with respect to various sectors from years 1990 to 2018.

Table 1. Distribution of Green House Gases (GHG) from the year 1990 to 2018.

Sl. No	Emission From (Sector Wise)	Type of Gas	Global Distribution (Approx.)
1	Transportation (Trucks, cars, planes, ships, etc.)	CO2, CO, CH4, N2O	28%
2	Electricity	CO2	24%
3	Industries	CO2, CH4 and N2O	20%
4	Agriculture	CH4 and N2O	16%
5	Residence and Commercial	CO2	12%

Case Study: Renewable energy integration into community systems for urban development.(Richardson, 2013)

Changing climate has to be mitigated in every aspect. Mitigating Green House Gas (GHG) emissions and adopting climate changes are the first priority.

A scenario-based and a survey-based study was conducted in certain areas to assess the emission of GHG and the Life Cycle Cost (LCC). The scenarios and results obtained were then compared with the stakeholders based on their costs and benefits. The survey-based study revealed a greater share of the renewable energy resources didn't significantly contribute for the reduction of the greenhouse gases. The other studies of various other mix provided beneficiary results of feasibility and acceptability of the Renewable Energy integration at other location. The community had developed a centralized 2MW power plant from the photovoltaics and had installed it at a household level. In another scenario, the geothermal heaters installed at household contributed to nearly about 60% of the household heating load. Also a installation of the centralized 300KW waste-to-energy (WTE) provided approximately 275-280KW of energy. Energy generators was being made aware of the investment they did in the type of the mix and the results obtained from them.

INTEGRATION

Integration of Electric Vehicles (EV) with the solar and wind energy induces a remarkable change in reduction of pollution, consuming the natural renewable resources and efficient results. The various integration methods with the renewable resources, their connectivity requirement, distribution management, etc. are discussed in the section below. The generation and utilization of various energies are as shown in the table 2.

Table 2. Summary of Source Contribution (Approx.) from various natural renewable energy resources

Sl. No	Energy Type	Energy Generation (Approx.)		Energy Consumption (Approx.)	
		1990 – 2015	Next 15 Years*	1990 – 2015	Next 15 Years*
1	Solar Energy	96.883 GW	80 - 90 GW	95 GW	75 – 80 GW
2	Wind Energy	66.98 GW	100 GW	65 GW	90 - 95 GW
3	Hydro Energy	35 GW	25 GW	32.43 GW	25 GW
4	Bio Gas Energy	9850.75 MW	6500 MW	9075.50 MW	6000 MW
5	Battery Storage	1900 GW	2500 GW	1860 GW	2410 GW

Extraction of energy from various resources can be done using various techniques and methods adopted by the respective harvester. Figure 1 depicts the energy extraction from various sources, their challenges and solutions.

Integration of Wind Energy with Electric Vehicles

Utilizing Electric Vehicles in energy systems making use of the wind energy has been noted to be obviously appropriate for the arrangement of auxiliary administrations. (Raviteja et al., 2018) Electric Vehicles should be facilitated for high-esteem administrations including auxiliary administrations to reduce the working expenses of EV proprietor in the transient time frame. (Almazrouei et al., 2018; Bampoulas &

Figure 1. Sources, Integration types, challenges and solutions.

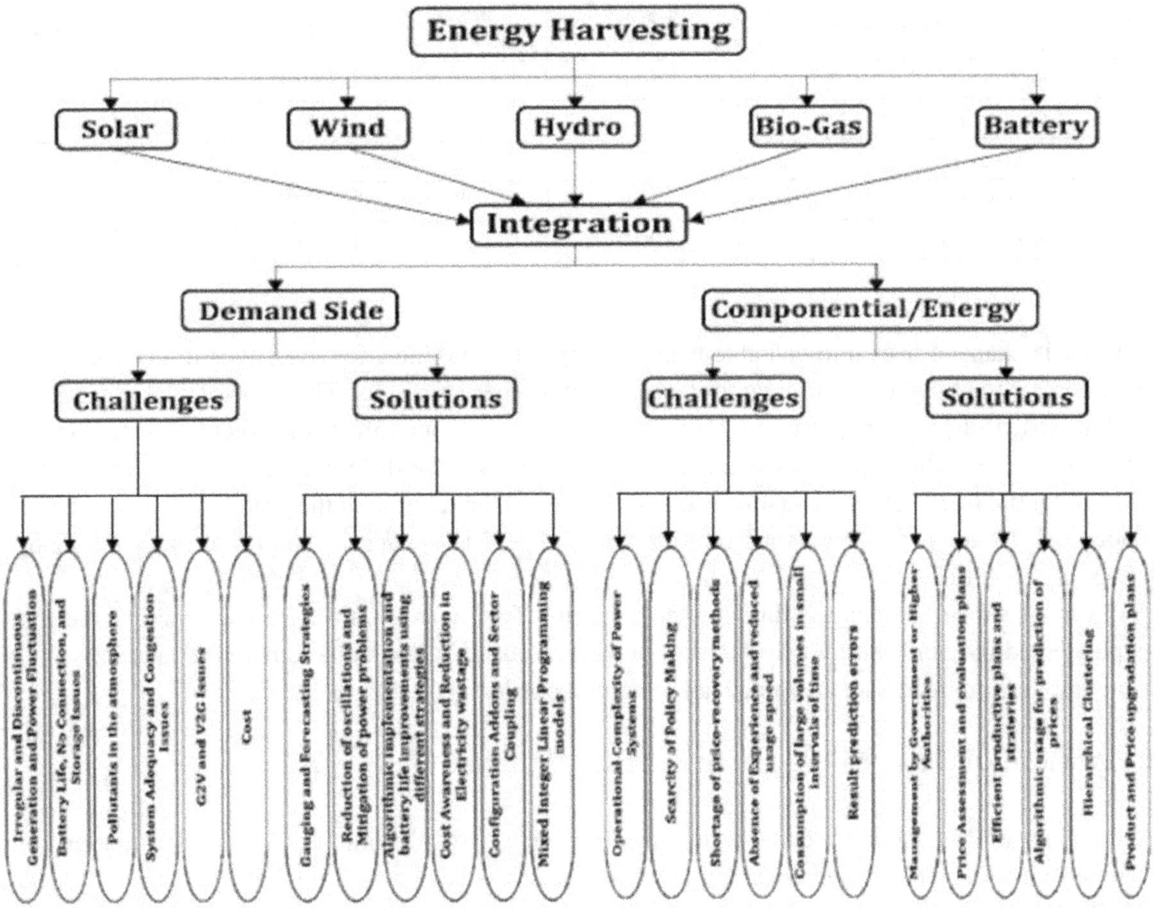

Karlis, 2017) The owners of the Electric Vehicles encounter lower cost regardless of a higher introductory worth contrasted with the ICE vehicles. The merits of auxiliary stores and load frequency control (LFC)are to be surveyed.(Vita & Koumides, 2019) EVs can proficiently control the mismatch of power resulting due to the inconstancy of wind power, thus eliminating conventional power plants. (Snehalika & Das, 2019) A lot of Electric Vehicles, can be utilized to offer auxiliary reserves and load frequency control in the power system.(Peças Lopes et al., 2010) The vehicle storage mechanisms are subjected to extensive energy extraction, that rises from void to full state of charge. (Richardson, 2013; Weigelt et al., 2018)(Patel et al., 2019) To diminish the unreasonable utilization of charge reserves and to restore the accessibility of these, it is fundamentally essential to regulate the power at regular time intervals. EV load coordination can be utilized in balancing out the frequency of the framework in micro-grid network with wind energy by utilizing a droop control technique. (Kempton & Tomic, 2005; Sensfus et al., 2008) Likewise, wind force penetration could be expanded significantly more by utilizing a planned EV load.

Case Study: Integration of Wind Energy in Africa.(Rahbar et al., 2018)

Table 3. Approximate values for Wind Energy Harnessing in Africa.

Sl. No	Countries	PF > 30% (TWh/Year)	PF> 40% (TWh/Year)	2017	2018	2019
1	South Africa	10,000	1800	105	217	364
2	Central Africa	1600	550	78	86	88
3	East Africa	31000	16500	51	68	90
4	West Africa	1700	60	62	81	375
5	North Africa	23000	7000	100	170	472

Wind is changed over into valuable energy using wind turbines, for use either to drive electrical generators or to straightforwardly control siphons and other apparatus. The hypothetical potential for wind in Africa surpasses request by significant degrees, and about 15% of the potential is described as a great resource. Table 3 presents wind energy potential over the different African divisions mulling over a Potential Factor of 30% and 40% (Potential Factor is the fraction of the year, the turbine generator is operated at rated power) into consideration. Table 3 depicts the approx. values of the Wind Energy harnessing in Africa

Given the expected manageability of wind assets in Africa, small-scale, excellent power frameworks supplemented by wind energy may forestall numerous energies lack situations in the landmass in the event that they are moderate and practically implementable.

Integration of Solar Energy with Electric Vehicles

Creating power PV at both low and medium inside the energy frameworks is possible. Furthermore, this option inspires the idea of joining the PV age with Electric Vehicles.. (Birnie, 2009). Usage of Electric Vehicles during the time invested in distributing-grid utilizing the photovoltaics is seen as an alternative for storing energy rather than controllable loads (Yadav & Yadav, 2018). As the solar radiation reaches the pinnacle during the daytime, the solar charge can be easily used to charge vehicle batteries for future utilization. During the day time, Electric Vehicles can be recharged at the parking places via a grid. Also, Electric Vehicles may be re-empowered altogether during the working duration to extract and utilize the Solar-to-Vehicle (SV2) approach (Etherden & Bollen, 2011).

Apart from the fact, that networks have high entrance limit of photovoltaic radiation, they may also have lower voltages. Under such situations, the need is to be connected to a variety of voltage sizes and feeders. In addition, similar errors may be visible especially within the time durations of high generation and low load (Birnie 2009; Yadav, & Yadav, 2018; Hashemi, & Østergaard, 2017). These occasions happen consistently and significantly on the zones that have exceptionally focused rooftop top photovoltaic radiation absorption. Various other options such as mollification of the voltage capacity such as reactive power control, (Yadav, & Yadav, 2018; Hashemi, & Østergaard, 2017; Jannesar et al., 2019; Liu, 2012; Seddig, et al., 2017) harmonized active power curtailment, (Fawzy et al., 2011) grid reinforcement (Hashemi & Østergaard, 2017) and permanent storage of energy (Klettke et al., 2018).

Case Study: Integration of Solar Energy in Africa (Rahbar et al., 2018).

Table 4. Solar Energy Harnessing in Africa.

Sl. No	Countries	2009-2014	2015	2016	2017	2018	2019
1	South Africa	1390	1066	1210	1421	1600	1750
2	Southem Africa (Excluding SA)	230	86	90	94	98	110
3	Central Africa	76	31	48	60	78	81
4	East Africa	375	276	384	462	510	600
5	West Africa	565	201	211	246	284	290
6	North Africa	715	263	291	366	400	450

Renewable energy utilization in sub-Sahara region of South Africa makes use of clean reliable and highly efficient energy sources such as solar, wind, bio-waste, thermal energy and hydrogen. The mandate for the utilization of the available natural renewable resources to the African population has been done on a large-scale entrusting state-owned utilities and funds. Though the dependency of the generation of energy on the energy resources is huge, sub-Sahara faces acute shortage of electricity. To increase the electricity generation, Africa majorly relies on the exceptional renewable resource. i.e. Solar Energy. The solar energy is excessively harnessed to generate power and is thus projected as growth market of the solar energy. Two types of solar energy may be generated. They are Solar Photovoltaic (PV) which can be used for various household and domestic utilities and Concentrated Solar Power (CSP) for higher utility scale. Indeed, the growth in the solar power harnessing is noticeable though the requirement for it dominatingly increases. The cumulative generation of solar energy resource in African countries is as shown in the Table 4 below.

Coordinating RE frameworks into the electric-power industry of the mainland will clearly help decrease the danger of a nonstop power lack in the whole landmass, just as change the mainland macroeconomic standpoint.

Connection Requirements

To guarantee appropriate co-ordination of all parts within the power system, protocols, features and specifications such as a specific network or a grid code needs to be created and followed to by all the stakeholders in the power sector. The grid or network codes spread across and envelopes various points such as association, planning, operating and market codes.(Cross & Hartshorn, 2016)

The grid or the network codes are highly particular for solar PV and wind energy plants and are technically advanced than the other class at generators. The impact of the VRE on the grid codes on the system is minimal and influential. With the increase in the generation of Variable Renewable Energy (VRE), the need to grow for VRE services such as reactive power, control of voltage(Neagu et al., 2019), frequency regulation, operating services and active power control also increases. Thus, more accurate and strict technical requirements are needed from the plants of Variable Renewable Energy that are grid-connected. (Cross & Hartshorn, 2016)

Methods to Match loads of Electric Vehicles with Renewables

Based on vehicle-type, the consumption of the electricity in the Electric vehicles vary. For lower- vehicles such as electric bike the utilization is less whereas for the higher vehicles the utilization is comparatively very much higher. The electric vehicles may be fitted with a pattern of the batteries being charged using the solar-energy in the mid-day and using the wind-energy at the night. Depending on the type of vehicle, size and its utilization, the demand of charging for some Electric Vehicles (EV) may be provided with much flexibility so as it can be shifted to adjust with the needs of the charging grid. For heavy-duty medium vehicles such as buses and trucks almost certainly, limits will be imposed on the capacity to alter the charging times since the vehicles are expected to meet the assistance commitments.(Tan et al., 2016)

Some approaches to match the Electric Vehicle (EV) charging demand with the renewable energy resource generations incorporate the methods such as empowering clients to get to renewables for EV charging through an organization charging course of action, offer renewables and empower charging during helpful occasions for the matrix grid, use time sensitive rates that coordinate sustainable power source flexibly, implement managed charging to adjust Electric Vehicle charging of the needs of the grid and couple the nearby renewables with charging framework of the Electric Vehicles (Cross & Hartshorn, 2016).

Charging Strategy Determination

The various charging strategies vary in flexibility and complexity. The different charging strategies are:

Immediate: This type of charging is a conventional charging and usually refers to the immediate EV charging when the vehicles arrive into the office in the morning or back to home in the evening hours. The shift in the charging times are not needed. The evaluation is done based on the increased or added load during the peak charging times.(Colmenar-Santos et al., 2017)

ToU Based: This type of charging considers the fixed type of tariff-based signals to deviate the charging behaviour of the customer EV and thus facilitate the integration of the electric vehicle to the grid or the power system. To improve the management of grid, requirements of the customer and the generation of electricity, the control and price signals were necessary. Thus, approach bears only a unidirectional communication. i.e. from the power system to the Electric Vehicle.(Colmenar-Santos et al., 2017)

RTP Based: Integrating the Renewable Energy Sources and limiting the power systems and reinforcing grids by charging it during durations of comparatively high generation from RES. (Colmenar-Santos et al., 2017)

Electric Network Integration with Distribution-Grid Management

The necessity for a progress to a more prudent energy framework prompts a significant change in the development, transport and energy segment. Power generation, extraction and utilization from the Renewable Energy Resources (RES) is turning out to be increasingly important. The best transition in the energy segment has resulted due to the improvement of distributed grid (DG) in a network. Distributed Grid plants exploits those primary and essential renewable energy sources distributed region-wise or territory-wise that couldn't be extracted in any case making use of a conventional unified plant. Also, a greatest advantage is that they gracefully supply and deliver the loads by working in a co-generative mode. Model for Distributed Grid (DG) boards include solar collectors and the photovoltaic (PV) panels mounted on head of structures.(Neagu et al., 2019)

A major downside of Distributed Grid is the high explicit venture cost for the most part because either they are little or medium energy plants and so the scale economy can't be applied. This can be confronted making use of a suitable incremental strategic system. The genuine issue is the trouble in anticipating and restricting the generated power and putting it on the appropriate distribution grid. So, the energy distribution making use of the renewable energy, along with the other distributed energy assets, i.e. Electrical vehicle batteries and reserves, is the fundamental drive power and motivation to move from a world of passage recharges to another world if smart grid recharging in the administration of the energy grid. (Hashemi & Østergaard, 2017)

Keen network is characterized as an advanced electric force matrix foundation that ensures the dependability of the framework and the security of flexibly, permitting to confront issues identified with the distributed power generation from Renewable Energy Resources (RES) load-control, promoting energy usage, its efficiency and effectiveness and including the detached last clients. To do so, the coordination of the electrical matrix with information and communication technology (ICT) is highly required.(Neagu et al., 2019)

Power generation, availability and accessibility is of critical significance for every human action. Thus, congruity and safety of the flexibly administration are important. Since, the energy can't be stored easily and in enormous amounts, electrical systems must ensure a steady balance among creation and utilization. This implies the power created needs to relate precisely and respond correspondingly to the any requirement at any interval of time. The electrical grid is governed in a manner to guarantee that this equalization is regarded in spite of any possible disturbances generating from the variations and fluctuations in the load to the issues determining the inaccessibility or unavailability of some of the framework components. (Kempton et al., 2001) (Hashemi & Østergaard, 2017)

In a customary organization arrangement, the generation of power occurs in a fundamentally huge, undistributed power plants. The energy generated is released on the transmission or HV network at high voltages (130–240–400 kV), and is then changed into medium voltages in the MV networks (15–20–23 kV) making use of the necessary components and inevitably it is changed over in optional substations or LV networks into Low Voltage (230–400 V) and is then distributed and redirected to the end users. Domestic customers usually get connected to the Low Voltage network and the business clients or industries are commonly associated with the MV network. High Voltage (HV) grid matrix is intended to transmit bulk energy from considerable amount of energy generators to regions of demand. It makes use of alternative pathways to deliver the energy to the huge organization structure in the event that a portion of its components are inaccessible or unavailable. In some countries, the transmission network has reached a highest automation level, prompting a decent unwavering quality deliverance and safety of supply. The transmission matrix is governed by the transmission system operator (TSO). (Neagu et al., 2019)

The ability of the distribution network to accept a specific measure of diffused energy generation is called hosting capacity. The usage of nodal hosting capacity (NHC) system to decide the amount DG that can be connected with a given network taking into consideration the performance limits of the components. Most often, the operating or performance limits are quick voltage changes, hamper, invert power stream and line warm cut-off points. The DSO manages the generator as a negative load, as it transfers the power on the grid as opposed to withdrawing it back. Thus, the distribution system operator overlooks it, since it can't control the generator during its activity. The generator can put on the force into the cross-section matrix at any moment contingent on the will of the producer or on the availability and accessibility of the energy resources. Thus, it is necessary to distinguish three primary issues identified with the actual distribution network. The issues may be as follows:

The distribution system operator is obliged to restrict distribution grid association so as to control the matrix, thus reducing the power that could be introduced through the grid. An approach to conquer this problem would permit the energy generators to team up for the grid-management.

The conduct of DG in the event of shortcomings or possibilities. On the off chance that there is an inconsistency in deliberate estimations of voltage and frequency at the association point. The distribution grid is detached regardless of whether the issue isn't identified with the distribution grid but to the transmission one, thus resulting in unexpected inaccessibility of the power generated by the grid increasing the probability of an occurrence of a dangerous situation.(Neagu et al., 2019)

Reverse power flows can happen in case the installed distribution grid grows. It implies that the power doesn't flow any longer just from the substations to the clients but the other way around, thus, increasing the risk.(Neagu et al., 2019)

The implementation of a new energy management system for the assurance and regulation frameworks involves a change to the electrical energy generation and distribution framework where the generation of power is unified and controllable while utilization is totally random, so the obligation regarding the harmony among generation of energy and utilization is completely on the production side where in some aspect of the structure is non-programmable. It can be counter balanced by a controllable part of utilization. (Kempton et al., 2001) (Hashemi & Østergaard, 2017)

Other than all the technical aspects with respect to protection, security and guidelines, so as to confront the progressions acquired by DG, the inclusion of the last client has a key job since he/she needs to get liable for controlling the load. While Electric Vehicles consume energy, they can moreover convey energy to the grid as "mobile energy" units. An extension in EV gathering means more adaptability for the system to respond to demand as well as supply. The energy suppliers offer the "vehicle-to-grid" arrangements to buyers of the vehicles. Under this circumstance, the reserves from the vehicle-to-grid organizations could charge with the yearly cost of charging an electric automotive. The owner of the auto will naturally exchange the energy with the grid at the peak and needed times so as the supplier will manage the auto's battery, except to charge with an exceptional charger at home. The energy suppliers also normally exchange the energy from the automotive vehicle's battery to the network grid at peak times when expenses are generally critical, potentially giving cash related return to the owner. The automotive vehicle's battery will be charged in the midst of off-top occasions when expenses are low. This enhancement depicts that vehicle-to grid arrangement could be used in various countries during peak hours to discharge the stored energy to the network and diminish charging prices. Changing the supply as well as demand of energy with electric cars could accomplish keeping up energy high from unnecessary ascensions to the grid, for instance, placing assets into new energy plants. Thus, transportation costs and energy bills could decrease with vehicle-to-grid arrangements. As the numbers of Electric Vehicles increase, the vehicle-to-grid arrangement need to be upheld as it helps in the levelling out the supply and the demand. This option may be especially useful in metropolitan cities where electric vehicles are used extensively for public transportation. These electric vehicles could offer capacity to the system when idle, thus reducing the costs of the city just as customers. An alerting or monitoring system should be in place to monitor the number of Electric Vehicles attached per grid. Attachment or addition of huge number of EVs to the grids could impact adequacy, efficiency, and operating costs of the grid. Thus, the impact of charging an electric vehicle is dependent upon its location on the grid network and the duration for which it is charged throughout the day.(Kempton et al., 2001) (Hashemi & Østergaard, 2017)

Electric Vehicles intend to use the assets such as sustainable energy source generation, storage reserves and demand response, to somewhat restrict the charging impacts of the Electric Vehicles. Progressed meter foundation embedded within a grid could show the merits and demerits in managing the charging and accommodation of EVs. Such devices license charging sub-stations to be aggregated with peak time rates that help off-peak charging.(Van der Kam & van Sark, 2015)

Coordination of Electrical Vehicles

Offering strengthening types of assistance from EVs is an entirely conceivable choice on the off chance that the cycle includes large fleet vehicles.(Kanchev et al., 2011) In every situation, the EV aggregator generates basic and essential signs for coordinating and maintaining the EV fleet dependent on the information shared between the generation unit, DSO and the TSO. There are different purposes for the implementation of EV aggregators. Initially, under current economic situations participation of small loads is restricted. Also, it licenses most effortless communication with the DSO for investigating. A proper methodology can help to decrease the risk of anticipated errors of the EV load. (Brenna et al., 2014; Rizvi et al., 2018)

As the entrance of the photovoltaics increments in low voltage grid, limiting the electric vehicle load can mark a progress in the operation of the feeder and a decrease to upgrade the technology and resources of the system. In the grids with very high penetration, there are various criterias to monitor. The user evaluates the modifications and addons based on the price, quality and reliability. A significant change in the voltage quality leads to the ascend in a decentralized RES generation. Also, the EV load coordination leads to the balance in between the production and consumption which has the capability to reduce over and under voltages. (Brenna et al., 2014; Rizvi et al., 2018)

Update Alternatives for More Noteworthy RES Infiltration

Various provisions and alternatives have been listed to infiltrate the Renewable Energy Sources and the reinforcement of the grid. Some of them are as follows:(Rehmani et al., 2018)

Diverting the route of the power to diminish circuit distance.
Improvement of MV or LV transformer with programmed control of voltage and the upstream transformer volume.
Improvement of the thermal capacity and voltage level of the grid circuit.
Setting up and additional circuitry for compensation of reactive power.
Installation of new substations for power distributions and modifying the architecture of the grid.

Automated power reduction is monitored in the DSOs, in cases where the RES instigated voltage variations were to surpass the permissable limits. To do so, it requires connectivity between all the units in the grid architecture, which again is a demerit to the proprietors. Extensive researches in reactive power techniques for low voltage grids, indicate limitations in the strength of diminishing voltage rise.

(Galus & Andersson, 2008; Harish & Kumar, 2014; O'Connell et al., 2014; Yang, 2017) Considering the instance of photovoltaics, most PV units associated with LV grid matrices work with unit power factor. A receptive force commitment by all the photovoltaic units result in the decrease of the voltage rise in a feeder.

An aggregator is indispensable prerequisite bridging between the network grid associated activities of a few EVs to accomplish the guidelines or norms by the TSO. The EV aggregator checks the driving requirements of the operators.(Rehmani et al., 2018) Consequently, the aggregator works by handling driver data. Considering the accommodation of all the activity profiles of the client, the aggregator delivers a "virtual force plant" in which the quantity of the autos anticipatedly reduced in addition to the accumulated power just as the degrees of energy. Besides, the assessment should be done every hour.

An alternate thought on the aggregator is that it is an elusive computational unit that screens and monitors the control of the grid. Also, it is an extremely smart interface amidst Electrical Vehicle and all of the electrical utilities. Harmonization of EV is very much easier if there is the probability of reacting to the propagated signs. Hence, the EV responds to the secondary service control and administrations as well as to technical inadequacies emerging from EV coordination processes.(Khodayar & Wu, 2015)

The capability of utilizing Renewable Energy Sources as option in contrast to customary powers, for example, petroleum derivatives, to control EVs can help lessen contamination with an extensive decarbonization impact and improve asset proficiency. Subsequently, there is a prerequisite to incorporate the two components in the best expense and compelling way.

DEMAND SIDE INTEGRATION

Demand-side participation, management and response are the various terminologies defined to describe the various initiatives from the demand-side. The demand-side integration depicts the relation between the energy sources, end-user loads and the energy-power systems. The Demand Side Integration includes significant involvement and deployment of high-end sensors, their communication systems and control devices having high energy exchanging functions. As the stress-level among the components varies, operational-complexity of the power-systems increases and thus the demand-controllability varies. To overcome which, there are very few methodologies to quantify the benefits and costs.(Khushalani et al., 2017; Pinson et al., 2008) The scarcity of policy-making in the large-scale electricity markets, the shortage of price-recovery methods in the grid-companies, the work of broad suspicions because of the absence of experience and the reduced usage need in ecological, energy and financial policies restrict the commercialization of Demand Side Integration. (Al-Shobaki & Mohsen, 2008; Siano, 2014; Strbac, 2008) Also, the anticipation in demand deteriorates due to the consumption of large volumes of electricity in small intervals of time, the unpredictability in the generation of the renewable and distributed resources, variation in the demands due to various factors on which it is dependent such as social, environmental and economic reasons. The result-prediction errors in the customer-participation levels and the forecasting models and inadequate historic electricity consumption data. (Šare et al., 2015) (Yesilbudak & Colak, 2018)

INTEGRATION CHALLENGES

Various uncertainties with respect to the design and supplies of a vehicle arise with less or deteriorated involvement of the renewable resources. Some uncertainties will arise from the capital investment part and proportions from the public and private sectors. Another issue is the involvement of the technology to transit the current conventional vehicles to Electrical Vehicles (EV) and to integrate the Renewable Energy Sources (RES) such as wind-energy, solar photovoltaics and diverse varieties of bio-energies to continually function as per the requirement. This section discusses the various challenges and strategies involved in the generation, integration and storage of the Renewable Energy Resources (RER) to the Electric Vehicles (EV).

Irregular and Discontinuous Generation

Due to irregular and discontinuous generation of electricity power from Renewable Energy Resources, the corresponding energy system frameworks should store abundant reserves as a source of conventional energy or critical limits of energy stockpiles. It prompts an expansion in the costs of electricity and the overall energy framework and thus limits an extensive conveyance of RES. (Mwasilu et al., 2014) The magnetic circuit saturation in power electronic gadgets, transformers and non-linear hardware likewise infuses harmonics into the network and results in voltage mutilation issues. (Sandy Thomas, 2009)

Battery Life

Due to varied incoming resource energy from the RES to the conversion and storage unit, an inappropriate and inefficient energy conversion happens resulting in the inefficient charge storage into the charge storage devices resulting in declining of the battery life. Depending upon the discharge depth, throughput of energy and overheating by charging or discharging by high flow of current for longer periods also reduces the life-span of the battery. (Sandy Thomas, 2009)

Pollutants in the Atmosphere

In case of wind energy harvesting to extract electricity, there are situations where the areas and territories are too much windy causing the mud and other waste materials to fly and deposit on the extraction system, leading to the reduction in the life-span of the extraction system and also in the reduction the efficiency of the system. Particularly, in the wind turbine, the turbine types, architecture and design, the wind turbulence and connection position from the site to the grid are mainly responsible for the reactive power consumption, low-power factor, interference from the electromagnetic components, power system transients and various synchronization issues. (Sandy Thomas, 2009)

In case of Solar Energy, the atmosphere may be cloudy restricting the generation of electric energy and therefore result in the reduction of the storage of the extracted energy. Also, the absorption of various chemical and harmful gases from the atmosphere into clouds resulting in rainfall causing damage to the equipment, that extracts the energy from it.

Difficulty in Reliability on Fossil Fuels

Petroleum derivative force age is as yet the prevailing power generation on the planet, among which coal is the primary fuel. Along these lines, charging EVs through the lattice clearly makes it hard to diminish the dependence on customary petroleum products except if power is given by inexhaustible sources.

Electricity Storage Issues

Compressed Air Energy storage (CAES) and Pumped Hydro Energy Storage (PHES) are site dependent and discharge times of supercapacitors and superconducting magnetic energy storages have very short operating lifetimes. (Sandy Thomas, 2009)

Replacement of Generators

The substitution of synchronous generators by PV with higher slope rates antagonistically influences dynamic and static voltage and frequency stability qualities of the framework and transient over voltage issues increment with the size of PV and the separation of Point of Interconnection from the main grid. (Vita & Koumides, 2019) Also, power and voltage instabilities arise due to the extensive loss of PV.

No connection During Swinging

Photovoltaic and wind power plants don't remain associated with the lattice or grid during voltage plunges and small deviations or disturbance in the signal little because of a group of generators in a one zone swinging against a group of generators in another zone. (Sandy Thomas, 2009)

Grid-to-Vehicle (G2V) and Vehicle-to-Grid (V2G) Issues

Thermal and transformer overloading of grid conductors by imbalance of the network, higher charging currents, loss of power, deflections and deviations in voltage while unregulated and arbitrary charging and usage of variety of charging loads due to various charging strategies and battery types during the hours of peak demand. Also, the vast majority of network incorporated battery chargers available create and add enormous measure of current harmonics to the power grid. Additionally, the number of discharge and charge cycles needed for Vehicle to Grid (V2G) are more which result in the faster draining and degradation of the battery and losses in energy conversion. (Sandy Thomas, 2009)

Case Study: Integration of the Generated energy into the Electrical Systems:

Matrix combination concerns have gone to the front lately as wind power entrance levels have expanded in various nations as an issue that may obstruct the broad organization of wind power frameworks. Two of the most grounded difficulties to wind force's future possibilities are the issues of irregularity and matrix reliability. The ordinary administration of transmission and dissemination activity is tested by power market rebuilding, security of flexibly concerns and the joining of more up to date age innovations, for example, wind and sun-oriented force. Seaward wind advancement presents a lower level of mechanical development and higher dangers than coastal breeze power developments. Internationally acknowledged guidelines for power execution, security, commotion and other climate related conditions are expected to diminish market hindrances, just as managerial and establishment costs.

Power frameworks must gracefully control in close equilibrium to request. The normal burden differs in unsurprising day by day and occasional examples, yet there is an unusual part because of irregular burden varieties and unexpected occasions. To make up for these varieties, extra age limit is expected to give guideline or put aside as stores. Generators inside an electrical framework have fluctuating working attributes: some are base-load plants; others, for example, hydro or ignition turbines, are more dexterous regarding reaction to vacillations and start-up occasions. The discontinuous profile can influence matrix unwavering quality and is another measurement that transmission and dissemination network administrators have not customarily needed to oversee on any critical scale. A few advancements of market rebuilding, for example, expanding network interconnections can improve matrix coordination and the executives of variable age sources. It can give occasions to advertise plan that diminish costs for adjusting power. While R&D advances and market learning are seen in the territories of turbines, base structures, activity and upkeep and correspondence transfers for seaward turn of events, all need further attention. It is relied upon to be High Voltage Direct Current (HVDC) innovation, which comes at expanded cost however offers extra usefulness for matrix incorporation and low line misfortunes.

Shortages in the Grid

The portability conduct of proprietors causes to sudden deficiencies in the lattice grid due to the inaccessibility of electric vehicles during top burden shaving. (Sandy Thomas, 2009)

System Adequacy

Lately, the introduction of the electric vehicles to the distributed, decarbonized and digitized power frameworks has exposed the systems to cyber-security issues and extraordinary climate occasions due to which the energy security has become a major issue. The dearth of framework adaptability has become the most prominent issue with the dissemination of the energy assets and rising Variable Renewable Energy (VRE), concurrent with an elimination of warm plants. Long haul planning of energy ought to be upheld and supported by appropriate adequacy evaluations to feature foreseen deficiencies of adaptability, and thus arises an utter need to discover market-based gadgets to address these.(Cross & Hartshorn, 2016)

Difficulty in Charging using Onsite Renewables

Multiple instantaneous and simultaneous connection of loads of charging EV makes it very difficult to scale up the coupling EV charging model for large public needs and for the customers with huge fleets. Thus, the PV panels might not be sufficient to produce the needed peak load and thus it becomes very much difficult to cater the charge to multiple EV's using onsite renewables. (Tan et al., 2016)

Fluctuating Power

Huge offshore wind farms generate huge electricity and tend to direct it to a grid of high capacity at a single location. In such cases, due to the variations in the wind speed, the power may fluctuate and may reach to extreme high values. This extremely high power may disrupt the charging grid (Peças Lopes, et al., 2010).

Congestion Issues

Uncoordinated charging of Electric Vehicles leads to a huge electric load equivalent to that of a higher power systems peak-load which in-turn results in distribution grid congestion issues (Pinson, et al., 2008).

Cost

Incorporation of new technologies and algorithms for effective utilization of the energy resources proves to be too expensive. Table 5 below depicts the Summary of Estimated Costs (Approx.) incurred for the wind energy generation.

Table 5. Summary of Estimated Costs (Approx.) incurred for the energy generation

Technology/ Infrastructure	Attribute	Wind Plants (Offshore)	Wind Plants (On Shore)
Grid Infrastructure	Grid-Connection expenses	30% of the total expenses of the offshore wind plants	14% of the total expenses of the onshore wind plants
	Upgradation expenses	4 EUR/MWh (Included in the total expenses)	5 EUR/MWh for 25% of Level of Energy Penetration
System Operation	System and Profile expenses	Integration Prices of about 25% of the level of energy penetration (About 10-15 EUR/ MWh)	Integration Prices of about 40% of the level of energy penetration (About 20-25 EUR/MWh)
	Balancing Expenses	About 4-5 EUR/MWh for Thermal and 2 EUR/MWh for Hydro Plants	About 5-6 EUR/MWh for Thermal and 2 EUR/MWh for Hydro Plants
Advanced Metering	Operation and Management Costs	Up to 225 USD/meter	Up to 250 USD/meter
Demand Response		220 USD / KW capacity of the wind plant	235-240 USD / KW capacity of the wind plant
Renewable Resource		2500 USD/month per wind plant	2500 USD/month per wind plant
Micro Grids		5 USD/Watt capacity of the wind plant	5 USD/Watt capacity of the wind plant

SOLUTIONS TO INTEGRATION ISSUES

Solutions are to be provided for various problems that occurs during the integration with the various components within the design, demand-side and renewable energy integration. Table 6 depicts the challenges and its corresponding solutions. The proposed solution for various issues may be as given in detail as below.

Solutions to Componential or Energy Integration Issues

There are various solutions to the issues occurred during the integration of the energy resources to the electric vehicle. Some of them are listed below as follows: (Sandy Thomas, 2009)

Gauging and Forecasting Strategies

Precise and intermittent gauging and forecasting strategies for load tracking and following, pre-load sharing, exchanging the energy, repair and maintenance in wind power plants and photovoltaics.

Reduction of Oscillations

Inter-area oscillations between wind power plants and photovoltaics can be compensated by implementing wide-area damping and power oscillation damping regulators.

Linear Arrangement

Arrangement in uniform lines with light hues for the improvement of aesthetic observations, acoustic protection and upgraded airfoils for decreasing clamors and radar frameworks for distinguishing migration paths of winged creatures in wind turbines.

Mitigation of Power Problems

Usage of Unified Power Flow Controller (UPFC), Static Synchronous Compensator (STATCOM) and Thyristor Controller Series Compensator (TCSC) devices and any other capacitor-reactor banks to mitigate reactive and active-power problems, frequency deviations, harmonic distortions and voltage instabilities.

Improving the Voltage Ride

Decomposition of Negative Sequence and Positive Sequence components and modification in the Grid-Side Converter (GSC) and Rotor-Side Converter (RSC) to improve the low-voltage ride-through capabilities of the wind-power and photovoltaic plants respectively.

Reduction in Electricity Wastage

Implement conversion strategies like Power-to-Hydrogen (P2H), Electricity-to-Thermal (E2T), Power-to-Gas (P2G) and Vehicle-to-Grid (V2G) to restrict electricity wastage in the curtailment of renewable energy.

Battery Life Maintenance

To maintain and restrict the life cycle of the battery within a tolerable range, the following parameters need to be taken care of and monitored. The temperature of the Battery (T) should be less than 35oC, Charge State (CS) to be less than 60% and Discharge Depth (DD) to be less than 60%.

Algorithmic Implementation

Implementation of algorithms like quadratic and linear programming, convex optimization, ant-colony, genetic and particle-swarm algorithms to optimize and enhance the unidirectional and bidirectional control strategies.

Cost-Awareness

Price and adaptability-aware electric vehicles communicating with sustainable power sources so as to lessen generation of electricity and costs of transmission, to limit charging and operational charging costs, to expand inexhaustible usage and to improve energy distribution.

Distributed Charging/Release

Addition of web empowered decentralized and versatile charging/releasing timetables so as to adjust and balance the demand/supply capability of interactions of V2G and G2V.

Configuration Addons

Adding configurations like full-bridge converter with phase-shifted ZVZCS (Zero Voltage Zero Current Switching) at the back-end DC-DC converter and active-rectifier with filter in the front-end AC-DC converter for Grid to Vehicle (G2V) operation helps to achieve galvanic isolation, DC-link voltage regulation, current harmonic elimination, primary current reset, reactive power compensation and unity power factor correction during freewheeling period. Addition of a configuration of three-phase, three-level diode clamped converter at the back end of AC-DC converter and three-phase transformer in front end AC-AC device for Vehicle to Grid (V2G) operation to accomplish a DC-connect voltage guideline, solidarity power factor revision, galvanic detachment, provides additional services and facilitates power flow in bidirectional to the grid.

System Adequacy

The System adequacy can be assessed and evaluated by exploring a hold edge in gracefully limit. (i.e. the installed supply capacity deducted by the peak load) (Cross & Hartshorn, 2016)

Sector Coupling

Sector coupling can altogether decrease essential energy request, through efficiencies and fuel exchanging, and empower adaptability in the demand-side of the power framework, while additionally supporting power sector income adequacy through electrification endeavors. Sector Coupling is a very brilliant technique of linkage between energy consumers and power sectors since the sensitivity, communication methodologies and control technologies of the devices is high. Also, the need to integrate Variable Energy Resources (VRE) the operational costs of the low-power system increases rapidly. (Cross & Hartshorn, 2016)

Available Charging Levels of Electric Vehicles

Parameters such as the time required for the complete charging of the battery of the EV, load management capability and the cost to customers may vary while designing the managed-charged utilities. Electric Vehicles (EV) can be charged using three-different levels. (Tan et al., 2016) They are

Level 1: A standard wall outlet fitted with a 120-V charger point for which the charging rate average vary in the range of 2-5 miles per hour.

Level 2: Common chargers fitted in the workplaces and households having a capability of disseminating 220-V charge which the average charging rate varying about 25 miles per hour.

DC Fast Charging: Electric Vehicles can be charged extremely fast by using DC Fast Charging which typically have the capability of charging the batteries up to 80 percent in a time duration of less than 30 seconds. Also, DC quick charging is less appropriate than different levels for management of the load onboard, provided that the average utilization of it is restricted and the duration of which ranges in between 5 and 45 minutes. Managed charging interferences could rattle transportation and fuelling prices.

Managed Charging

Designing the programs that control the timing of charging of the Electric Vehicles (EV) possibly when working at home or office can be often related the charging with the usage of the renewables due to no mobility of the vehicles, availability of energy from the renewable resources and limit the time duration. This method, also known as smart or controlled charging, allows any third party to control the time duration and voltage remotely by altering the rate of charging so as to match it closely with the needs of the grid. Wireless networking and cellular communication acts as a major interface in between the Electric Vehicle (EV), managed charger and the utility. This charging is of benefit to the utility or third party as well as customer since the charging duration of the EV can be altered with respect to the grid-needs, time-match of the generation of the Renewable Energy (RE). Also, managed charging programs can be a mixture of the level 1 and level 2 charging strategies. i.e. dropping and rising the users to the charging needs of their Electric Vehicles (EV) based on the grid-conditions rather utilizing it all the time. Implementation of Vehicle-to-Grid (V2G) strategy where a bidirectional energy flow occurs. i.e. The energy not only flows from the grid to the vehicle batteries but also flows from the batteries to grid under certain controlled situations (Examples include batteries charged with wind and Solar energy). The merit of Vehicle-to-Grid (V2G) is that the energy can not only be purchased by the system operators for transmission or other purposes but also, the EV charged batteries can themselves be used for any other additional services such as voltage or frequency control and so this technology can be used to address variability in renewable energy and make it easily accessible to the public. (Tan et al., 2016)

Renewables and Electric Vehicle Charging

Typically, EV charging is integrated with the onsite renewables such as solar energy. Pairing an EV charging with an onsite renewable provides several approaches such as placing the system at the same place with no direct control of the load so as not to influence the charging times, utilizing controlled charging system to control the load and match the renewable generation and battery storage integration into the system. Here, a level-2 charging-EV system is integrated with a software that manages the charging of the EV and align it to the output of the PV system. (Tan et al., 2016)

Mixed Integer Linear Programming Model

An optimized model of an infrastructure of smart energy is proposed where the electricity generation will be done by using the renewable resources and will be provided to the city or district taking into the consideration the number of storage batteries and PV panels to provide the charging points in the buildings and for the Electric Vehicles (EV). This type of modelling reduces the overall costs. This type of modelling facilitates the connection to the general public and also provides the electricity distribution to the grid which models the electricity load profiles and also the transportation demand of the users owning EV. (Snehalika & Das, 2019) Considering electric mobility in the grid, the release back of charged battery energy to the grid when the transportation is not needed and its utilizing it for charging of the EV's which need to travel.

Solutions to Demand Side Integration

The problems arising from the demand side integration can be resolved by providing one of the solutions below: (Sandy Thomas, 2009)

Management by Government Bodies

The foundation of the demand side integration and coordination industry and the strategem of demand side coordination portfolios by the governments.

Price Assessments

Creation price simulations and load limit value calculations for deciding and assessing incorporation expenses and advantages.

Efficient Productive Plans

Usage of long-haul limit markets, energy saving certificates and energy execution and performance contracts for framework administrator, stakeholder, load based energy productivity plans of action and efficient business plans.

Algorithm Usage

Usage of algorithms such as wavelet-transform, particle swarm optimization and genetic algorithm based on the artificial neural networks, generalized regression neural networks, integrated and auto-regressive moving average moving models, adaptive neuro-fuzzy interference systems for high demand conjecture exactness related with scheduling, regulation, dispatching and unit commitments of the power-grids. (Rizvi et al., 2018)

Hierarchical Clustering

Usage of algorithm involving different methods involving hierarchical clustering, self-organizing maps, fuzzy c-means algorithms, k-means algorithm, ant and bee-colony optimization for deriving the important information from the massive data obtained from the load and finally differentiating the types of load in the smart-grid environment.(Tan et al., 2016)

Upgradation Plans

Combinatoric stimulus-based and cost-based plans of action for upgrading load control attributes, for example, response span, response speed, advance warning or notification, area sensitivity and real utilization rate.

Providing Funds

Providing financial support by providing subsidiaries, loans and grants for various proposals applied by the agencies and companies. Facilitating various mechanisms for bidding in legislative support, exemption of taxes in investments and providing an initiative of environmental taxes for fiscal support.

Merits of Renewable Resources in Electric Vehicles

Usage of the renewable resources promotes non-pollutable and energy-conservative environment and has the following merits mentioned below:

Reduced Pollution

The harmful gas emissions such as CO2, SO2, CO and other gases from the industries and the exhaust of the vehicles to the atmosphere is reduced the usage of Electrical Vehicles making utilization of the renewable energy resources An Electrical Vehicles has null exhaust emissions. (Patel et al., 2019)

Recharge Anytime

Usage of renewable energy resources to recharge the electrical vehicles enables the customer to ease of recharging his vehicles from any grid. Also, the recharge of the electric vehicle is possible from the natural resources such as solar energy directly. Another solution for the electrical energy recharging is to purchase the electric power from the electricity retailer. Thus, this method eases the recharge of the electric vehicles and hence greenhouse gas emissions reduce.

Usage of Environment-Friendly Materials

Usage of eco-friendly production and materials such as making use of recycled materials and bio-based materials for the manufacture of the components of the Electric Vehicle components such as padding, bodywork and interior reduces the waste as well as produces a brand-new vehicle. (Liu et al., 2015)

Table 6. Issues, Challenges and Solutions for the Integration of Renewable Energy Resources to the system or grid.

Sl. No	Energy Type	Challenges	Solutions / Techniques to overcome challenges
1	Solar	Inefficient Photovoltaic Cells Pollutants in the atmosphere Active and Passive Thermal Voltage and Power stabilities rise due to loss of PV	Gauging and forecasting strategies may be used. Design of efficient Photovoltaic cells. Usage of stabilizer to stabilize the voltage generated from the solar reduces the loss of PV.
2	Wind	Variable wind energy at wind farms lead to the blade swinging in a random direction and no power is generated.	Reduction of oscillation by usage of wide area and power oscillation dampers. Linear arrangements improve aesthetic observations and acoustic protection to decrease radar frameworks.
3	Hydro	Dry out of the water source. Run length/ force exerted by the river or any other water source on the turbines.	Designing of turbines that generate maximum electricity with minimum hydro resource.
4	Bio Gas	Limited fossil-fuels. Inadequate combustion, fermentation and gasification	Implement conversion strategies like P2H, E2T, P2G, and V2G to avoid wastage.
5	Battery Storage / Power Fluctuation	Varied incoming energy generation leading to inefficient charge storage. Energy storage schemes are site dependent. Discharge times have short life cycle operated lifetimes. Power Conversion issues during V2G and G2V System inadequacy. Congestion Issues	UPFC, STATCOM and TCSC may be used to mitigate the power problems. Implementation of Grid-Side Converter (GSC) and Rotor-Side Converter (RSC) to improve the low-voltage ride-through capabilities. Maintaining and monitoring the status of various parameters of a battery to improve its life. Distributed charging/release, managed charging balances the demand-supply capability of the battery. System inadequacy can be addressed by exploring the limits of the system.

Human Health Benefits

The reduction in the harmful exhaust emissions improves the human health as the better quality of leads to better health and its costs.

Improvements in Safety

Numerous features such as lower center of gravity, fire safety, body durability and construction and of Electric Vehicles (EV) improve human safety from vehicle roll over, major fire-explosions and collisions.(Cross & Hartshorn, 2016)

Reduced Sound Pollution

The usage of electricity in the EVs produces very little sound than the Petrol or Diesel Vehicles and almost very little or no vibration at all.

Battery Usage Enhancement and Safety

Usage of electricity to store the energy in the batteries is much reliable, efficient and durable with the reduction of battery, volume and weight, and increasing their safety. The major advantage was the reduction in the cost.

FUTURE SCOPE

The usage of the renewable energy resources to charge the electric vehicles is partially a success. Though the distance covered by the charged vehicles is satisfactory, it is less efficient with respect to the rechargeable strategies and techniques. Thus, there arises a need to find an efficient method of recharging a battery such as on-the-go recharge making use of the solar and wind energy from the environment. Also, the battery protection and storage mechanisms may be improved. Implementation of the charging strategies may be developed from the reuse or non-reusable waste. Biowaste may also be considered a method to improvise the charging of the battery thus making the electric vehicle to cover more distance. Also, charging strategies such as turbo charging of the batteries may be included with which more charging in less time duration. Implementation of such ideas results in the increase of the efficiency of the electric vehicle with respect to the distance covered, charging strategies, battery reserves, utilization of reusable and renewable resources and reduction of pollution.

CONCLUSION

Investing great energies embeds great potential. Making use of the renewable resources such as solar and wind energy tackles not only the problems of usage non-renewable energy but also reduces the pollution. Various challenges such as energy conversion, storage of energy into the batteries, integration of the wind and solar energy, the necessities of the connections with the grid arising during the charge requirement to the current vehicle or to other vehicle, interconnection of vehicles via an interface such as grid to exchange the charge, load matching for effective utilization of the stored energy and battery charging. Making use of different charging strategies such as Immediate, ToU and RTP based charging increases the efficiency of the storage of the charge in the batteries. Extraction of the energy from the solar radiation making use of the photovoltaic panels is an efficient method to recharge the batteries on the go. The voltage management and conversion within the grid by the DSO and TSO to satisfy the charge requirement of the end user vehicle is being discussed. The integration strategies, challenges and the solutions with respect to the demand and the supply of the energy paved a way for a strategic business management for satisfying the user need from a business perspective. Various integration challenges and solutions were discussed. Considering the overall scenario of the utilization of the renewable resources, various merits were discussed that may pave the way for an incorporation and implementation of advanced featured vehicle which are pollution-free, energy and utilization-aware and efficient.

REFERENCES

Al-Shobaki, S., & Mohsen, M. (2008). Modeling and forecasting of electrical power demands for capacity planning. *Energy Conversion and Management*, *49*(11), 3367–3375. doi:10.1016/j.enconman.2008.05.005

Almazrouei, S., Hamid, A.-K., & Shamsuzzaman, M. (2018). Predictive energy management in large-scale grid connected PV-batteries system. *2018 5th International Conference on Renewable Energy: Generation and Applications (ICREGA)*, 315–318. 10.1109/ICREGA.2018.8337601

Bampoulas, A., & Karlis, A. (2017). Provision of frequency regulation by a residential microgrid integrating PVs, energy storage and electric vehicle. *2017 IEEE International Conference on Environment and Electrical Engineering and 2017 IEEE Industrial and Commercial Power Systems Europe (EEEIC / I&CPS Europe)*, 1–6. 10.1109/EEEIC.2017.7977472

Birnie, D. P. III. (2009). Solar-to-vehicle (S2V) systems for powering commuters of the future. *Journal of Power Sources*, *186*(2), 539–542. doi:10.1016/j.jpowsour.2008.09.118

Brenna, M., Foiadelli, F., & Longo, M. (2014). The Exploitation of Vehicle-to-Grid Function for Power Quality Improvement in a Smart Grid. *IEEE Transactions on Intelligent Transportation Systems*, *15*(5), 2169–2177. doi:10.1109/TITS.2014.2312206

Cetinkaya, H. B., Kurucay, B., Stamoulis, P., Turhan, T., Erisik, O., Ozkan, M., & Ozderli, E. (2020). A Case Study on the Use of Energy Storage in Industrial Plants with a Renewable Energy Plan. *2020 7th International Conference on Electrical and Electronics Engineering (ICEEE)*, 170–175. 10.1109/ICEEE49618.2020.9102573

Colmenar-Santos, A., Linares-Mena, A.-R., Borge-Diez, D., & Quinto-Alemany, C.-D. (2017). Impact assessment of electric vehicles on islands grids: A case study for Tenerife (Spain). *Energy*, *120*, 385–396. doi:10.1016/j.energy.2016.11.097

Cross, J. D., & Hartshorn, R. (2016). My Electric Avenue: Integrating Electric Vehicles into the Electrical Networks. *6th Hybrid and Electric Vehicles Conference (HEVC 2016)*, 1–6. 10.1049/cp.2016.0972

Etherden, N., & Bollen, M. H. J. (2011). Increasing the hosting capacity of distribution networks by curtailment of renewable energy resources. *2011 IEEE Trondheim PowerTech*, 1–7. doi:10.1109/PTC.2011.6019292

Fawzy, T., Premm, D., Bletterie, B., & Goršek, A. (2011). Active contribution of PV inverters to voltage control – from a smart grid vision to full-scale implementation. *E & I Elektrotechnik Und Informationstechnik*, *128*(4), 110–115. doi:10.100700502-011-0820-z

Figueiredo, R., Nunes, P., & Brito, M. C. (2017). The feasibility of solar parking lots for electric vehicles. *Energy*, *140*, 1182–1197. doi:10.1016/j.energy.2017.09.024

Galus, M. D., & Andersson, G. (2008). Demand Management of Grid Connected Plug-In Hybrid Electric Vehicles (PHEV). *2008 IEEE Energy 2030 Conference*, 1–8. doi:10.1109/ENERGY.2008.4781014

Güneralp, B., Lwasa, S., Masundire, H., Parnell, S., & Seto, K. C. (2017). Urbanization in Africa: Challenges and opportunities for conservation. *Environmental Research Letters, 13*(1), 1–8. doi:10.1088/1748-9326/aa94fe

Harish, V. S. K. V., & Kumar, A. (2014). Demand side management in India: Action plan, policies and regulations. *Renewable & Sustainable Energy Reviews, 33*, 613–624. doi:10.1016/j.rser.2014.02.021

Harnessing Renewable Energy for Sustainable Development. (2018). *International Centre for Trade and Sustainable Development*. https://ictsd.iisd.org/bridges-news/bridges-africa/news/harnessing-renewable-energy-for-sustainable-development

Hashemi, S., & Østergaard, J. (2017). Methods and strategies for overvoltage prevention in low voltage distribution systems with PV. *IET Renewable Power Generation, 11*(2), 205–214. doi:10.1049/iet-rpg.2016.0277

Honarmand, M., Zakariazadeh, A., & Jadid, S. (2014a). Integrated scheduling of renewable generation and electric vehicles parking lot in a smart microgrid. *Energy Convers Manag, 86*, 745-55. . doi:10.1016/j.enconman.2014.06.044

Honarmand, M., Zakariazadeh, A., & Jadid, S. (2014b). Optimal scheduling of electric vehicles in an intelligent parking lot considering vehicle-to-grid concept and battery condition. *Energy, 65*, 572–579. doi:10.1016/j.energy.2013.11.045

Hutchinson, N., & Bird, L. (2019). Using Renewables for Electric Vehicle Demand: A Review of Utility Program Designs & Implementation Strategies. In *Working Paper*. World Resources Institute. Available online at https://www.wri.org/publication/renewables-for-ev-demand

Igbinovia, F. O., & Krupka, J. (2018). Renewable Energy Integration in Africa: A Case Study of the Adoption of New Technology by the Electricity Transmission Company of Nigeria. *2018 International Conference on Power System Technology (POWERCON)*, 2034–2042. 10.1109/POWERCON.2018.8602197

Introduction to System Integration of Renewables – Analysis. (2020). *IEA*. https://www.iea.org/reports/introduction-to-system-integration-of-renewables

Jannesar, M. R., Sedighi, A., Savaghebi, M., & Guerrero, J. M. (2018). Optimal placement, sizing, and daily charge/discharge of battery energy storage in low voltage distribution network with high photovoltaic penetration. *Applied Energy, 226*, 957–966. doi:10.1016/j.apenergy.2018.06.036

Kanchev, H., Lu, D., Colas, F., Lazarov, V., & Francois, B. (2011). Energy Management and Operational Planning of a Microgrid With a PV-Based Active Generator for Smart Grid Applications. *IEEE Transactions on Industrial Electronics, 58*(10), 4583–4592. doi:10.1109/TIE.2011.2119451

Karunathilake, H., Perera, P., Ruparathna, R., Hewage, K., & Sadiq, R. (2018). Renewable energy integration into community energy systems: A case study of new urban residential development. *Journal of Cleaner Production, 173*, 292–307. doi:10.1016/j.jclepro.2016.10.067

Kempton, W., & Tomic, J. (2005). Vehicle to grid fundamentals: Calculating capacity and net revenue. *Journal of Power Sources, 144*(1), 268–279. doi:10.1016/j.jpowsour.2004.12.025

Kempton, W., Tomic, J., Letendre, S. E., Brooks, A., & Lipman, T. (2001). *Vehicle to grid power: Battery, hybrid, and fuel cell vehicles as resources for distributed electric power in California*. Working Paper Series ECD-ITS-RR-01-03.

Khodayar, M. E., & Wu, H. (2015). Demand Forecasting in the Smart Grid Paradigm: Features and Challenges. *The Electricity Journal*, *28*(6), 51–62. doi:10.1016/j.tej.2015.06.001

Khushalani, B., Alluri, P., Kasani, V. S., & Solanki, J. (2017). Coordinated charging strategies for plug-in hybrid electric vehicles. *2017 North American Power Symposium (NAPS)*, 1–5. 10.1109/NAPS.2017.8107264

Klettke, Moser, Bossmann, Barberi, & Fournié. (2018). *Effect of Electromobility on the Power System and the Integration of the RES-S13 Report*. European Union, METIS Studies.

Li, Y., Chi, Y., Wang, Z., Tang, H., Li, G., Sun, S., & Cheng, Y. (2019). Study on Fault Control Requirement and Strategy of Large-Scale Renewable Energy VSC-HVDC Integration. *2019 IEEE 3rd Conference on Energy Internet and Energy System Integration (EI2)*, 1438–1441. 10.1109/EI247390.2019.9062065

Li, Y., Tian, X., Liu, C., Su, Y., Li, L., Zhang, L., Sun, Y., & Li, J. (2017). Study on voltage control in distribution network with renewable energy integration. *2017 IEEE Conference on Energy Internet and Energy System Integration (EI2)*, 1–5. 10.1109/EI2.2017.8245755

Liu, L., Kong, F., Liu, X., Peng, Y., & Wang, Q. (2015). A review on electric vehicles interacting with renewable energy in smart grid. *Renewable & Sustainable Energy Reviews*, *51*, 648–661. doi:10.1016/j.rser.2015.06.036

Liu, X., Aichhorn, A., Liu, L., & Li, H. (2012). Coordinated control of distributed energy storage system with tap changer transformers for voltage rise mitigation under high photovoltaic penetration. *IEEE Transactions on Smart Grid, 3*(2), 897-906. DOI: .2177501 doi:10.1109/TSG.2011

Mwasilu, F., Justo, J. J., Kim, E.-K., Do, T. D., & Jung, J.-W. (2014). Electric vehicles and smart grid interaction: A review on vehicle to grid and renewable energy sources integration. *Renewable & Sustainable Energy Reviews*, *34*, 501–516. doi:10.1016/j.rser.2014.03.031

Neagu, B. C., Gavrilas, M., Grigoras, G., & Ivanov, O. (2019). Voltage Control in Microgrids in the Presence of Small-Scale Renewable Energy Source. *2019 International Conference on Electromechanical and Energy Systems (SIELMEN)*, 1–4. 10.1109/SIELMEN.2019.8905811

O'Connell, N., Pinson, P., Madsen, H., & O'Malley, M. (2014). Benefits and challenges of electrical demand response: A critical review. *Renewable and Sustainable Energy Reviews, 39*, 686–699. doi:10.1016/j.rser.2014.07.098

Patel, K., Das, N., & Khan, M. M. K. (2019). Optimization of Hybrid Solar, Wind and Diesel Energy Systems from Economic Point of View. *2019 29th Australasian Universities Power Engineering Conference (AUPEC)*, 1–6. 10.1109/AUPEC48547.2019.211861

Peças Lopes, J. A., Polenz, S. A., Moreira, C. L., & Cherkaoui, R. (2010). Identification of control and management strategies for LV unbalanced microgrids with plugged-in electric vehicles. *Electric Power Systems Research, 80*(8), 898–906. doi:10.1016/j.epsr.2009.12.013

Pinson, P., Christensen, L.E.A., Madsen, H., Sørensen, P.E., Donovan, M.H., & Jensen, L.E. (2008). Regime switching modelling of the fluctuations of offshore wind generation. *Journal of Wind Engineering & Industrial Aerodynamics, 96*(2), 2327-2347. DOI: .03.010 doi:10.1016/j.jweia.2008

Puscasiu, A., Fanca, A., Gota, D.-I., & Valean, H. (2019). Dynamic Routing of Traffic Based on Greenhouse Gases Measurements. *2019 23rd International Conference on System Theory, Control and Computing (ICSTCC),* 549–553. 10.1109/ICSTCC.2019.8886113

Rahbar, K., Chai, C. C., & Zhang, R. (2018). Energy Cooperation Optimization in Microgrids With Renewable Energy Integration. *IEEE Transactions on Smart Grid, 9*(2), 1482–1493. doi:10.1109/TSG.2016.2600863

Rahbari, O., Vafaeipour, M., Omar, N., Rosen, M. A., Hegazy, O., Timmermans, J.-M., Heibati, S., & Bossche, P. V. D. (2017). An optimal versatile control approach for plug-in electric vehicles to integrate renewable energy sources and smart grids. *Energy, 134*, 1053–1067. doi:10.1016/j.energy.2017.06.007

Raviteja, K., Kar, P. K., & Karanki, S. B. (2018). Renewable Energy Resources Integration To Grid With Improved Power Quality Capabilities And Optimal Power Flows. *2018 IEEE International Conference on Power Electronics, Drives and Energy Systems (PEDES)*, 1–6. 10.1109/PEDES.2018.8707646

Rehmani, M. H., Reisslein, M., Rachedi, A., Erol-Kantarci, M., & Radenkovic, M. (2018). Integrating Renewable Energy Resources Into the Smart Grid: Recent Developments in Information and Communication Technologies. *IEEE Transactions on Industrial Informatics, 14*(7), 2814–2825. doi:10.1109/TII.2018.2819169

Richardson, D. B. (2013). Electric vehicles and the electric grid: A review of modeling approaches, Impacts, and renewable energy integration. *Renewable & Sustainable Energy Reviews, 19*, 247–254. doi:10.1016/j.rser.2012.11.042

Rizvi, S. A. A., Xin, A., Masood, A., Iqbal, S., Jan, M. U., & Rehman, H. (2018). Electric Vehicles and their Impacts on Integration into Power Grid: A Review. *2018 2nd IEEE Conference on Energy Internet and Energy System Integration (EI2),* 1–6. 10.1109/EI2.2018.8582069

Sandy Thomas, C. E. (2009). Transportation options in a carbon-constrained world: Hybrids, plug-in hybrids, biofuels, fuel cell electric vehicles, and battery electric vehicles. *International Journal of Hydrogen Energy, 34*(23), 9279–9296. doi:10.1016/j.ijhydene.2009.09.058

Šare, A., Krajačić, G., Pukšec, T., & Duić, N. (2015). The integration of renewable energy sources and electric vehicles into the power system of the Dubrovnik region. *Energy, Sustainability and Society, 5*(1), 27–34. doi:10.118613705-015-0055-7

Schuller, A., Flath, C. M., & Gottwalt, S. (2015). Quantifying load flexibility of electric vehicles for renewable energy integration. *Applied Energy, 151*, 335–344. doi:10.1016/j.apenergy.2015.04.004

Seddig, K., Jochem, P., & Fichtner, W. (2017). Integrating renewable energy sources by electric vehicle fleets under uncertainty. *Energy, 141*, 2145–2153. doi:10.1016/j.energy.2017.11.140

Sensfus, F., Ragwitz, M., & Genoese, M. (2008). The merit-order effect: A detailed analysis of the price effect of renewable electricity generation on spot market prices in Germany. *Energy Policy, 36*(8), 3086–3094. doi:10.1016/j.enpol.2008.03.035

Siano, P. (2014). Demand response and smart grids—A survey. *Renewable & Sustainable Energy Reviews, 30*, 461–478. doi:10.1016/j.rser.2013.10.022

Simaremare, A. A., Tambunan, H. B., Pramana, P. A. A., & Munir, B. S. (2018). Optimization of Renewable Energy Penetration for Microgrid System – Case Study of Tomia Island. *2018 3rd International Conference on Information Technology, Information System and Electrical Engineering (ICITISEE),* 180–183. 10.1109/ICITISEE.2018.8720965

Snehalika, & Das, S. (2019). Emerging trends in Electric Vehicle in Indian market. *2019 3rd International Conference on Recent Developments in Control, Automation & Power Engineering (RDCAPE),* 514–518. doi:10.1109/RDCAPE47089.2019.8979012

Strbac, G. (2008). Demand side management: Benefits and challenges. *Energy Policy, 36*(12), 4419–4426. doi:10.1016/j.enpol.2008.09.030

Tan, K. M., Ramachandaramurthy, V. K., & Yong, J. Y. (2016). Integration of electric vehicles in smart grid: A review on vehicle to grid technologies and optimization techniques. *Renewable & Sustainable Energy Reviews, 53*, 720–732. doi:10.1016/j.rser.2015.09.012

Van der Kam, M., & van Sark, W. (2015). Smart charging of electric vehicles with photovoltaic power and vehicle-to-grid technology in a microgrid; a case study. *Applied Energy, 152*, 20–30. doi:10.1016/j.apenergy.2015.04.092

Vita, V., & Koumides, P. (2019). Electric Vehicles and Distribution Networks: Analysis on Vehicle to Grid and Renewable Energy Sources Integration. *2019 11th Electrical Engineering Faculty Conference (BulEF)*, 1–4. 10.1109/BulEF48056.2019.9030787

Weigelt, M., Mayr, A., Masuch, M., Batz, K., Franke, J., Bican, P. M., Brem, A., Russer, J., & Russer, P. (2018). Techno-Economic Evaluation of Strategic Solutions to Extend the Range of Electric Vehicles. *2018 8th International Electric Drives Production Conference (EDPC),* 1–7. 10.1109/EDPC.2018.8658329

Yadav, T., & Yadav, D. (2018). Integration of Electric Vehicle and Renewable Generation to Improve the Power Quality Problems in Distribution System. *2018 5th IEEE Uttar Pradesh Section International Conference on Electrical, Electronics and Computer Engineering (UPCON),* 1–5. 10.1109/UPCON.2018.8596941

Yan, L., Yongning, C., Xinshou, T., Zhankui, Z., & Haoyong, C. (2019). The Experiences and Practices on Market with Large-Scale Renewable Energy Grid Integration. *2019 IEEE Sustainable Power and Energy Conference (ISPEC)*, 68–71. 10.1109/iSPEC48194.2019.8975174

Yang, C.-J. (2017). Opportunities and barriers to demand response in China. *Resources, Conservation and Recycling, 121*, 51–55. doi:10.1016/j.resconrec.2015.11.015

Yesilbudak, M., & Colak, A. (2018). Integration Challenges and Solutions for Renewable Energy Sources, Electric Vehicles and Demand-Side Initiatives in Smart Grids. *2018 7th International Conference on Renewable Energy Research and Applications (ICRERA)*, 1407–1412. 10.1109/ICRERA.2018.8567004

KEY TERMS AND DEFINITIONS

Algorithm: A method of properly defining a process according to which the execution occurs.

Electric Vehicles: Vehicles that run with the help of electrically stored energy.

Greenhouse Gas: Gases that increase the global temperature such as carbon dioxide, carbon monoxide, sulphur gases, etc.

Grid: A single point of contact where all the recipients gather to share a same resource.

Photovoltaics: Generation of electricity or voltage making use of solar-radiation which is a result of the electron liberation from a lower level to a higher level from the manufacture material.

Regulation: Moderating any parameter or bringing it to a predefined level.

Renewable Energy Resource: An abundant component making use of which the energy can be repeatedly generated.

Chapter 5
Mathematical Modeling of DC–DC Converters and Li Ion Battery Using MATLAB/Simulink

Sumukh Surya
KPIT, Bangalore, India

ABSTRACT

In the present work, three different methods for obtaining the DC response for modeling practical DC-DC buck and boost converters operating in continuous conduction mode (CCM) are demonstrated using MATLAB/Simulink. The method of selection for inductor, diode, and MOSFET for a DC-DC converter is discussed in detail. The governing equations for the non-ideal converters were derived using volt-sec and amp-sec balance equations. Mathematical modeling of basic converters was carried out using 'commonly used blocks' reducing the dependence on SimPower System tool box in Simulink. The non-ideal parameters in the converter caused a drastic variation in the duty cycle and output voltage which in turn had an adverse effect on the efficiency. The transients in output voltages and inductor currents were observed. In addition, a Li ion polymer battery was mathematically modeled. Accurate battery modeling for pulse charging was proposed. A comparative analysis of 1, 2, ..., 5 RC pair/s modeling of the battery was presented.

INTRODUCTION

A converter is a device used to transform electrical power from one form to another. Converters can be of various types, such as AC to AC, AC to DC, DC to DC and DC to AC. Among DC-DC converters, switched mode converters and linear regulators are more commonly used. Since the latter has the increased I^2R losses, the former is preferred. Among switched mode converters, several types, categorized as isolated and non-isolated can be used as LED-drivers. These converters are differentiated based on voltage levels and isolation at the input. Buck, Boost, Buck-Boost, CUK and SEPIC are some of the non-isolated / Transformerless converters. Isolated converters are further classified as Unidirectional (e.g., flyback and forward) and Bi-directional (e.g., Push-pull, full bridge and half-bridge) Isolated converters possess

DOI: 10.4018/978-1-7998-7626-7.ch005

a transformer at the input side, which offers the advantage of isolation against short circuits and large surges. Hence, the load which is downstream of the transformer would be protected. These transformers operate at high frequencies (typically kHz) and hence the design of such transformers is a challenge. The flyback and forward converters are the isolated versions of buck-boost and buck converters.

The methodology for obtaining the average current control for DC-DC converters in ideal case operating in CCM is shown using State Space Averaging Technique. It was observed from the frequency response of the transfer functions that ideal buck and boost converters showed high stability. The analysis was carried out using MATLAB software and the code is presented. The basics of peak current mode (PCM) control and its advantages are discussed. The importance of estimating core temperature (Tc) in a Battery Management System (BMS) is shown.

DC-DC converters play a vital role in the design of Electric Vehicle (EV) charging systems, LED drivers and power supply units for critical loads like micro-processors. The converters show different behaviors during ideal and non-ideal scenarios.

The analyses for variety of DC –DC converters have been performed in the recent years. In (Canalli, 1996), a generalized approach for obtaining the large signal averaged model for converters is proposed. The behavioral model proposed provides results independent of the operation of the converter (CCM / DCM). The paper provides the modeling using LTSpice and experimental validations for CCM and DCM operations of buck, boost and SEPIC converters. The factor used to differentiate the operations was duty cycle, μ. μ for DCM was higher than that of CCM.

The behavior of the AC response to the converters can be modeled using various methods viz,. (a) State Space Averaging (b) Small signal model and (c) Circuit Averaging Techniques.

In state space averaging technique, the volt sec and amp sec balance equations for inductors and capacitors are used to obtain the open loop transfer function for the converters. In (Moussa & Morris, 1990), the necessary equations required for the modeling are provided for second order converters. The advantage is that the DC response in addition to small signal modeling equations can be obtained.

The small signal model provides the AC equivalent circuit in which the non-linear equations are converted into linear equations. The volt-sec and amp-sec balance equations for the converter are derived and are perturbed. Later, the system is linearized by re-arranging the terms and the transfer functions of output voltage to the duty ratio (G_{vd}) is derived.

The circuit averaging of any converter involves three major steps viz,. (a) Separate the switch network from the converter and define the ports (b) Sketch the waveform of the switch current and voltage waveforms and average it and (c) Simplification of the equations and draw the equivalent switch network. In (Cuk & Middlebrook, 1979), a generalized method for modeling any three-state (Input voltage, output voltage & duty cycle) switching converter operating in the DCM was presented and the experimental validation for a Buck Boost converter as carried out. This also provided the design of closed loop transfer functions using P, PI, PD and PID controllers for constant voltage, constant current, peak current and average current control techniques for CCM and DCM operations

In reality, the converters possess ESR (Equivalent Series Resistance) for the inductor and capacitor. This causes increase in the conduction losses. To achieve desired voltage at load, increased duty ratio is essential. In (Erickson & Dragon, 2007), the authors showed that the switching losses are much higher than that of the conduction. In (Czarkowski & Kazimierczuk, 1993), Bilinear large-signal, linear, dc, and small-signal circuit models of the PWM buck converter for a buck converter operating in CCM involving switching losses were shown.

In (Bertoldi, 2018), a SEPIC operating in DCM was selected to drive a LED for constant voltage application. An average and a switching model was developed, modeled in MATLAB / Simulink and validated against the experimental results. The transfer functions G_{vd} and G_{vg} (Output voltage to input voltage) were derived. It was shown that the SEPIC provided lower input current harmonics.

In (Eng & Bunlaksananusorn, 2009), SEPIC modeling for DCM operation was carried out using State Space averaging technique using MATLAB and LTSpice simulation tools. It was shown that the Bode plots obtained from the tools closely matched at frequencies below 10 kHz. At higher frequencies, the LTSpice plots diverged from the other as it used reduced order matrix.

In (Lee & Lai, 2007), an averaged model in LTSpice based on converters like ideal Buck and Boost operating in CCM was constructed using CCM1 block available under average.lib

In (Niculescu & Iancu, 2000), a mathematical model for Cuk converter operating in CCM was derived and modeling using Simulink. The importance of step size while capturing the transients was shown. The closed loop operation was carried out using a PI controller.

In (Hart, 2011), a generic state space modeling approach for fourth order converter using MATLAB / Simulink was shown for four topologies viz,.: Non-isolated Cuk, SEPIC and Zeta PWM converters and the Ripple-free input-current PWM boost converter.

In (Surya et al., n.d.), the steady state equivalent circuits and operations of ideal Buck, Boost, Buck Boost, Interleaved Boost, Cuk & SEPIC were discussed in detail.

In (Plett, 2015), the frequency response of G_{vd} (Duty Cycle to Output Voltage) for fourth order converters was plotted using LTSpice software tool. It was assumed that the non –ideal converters operated in DCM. G_{vd} for converters showed high stability in the open loop operation.

Electric Vehicles (EV's) and Hybrid Electric Vehicles (HEV's) are considered more superior than the conventional vehicles. EV's majorly use Lithium (Li) ion batteries to drive them. Some of the important Energy Storage Devices include (a) Battery (b) Fuel Cell (c) Ultra Capacitor and (d) Pumped Hydro Storage

A battery is a device that converts chemical energy into electrical energy. Li ion batteries are preferred as they possess high specific density and energy density. Different types of batteries are (a) primary (irreversible chemical reaction) and (b) secondary (reversible chemical reaction). Alkaline battery, Zinc battery are examples of primary battery. Nickel Cadmium and Lithium ion batteries are examples of secondary battery (Williamson, 2013). One of the major disadvantages of Li ion batteries is that they are highly sensitive to temperature. Hence, they must be operated under controlled conditions to achieve best performance and improved life (Surya et al., 2020).

A fuel cell (FC) is an electrochemical cell that converts the chemical energy of a fuel (often hydrogen) and an oxidizing agent (often oxygen) into electricity through a pair of redox reactions. Different types of fuel cells are (a) Alkaline FC (b) Phosphoric acid FC (c) Polymer electrolyte membrane FC (d) Direct methanol FC etc,. Fuel cells are inherently modular and therefore lend themselves to a wide range of applications, from large stationary power plants to small portable power packs.

Ultra capacitors possess two metal plates coated with a soft material which is porous and is known as Activated Carbon. The plates are immersed in liquid containing positive and negative charged particles. They are also known as EDLC (Electric Double-Layer Capacitor). The energy density of UC is much higher than the electrolytic capacitor, No chemical reactions take place in a UC and hence they can run efficiently for many cycles and hence are used in HEV's. Some of disadvantages of UC are (a) low voltage per individual cell (b) Self-discharge is high in comparison with battery (d) Issues during balancing when more than three UC's are connected in series.

In Pumped Hydro Storage, behind dams water is stored in reservoirs and dispatched when required to generate electricity. Pumped hydro storage preserves fossil fuel resources. It also significantly reduces the greenhouse gases. Some of the disadvantages are geographic restrictions prevent the installation of pumped hydro storage. Initial investment is high and construction time is long.

Compared to all energy storage devices, batteries are considered superior, as they are portable and have low self-discharge rates. In order to have protection against overcharge or over temperature, a Battery Management System (BMS) is mandatory.

Using the battery model, polynomial regression model and the Gaussian process (GP) models were proposed to compute SOC for higher accuracy (Surya et al., 2020) . It was shown that GP regression model generally performs much better than the polynomial regression model. The SOC computation plays a critical role in deciding the range of the EV.

A virtual sensor for the computation of SOC was used for a small EV with two passenger capacity (Xi et al., 2019). For a battery capacity of 150Ah and OCV of 72V, SOC was computed using experimental data of vehicle speed, pedal position and acceleration and battery bank voltage. Thus, obtained SOC was compared with that of the 2RC pair battery model. The prediction was 91.8% accurate.

In this paper, practical Buck and Boost converters were modeled in three different methods using MATLAB/ Simulink and its DC response was analyzed. The design and procedure for selection of every component in the DC-DC converter are discussed in detail. In addition, n RC pair battery modeling for pulse charging was carried out and the results are compared.

SPECIFICATIONS OF BUCK AND BOOST CONVERTERS

Table 1. Buck Converter

Parameter	Value
Input Voltage, V_g	10V
Output Voltage, V_0	5V
Output Current, I_0	2A
Current Ripple, $\Delta i_L/I_L$	15%
Voltage Ripple, $\Delta V_0/V_0$	5%
Switching Frequency	20kHz

DESIGN OF PRACTICAL CONVERTERS

The following steps were carried out to design practical DC-DC converters.

1. Design inductor and capacitor values for ideal scenario (Surya et al., n.d.)
2. Choose a proper inductor value from the datasheet provided by the manufacturer
3. Select the non-ideal parameters for diode (Forward Voltage and Dynamic Resistance) and MOSFET (Switch resistance) based on reverse voltage and drain to source voltage respectively

Table 2. Boost Converter

Parameter	Value
Input Voltage, V_g	5V
Output Voltage, V_0	10V
Output Current, I_0	1A
Current Ripple, $\Delta i_L/i_L$	15%
Voltage Ripple, $\Delta V_0/V_0$	5%
Switching Frequency	20kHz

Figure 1. Buck Converter

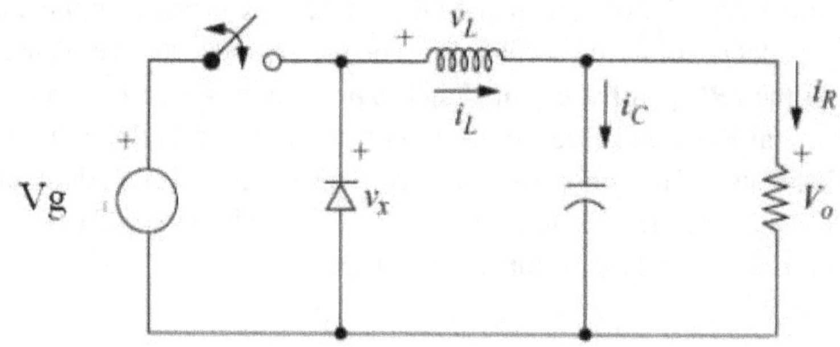

4. Derive the equation for the ratio of output voltage to input voltage considering the non-idealities
5. Re-calculate the value of duty cycle, D based on values obtained in steps (2) and (3)

MODELING BUCK CONVERTER

A schematic of Buck converter is shown in Figure 1. Figs. 2 and 3 show the equivalent circuits when the switches are closed and opened respectively. The diode gets reverse biased during switch ON condition. The inductor current i_L flows through the inductor L_1 and charges it. In the next cycle, the inductor

Figure 2. Switch Closed

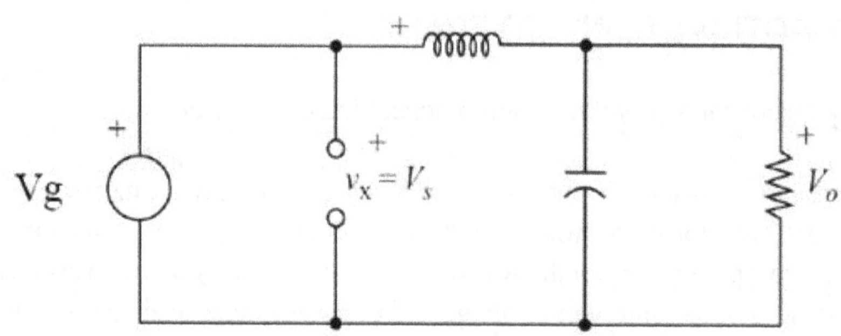

Figure 3. Switch Opened

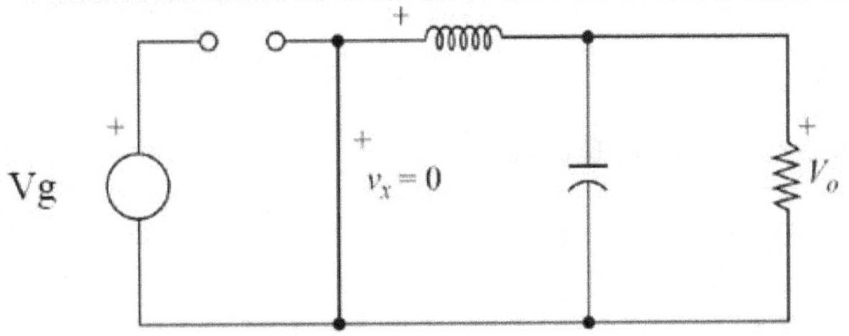

polarity gets reversed as the switch is turned OFF. The diode gets forward biased. The equations for different voltages and currents are shown in the equations (1) and (2). Using the equations in (Surya et al., n.d.), the values of D, L and C are computed. D=0.5, L=416.66μH and C=7.5μF

Figure 4 shows the non-ideal buck converter. The equation for V_0/V_g is derived by applying KVL and KCL to Figure 4

Figure 4. Non Ideal Buck Converter

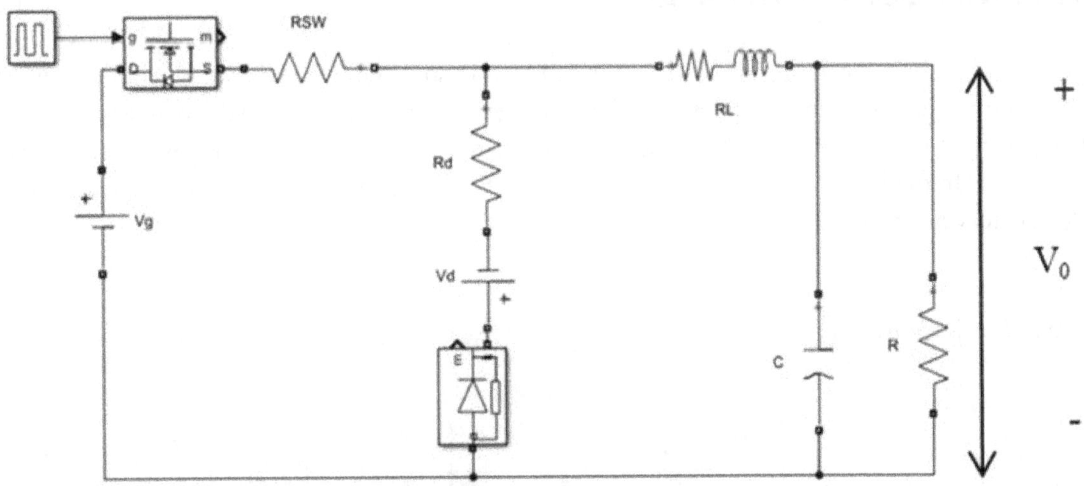

When the switch is closed,

$$V_g - i_L(R_{sw} + R_L) - V_0 = V_L \tag{1}$$

$$i_c = i_L - V_0/R \tag{2}$$

When the switch is opened,

$$-i_L R_L - V_0 + i_L R_d - V_d = V_L \quad (3)$$

$$i_c = i_L - V_0/R \quad (4)$$

By applying volt-sec and amp-sec balance,

$$DV_g - i_L(DR_{sw} + R_L - R_d(1-D)) - V_0 - V_d(1-D) = 0 \quad (5)$$

$$i_L = I_0 = V_0/R \quad (6)$$

Substituting (6) in (5),

$$V_0/V_g = (D - (V_d(1-D)/V_g))/((R + DR_{sw} + R_L - R_d(1-D))/R) \quad (7)$$

Where $R_{SW}(\Omega)$ is the MOSFET Switch resistance, V_d (V) & $R_d(\Omega)$are the forward voltage and dynamic resistance of the diode respectively and $R_L(\Omega)$ is the Equivalent Series Resistance (ESR) of the inductor.

MODELING BOOST CONVERTER

A schematic of boost converter is shown in Fig 5. Figs. 6 and 7 show the equivalent circuits when the switches are closed and opened respectively. The diode gets reverse biased when the switch is closed. The inductor current i_L flows through the inductor L and charges it. The load current I_0 discharges through the capacitor. In the next cycle, the inductor polarity gets reversed when the switch is opened, and the diode gets forward biased.

Figure 5. Boost Converter

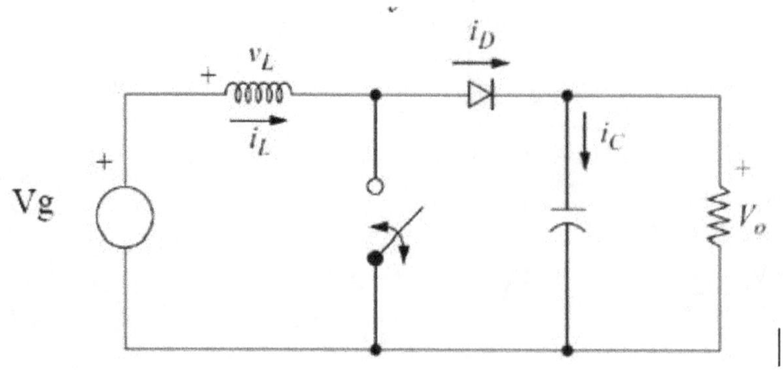

Figure 6. Switch Open

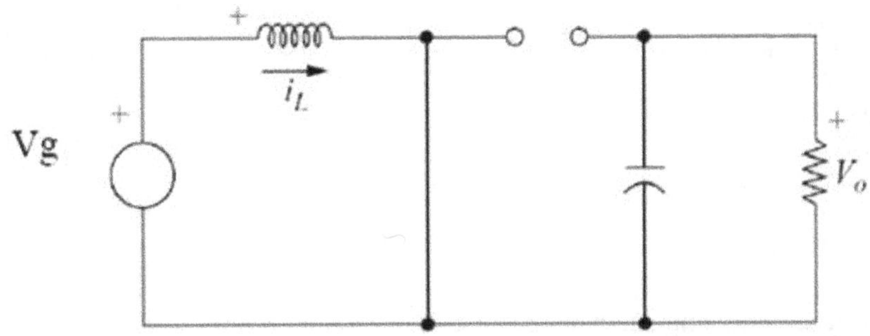

Figure 7. Switch closed

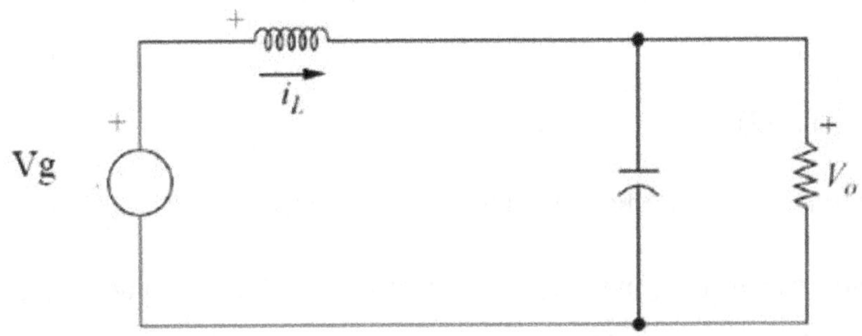

Using the equations shown in (Surya et al., n.d.), the values of D, L and C are computed. D=0.5, L=416.6μH and C=50μF

Figure 8 shows the non-ideal boost converter. The equation for V_0/V_g is derived by applying KVL and KCL to Figure 9

When the switch is closed,

$$V_g - i_L R_L - i_L R_d = V_L \tag{8}$$

Figure 8. Non Ideal Boost Converter

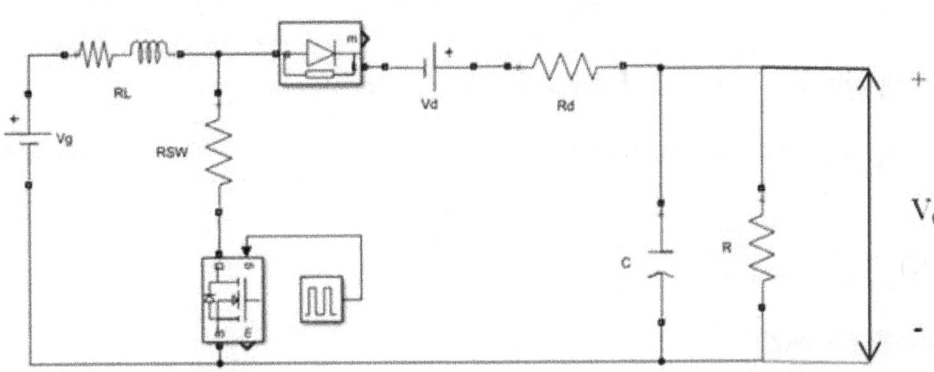

$$i_c=-V_0/R \tag{9}$$

When the switch is opened,

$$V_g-i_LR_L-V_d-i_LR_D-V_0=V_L \tag{10}$$

$$i_c=i_L-V_0/R \tag{11}$$

By applying volt-sec and amp-sec balance,

$$V_g-i_L(DR_d+R_L+R_d(1-D))-V_0(1-D)-V_d(1-D)=0 \tag{12}$$

$$i_L=V_0/(R(1-D)) \tag{13}$$

Substituting (13) in (12),

$$V_0/V_g=(V_g-V_D(1-D))/V_g) * (R(1-D)/R(1-D)^2+DR_d+R_L+R_d(1-D)) \tag{14}$$

Using the equations shown in (Eng & Bunlaksananusorn, 2009), the values of D, L and C are computed. D=0.66, L=375μH and C=132μF

When the switch is closed,

$$V_g-i_L(R_{sw}+R_L)=V_L \tag{15}$$

$$i_c=-V_0/R \tag{16}$$

When the switch is opened,

$$i_LR_L+V_0+i_LR_d-V_d=V_L \tag{17}$$

$$i_c=i_L-V_0/R \tag{18}$$

By applying volt-sec and amp-sec balance,

$$DV_g-i_L(DR_{sw}+R_L-R_d(1-D))-V_0(1-D)+V_d(1-D)=0 \tag{19}$$

$$i_L=V_0/R(1-D) \tag{20}$$

Substituting (19) in (20),

$$V_0/V_g=(DV_g+V_d(1-D))/V_g * (R_{sw}D+R_d(1-D)+R_L+R(1-D)^2)/R(1-D) \quad (21)$$

SIMULINK MODELING OF NON IDEAL CONVERTERS

The modeling in Simulink can be done in three different ways.

1. Method 1 is using Simscape components
2. Method 2 is using Commonly Used Blocks containing basic mathematical operations with integrator and hence, reduces the dependence on SimPowerSystem tool box. This method gets complicated when the circuit has more than one energy storing element.
3. Method 3 is using a Mux, a function block (Fcn) and an integrator. The inputs to the mux are defined in the form of u(1), u(2) etc;

Methods (1), (2) & (3) would yield similar solutions. Illustration of different methods for buck converter is shown in Figs. 10, 11 and 12. Figs. 12, 13 and 14 represent the output voltage V_0, output current, I_0 and inductor current, (i_L)

Figure 9. Simscape model (Method 1)

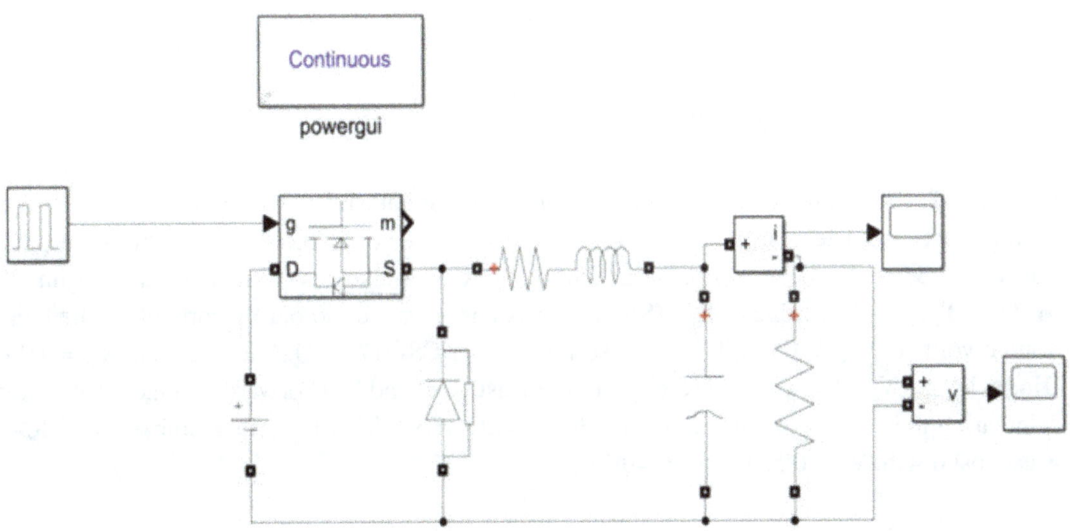

Illustration of different methods for boost converter is shown in Figs. 15, 16 and 17. Transients in output voltage, output current and the inductor current are shown in Figs. 18, 19 and 20.

Figure 10. Mathematical Model (Method 2)

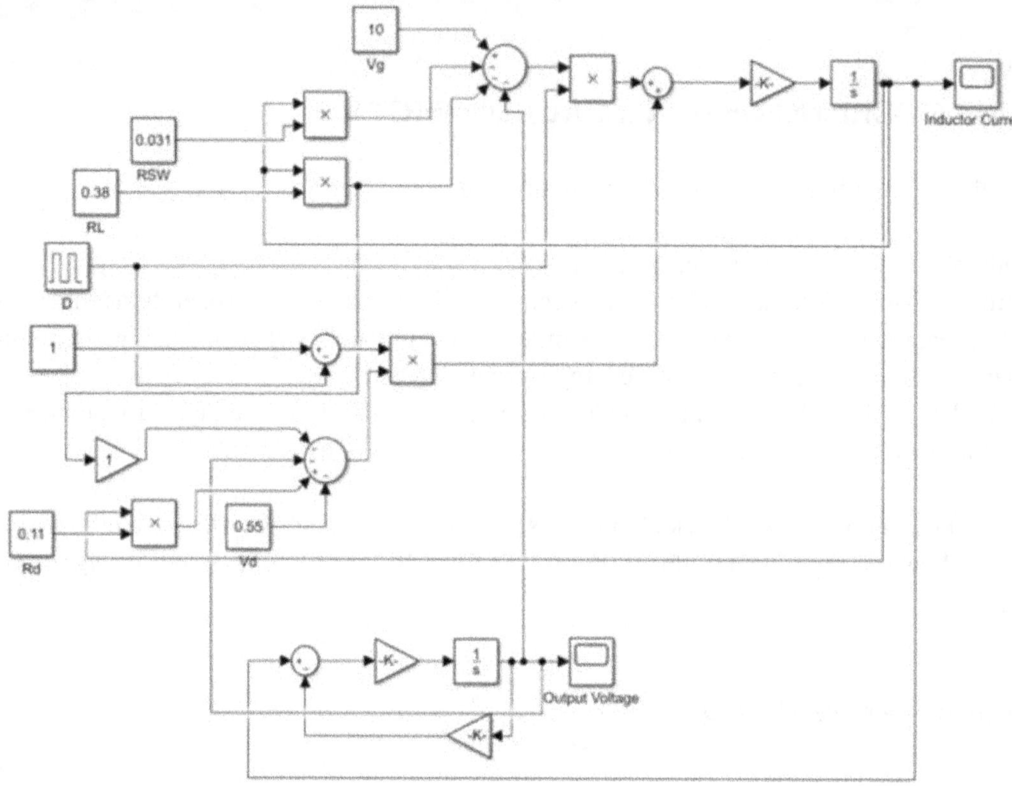

SELECTION OF COMPONENTS

1. **Inductor**, L – It was selected based on standard inductor datasheet. The value of L was computed during ideal case. The nearest value of L was selected based on 1482 series and the corresponding Equivalent Series Resistance (ESR) was found out. R_L was taken as 0.38Ω (Surya & Arjun, 2020)
2. **MOSFET**, r_{SW} – The value of R_{SW} (Switch Resistance) was found out by computing the Drain to Source voltage, V_{DS}. MOSFET from TI series number CSD17313Q2Q1 is chosen. R_{SW}=0.031Ω
3. **Diode**, R_d and V_d – The value of R_d (Dynamic Resistance) and V_d (Forward voltage) were found by calculating the reverse voltage across the diode. DIODES INC with series number – B530C-13-F was chosen with V_d = 0.55V and R_d=0.11 Ω

AVERAGE CURRENT CONTROL (BOOST CONVERTER)

Figure 21 shows the block diagram of average current control in a DC-DC converter. V_c is the control voltage (V), i is the switch current (A). R_f is the sensing resistance (Ω). Hence $V_c = R_f i_c$ is the control input or the reference to the control loop involving current. G_{ci} is the compensator transfer function and the compensator output feeds the PWM generator which controls the switch operation. V_m is the peak

Figure 11. Mathematical Model (Method 3)

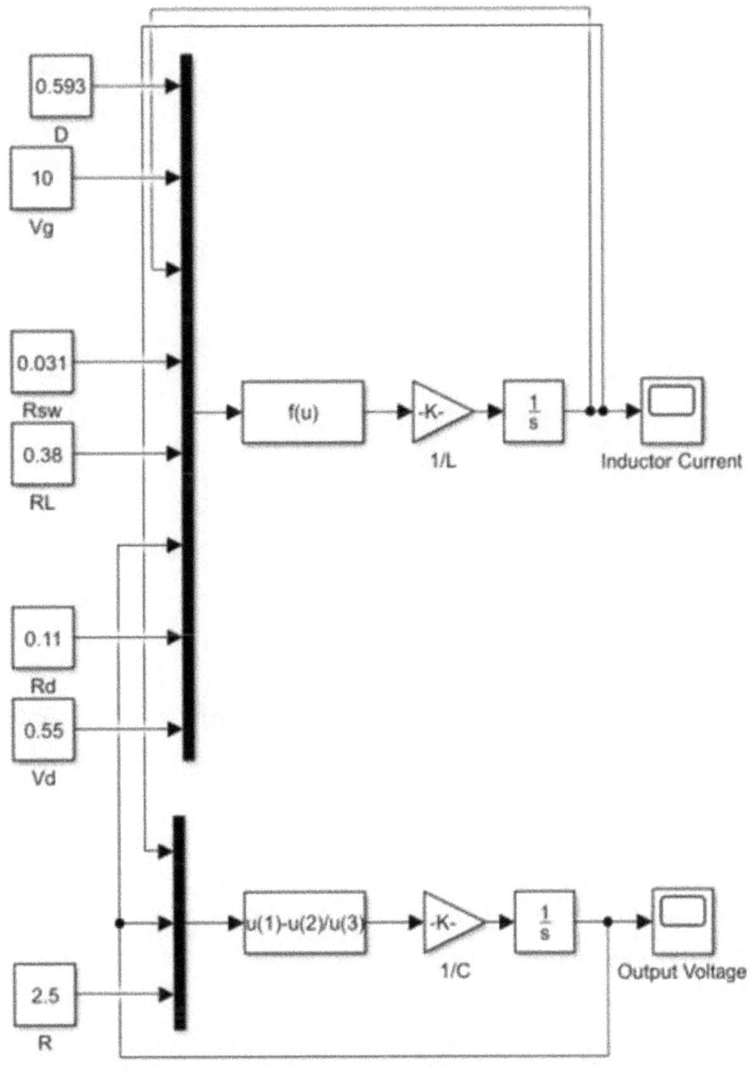

value of the saw-tooth signal which is required to produce PWM signal. V_g and V_0 are the input and output voltage (V).

Figure 22 shows the voltage loop around an average current controlled DC-DC converter. This configuration ensures output voltage control and average current control. H represents the sensor gain. G_{cv} represents the compensator required to control the output voltage.

The current control in DC-DC converters is classified as (a) Average Current Control (ACC) and (b) Peak Current Mode (PCM) control.

Advantages of ACC

1. Improved immunity towards noise
2. Performed on wide range of applications

Disadvantages of ACC

Figure 12. Output Voltage Vs. Time

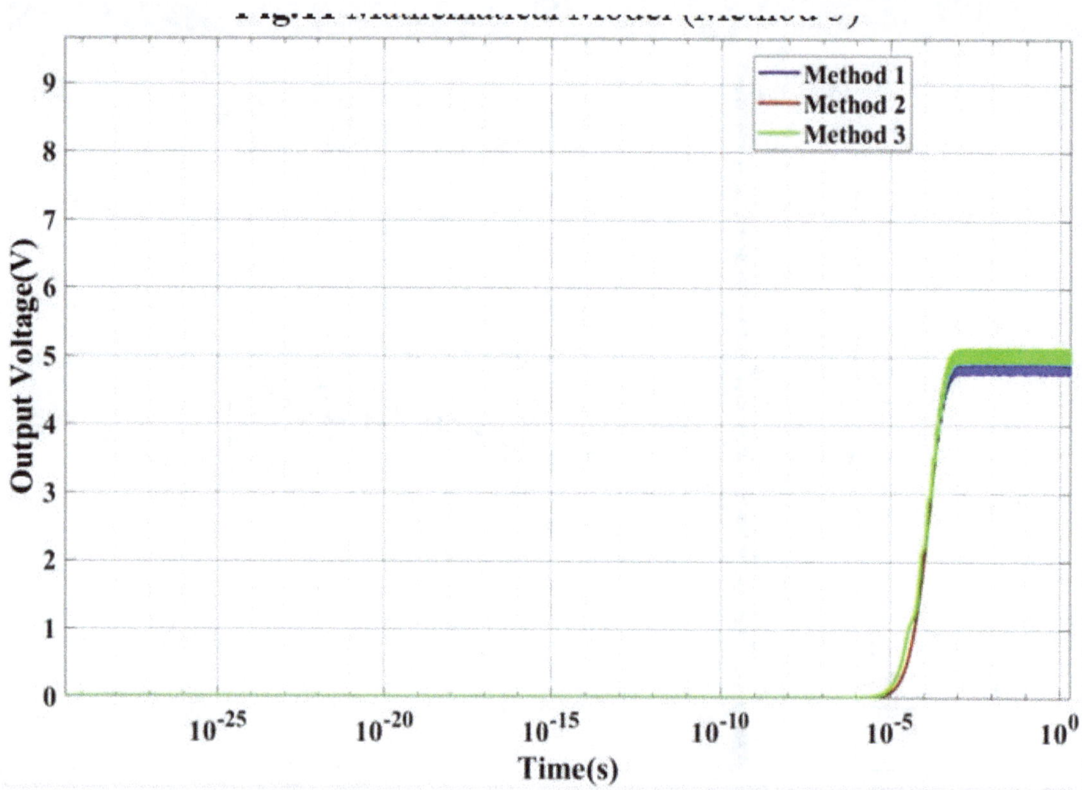

1. Cannot provide instantaneous control on switch current
2. Does not mitigate transformer saturation problems in push-pull converters

Applications of average current control

1. Inverters
2. EV On-board chargers
3. PFC converters

An EV Onboard charger consists of AC-DC converter, DC-DC converter and a Power Factor Correction (PFC) control circuit. PFC creates sinusoidal input currents and maintains the distortions under limit.

In order to evaluate the feasibility of PFC control, average inductor current is controlled, enabling the possibility of switch current control.

In this work, the ideal buck and boost converters are considered and the open loop transfer function of duty cycle to inductor current (G_{id}) is studied. G_{id} is found using the state space averaging technique using the governing equations shown in (22) and (23)

$$\dot{X}=AX+BU \tag{22}$$

$$Y=CX+DU \tag{23}$$

Figure 13. Output Current Vs. Time

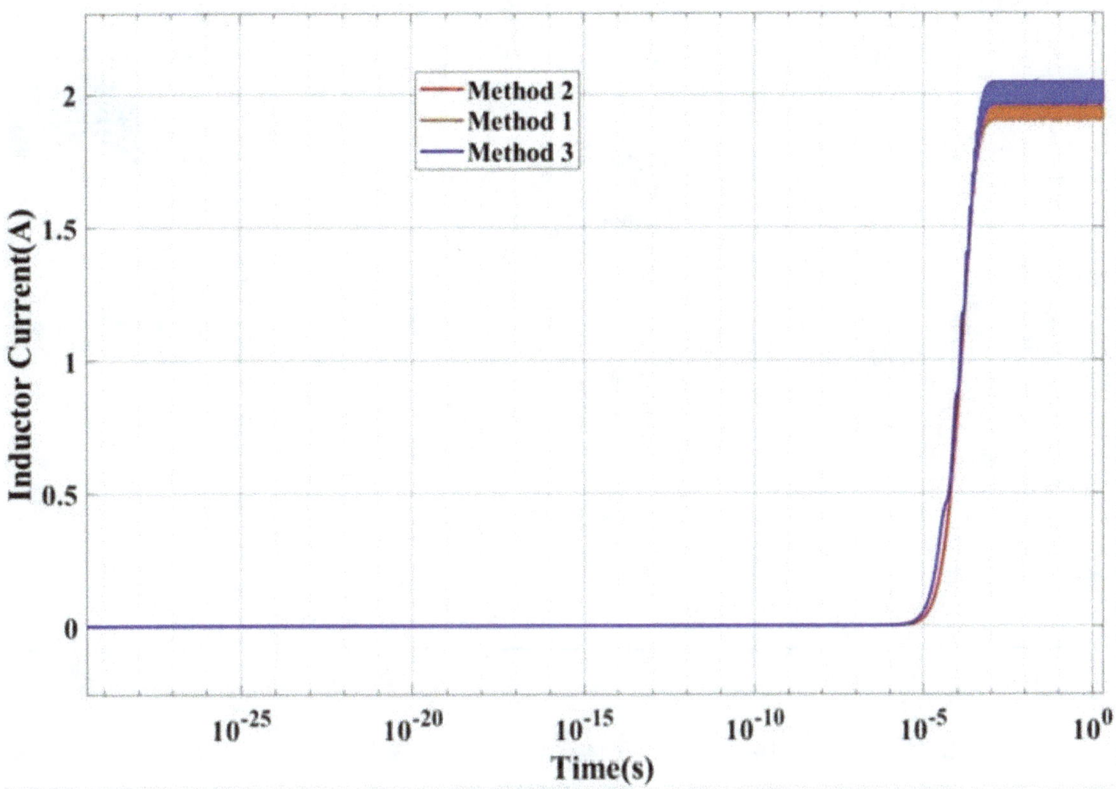

Where A, B, C and D are the state space matrices. The state variables are considered as i_L and V_0, input as V_g and output as i_L. The operation of the converter is determined by the switch positions.

When the switch is closed,

$$V_L=V_g \tag{24}$$

$$i_c=-V_0/R \tag{25}$$

When the switch is opened,

$$V_L=V_g-V_0 \tag{26}$$

$$i_c=i_L-V_0/R \tag{27}$$

The state space equations can be written as shown below for switch closed and opened by combining (24) with (26) and (25) with (27)

Figure 14. Inductor Current Vs. Time

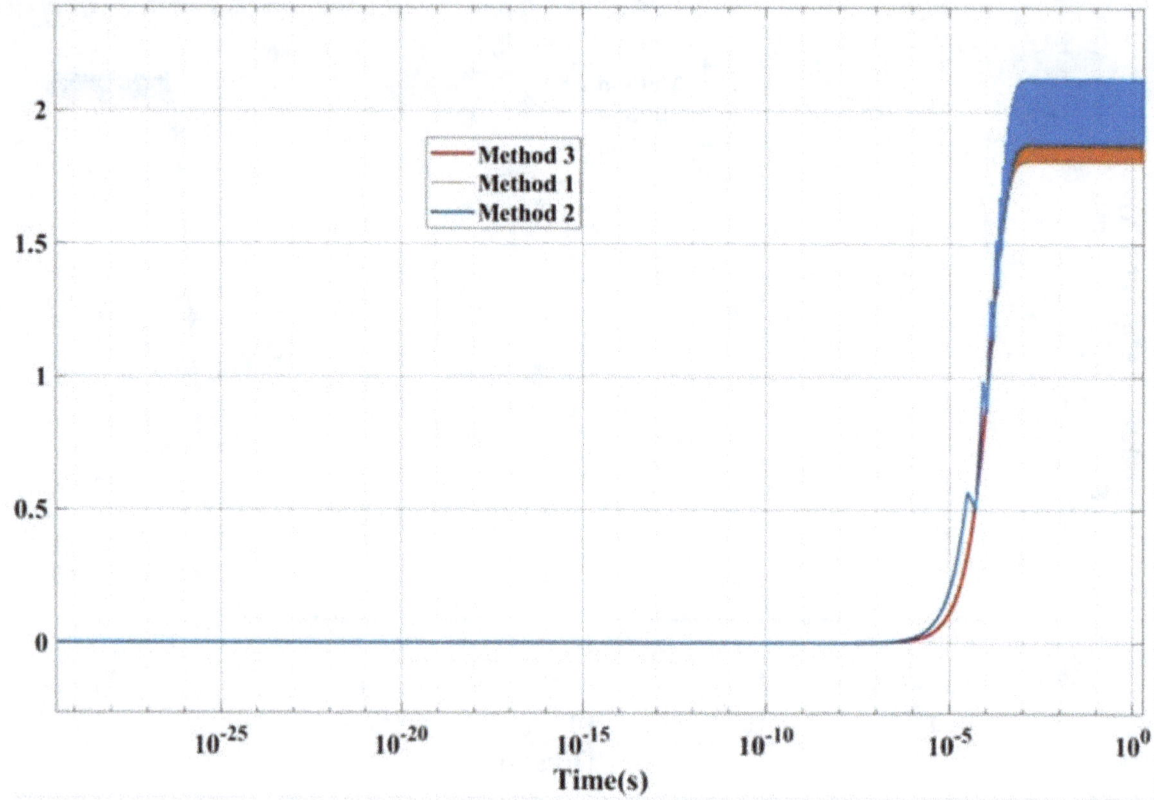

$$\begin{bmatrix} diL/dt \\ dv0/dt \end{bmatrix} = \begin{bmatrix} 0 & 0 \\ 0 & -1/RC \end{bmatrix} \begin{bmatrix} iL \\ v0 \end{bmatrix} + \begin{bmatrix} 1/L & 0 \end{bmatrix} \begin{bmatrix} Vg \\ 0 \end{bmatrix} \tag{28}$$

$$\begin{bmatrix} diL/dt \\ dv0/dt \end{bmatrix} = \begin{bmatrix} 0 & -1/L \\ 1/C & -1/RC \end{bmatrix} \begin{bmatrix} iL \\ v0 \end{bmatrix} + \begin{bmatrix} 1/L & 0 \end{bmatrix} \begin{bmatrix} Vg \\ 0 \end{bmatrix} \tag{29}$$

Averaging A, we get $A=A_1D+A_2*(1-D)$

$$A=\begin{bmatrix} 0 & -(1-D)/L \\ (1-D)/C & -1/RC \end{bmatrix} \tag{30}$$

$$(A_1\text{-}A_2)X + (B_1\text{-}B_2)U = \begin{bmatrix} 0 & 1/L \\ -1/C & 0 \end{bmatrix} \begin{bmatrix} iL \\ V0 \end{bmatrix} = \begin{bmatrix} V0/L \\ -IL/C \end{bmatrix} \tag{31}$$

Figure 15. Simscape model of a Boost converter (Method 1)

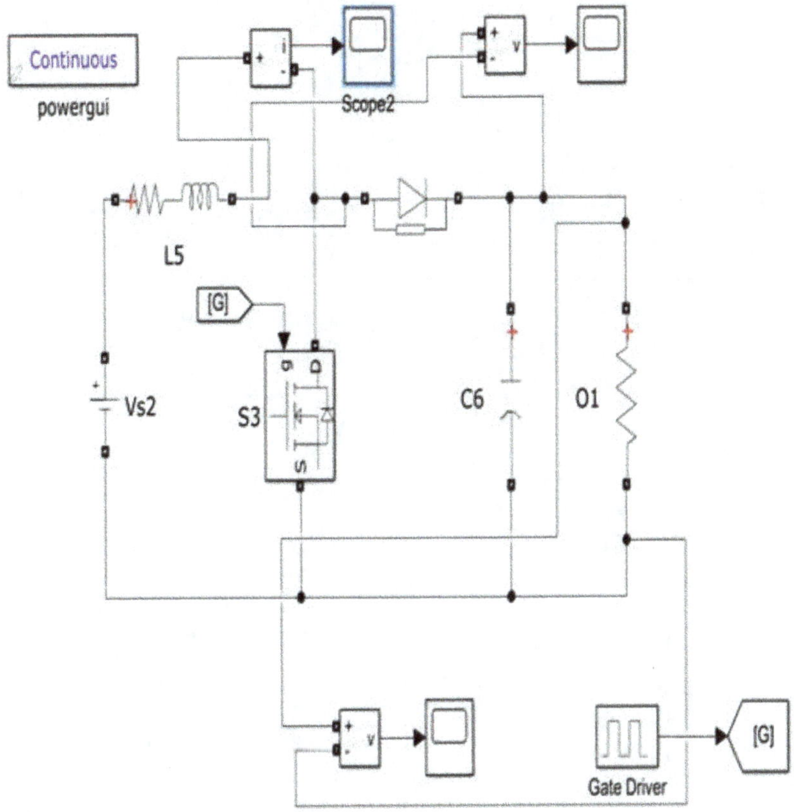

$$\hat{i}_L/\hat{d} = C[sI-A]^{-1}*((A_1-A_2)X + (B_1-B_2)U) \quad (32)$$

$$\text{Where } CX=\begin{bmatrix}1 & 0\end{bmatrix}\begin{bmatrix}iL \\ v0\end{bmatrix} \quad (33)$$

$$sI\text{-}A=\begin{bmatrix} s & (1-D)/L \\ -(1-D)/C & s+1/(RC)\end{bmatrix} \quad (34)$$

$$\hat{i}_L/\hat{d} = ((s+1/RC)V_0/L+(V_0/\ RLC))/\Delta \quad (35)$$

Where $\Delta = s^2+s/RC+(1-D)^2/LC$ (36)

The specifications of the converter are shown in Table 3.

Figs. 23 and 24 show the frequency response and root locus of G_{id} for an ideal boost converter. The system should high stability of gain margin of infinity and phase margin of 89.8.

Figure 16. Mathematical model (Method 2)

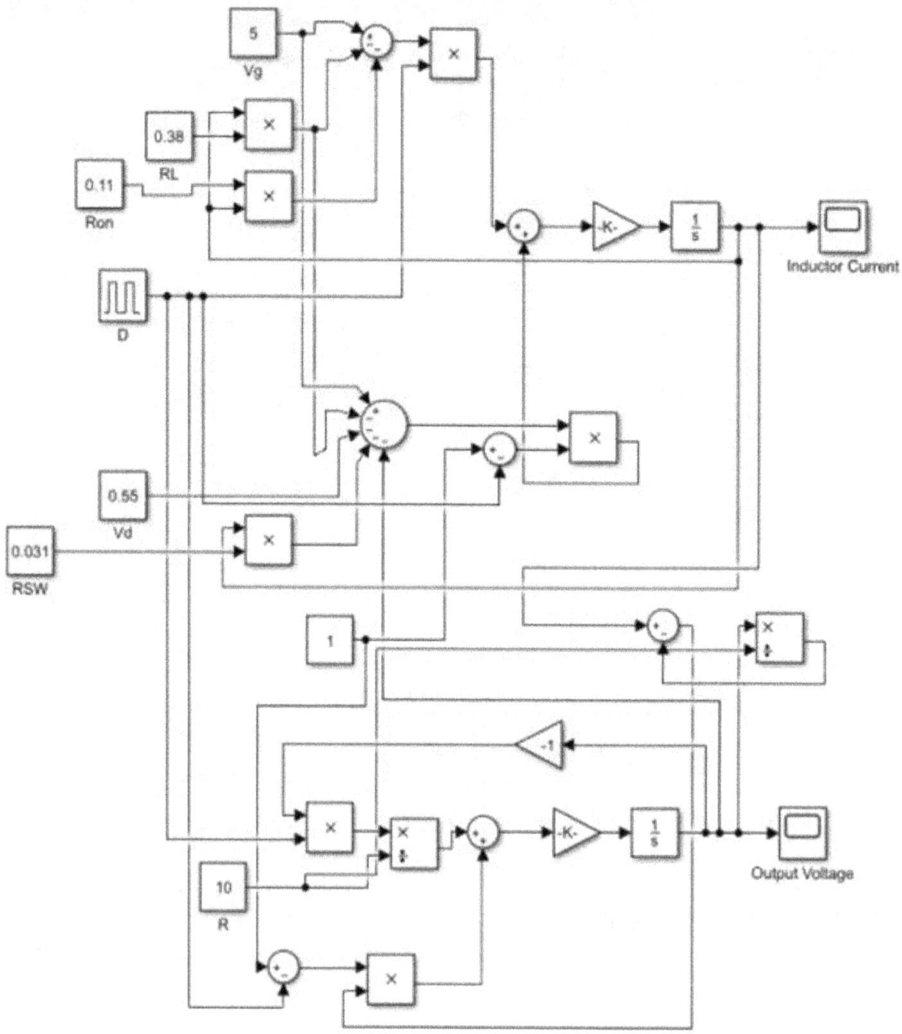

$$\hat{i}_L/\hat{d} = (2.1938*10^6*(s+1.725*10^4))/(s^2 + 8627s + 3.286*10^9) \quad (37)$$

MATLAB CODE (BOOST CONVERTER)

```
clc;clear;
    vg=5;% Define Input Voltage
    l=4.7e-6;% Define L value
    c=9.66e-6;% Define C value
    d=0.5833;% Define duty cycle
    r=12;% Define load resistance
    v0=vg/(1-d);% calculation of output voltage
```

Figure 17. Simulink model (Method 3)

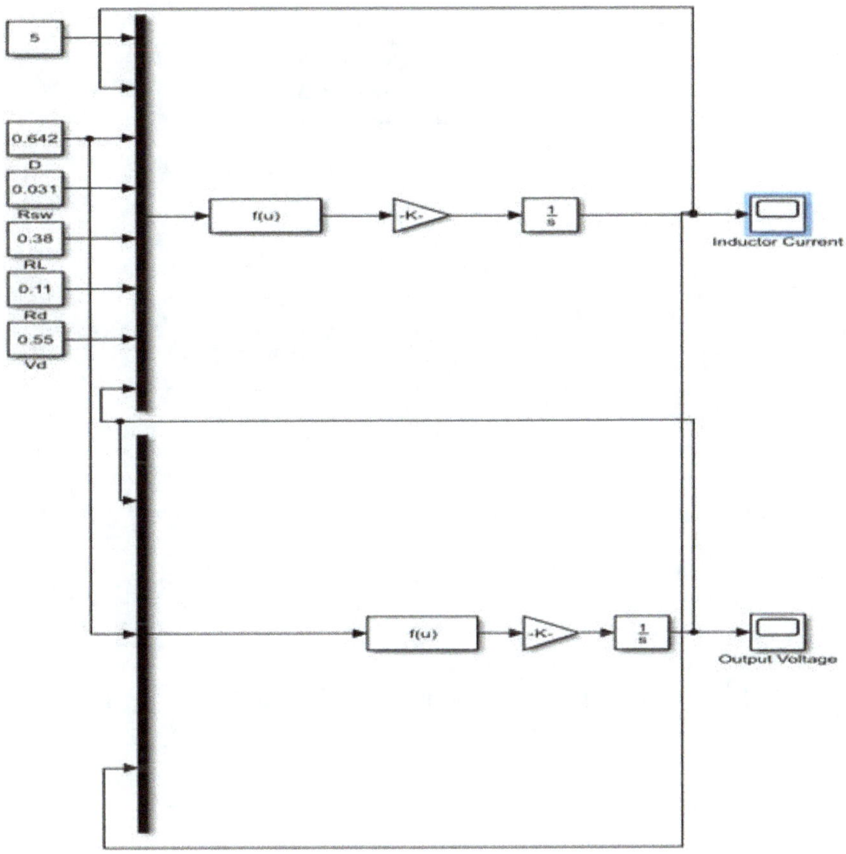


```
il=(v0/(r*(1-d)))% average inductor current
s=tf('s')% To define the transfer function in laplace domain
%% Define the transfer function %%
deltai=s*s+(s/(r*c))+(ddashi^2/(li*c))
sys1=((v0/li)*(s+1/(r*c))+(v0/(r*li*c)))/deltai
zpk(sys1)% To obtain the transfer function in zero and pole form
rlocus(sys1)% To plot the root locus of the system defined (sys1)
margin(sys1)% To plot the frequency response and the phase plot of the system defined. This command directly calculates the gain and phase margins.
```

AVERAGE CURRENT CONTROL (BUCK CONVERTER)

The specifications of the boost converter are shown in Table 4.

$$A_1 = \begin{bmatrix} 0 & -1/L \\ 1/C & -1/RC \end{bmatrix} \tag{38}$$

Figure 18. Output Voltage Vs. Time

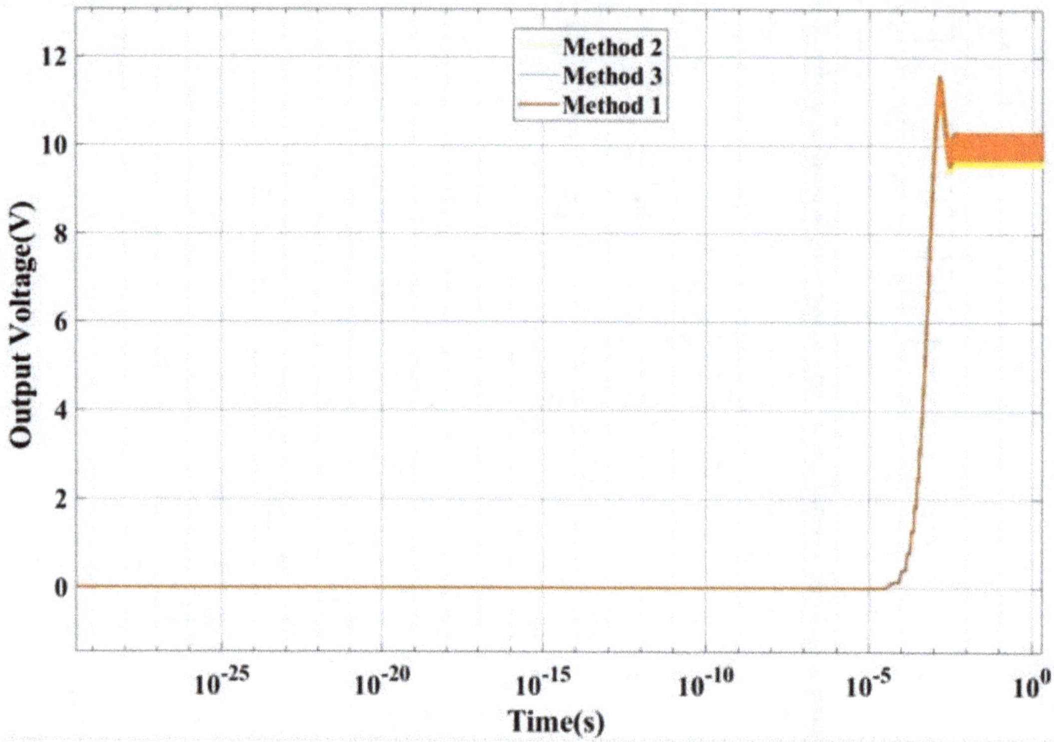

$$A_2=\begin{bmatrix} 0 & -1/L \\ 1/C & -1/RC \end{bmatrix} \tag{39}$$

$$B_1=\begin{bmatrix} 1/L \\ 0 \end{bmatrix} \tag{40}$$

$$B_2=\begin{bmatrix} 0 \\ 0 \end{bmatrix} \tag{41}$$

$$A=A_1D+A_2(1\text{-}D)=\begin{bmatrix} 0 & -1/L \\ 1/C & -1/RC \end{bmatrix} \tag{42}$$

Figure 19. Output Current Vs. Time

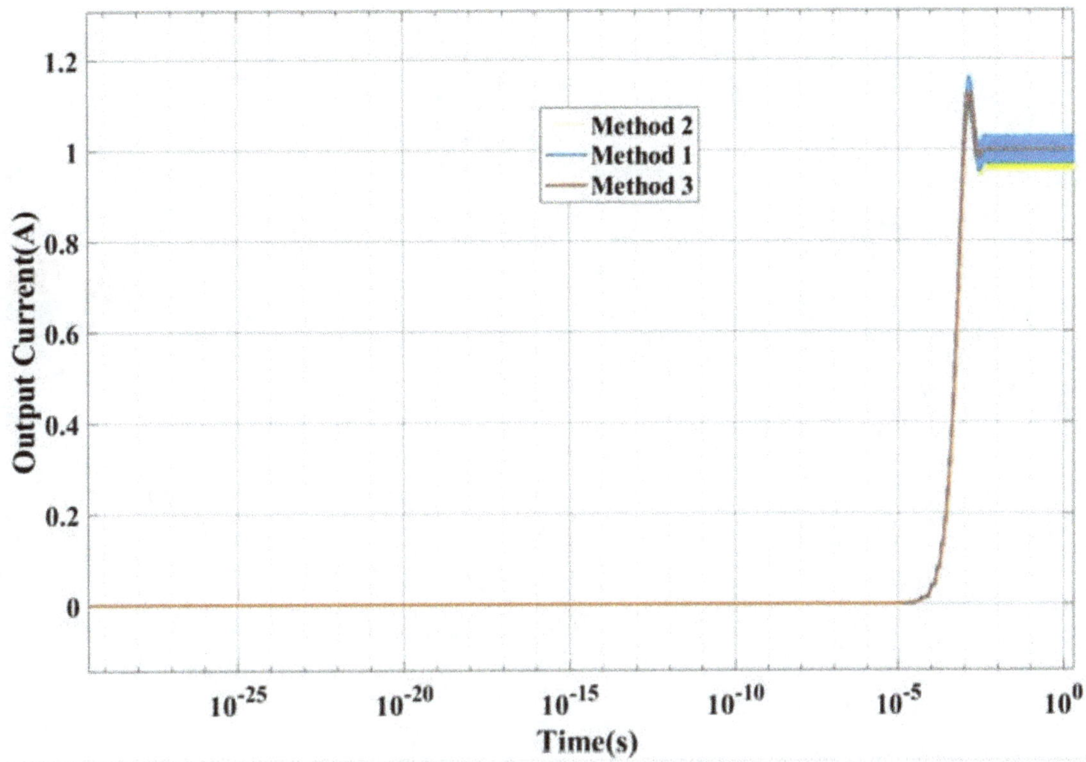

$$\text{sI-A} = \begin{bmatrix} s & 1/L \\ -1/C & s+1/RC \end{bmatrix} \tag{43}$$

$$\hat{i}_L/\hat{d} = ((V_g/L)*(s+1/RC))/\Delta \tag{44}$$

$$\Delta = s^2+s/RC+1/LC \tag{45}$$

$$G_{id} = 14545(s+1082)/(s^2+1082s+1.082*10^7) \tag{46}$$

Figs. 25 and 26 show the frequency response and the root locus of G_{id}. The system should high stability having gain margin and phase margin of infinity and 90.2.

MATLAB CODE (BUCK CONVERTER)

```
clc;clear;
```

Figure 20. Inductor Current Vs. Time

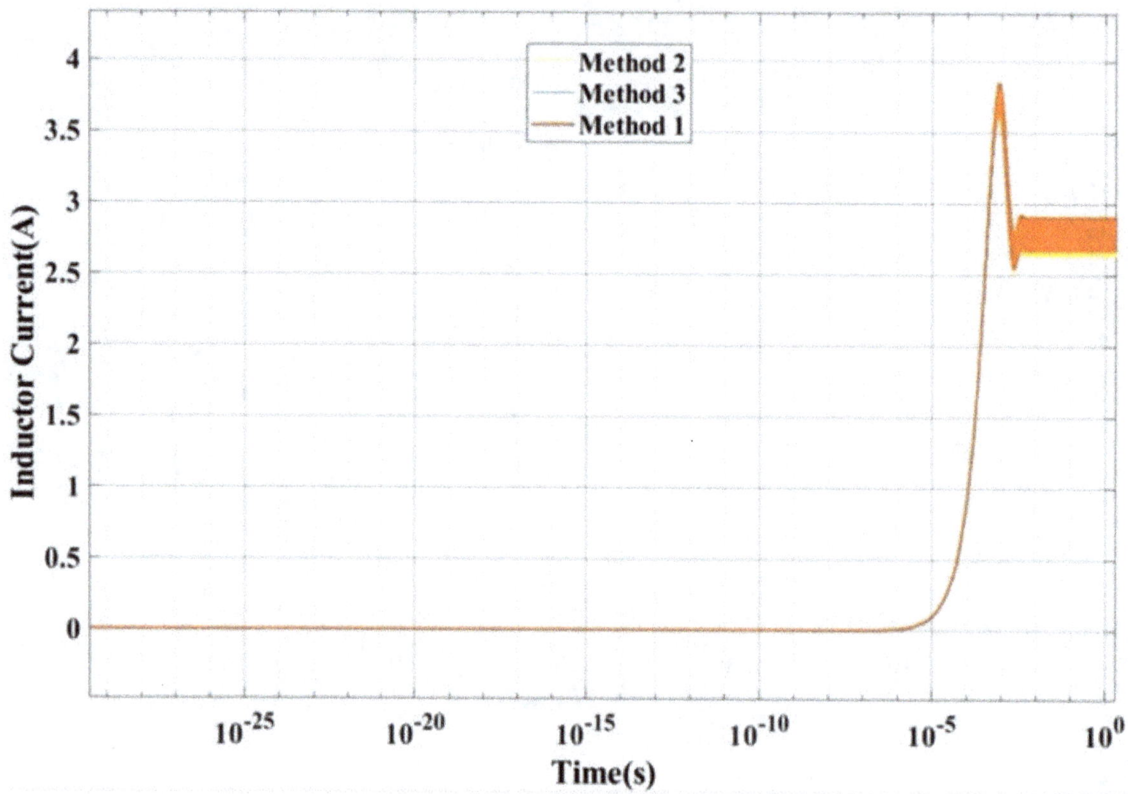

Figure 21. Block diagram of average current control

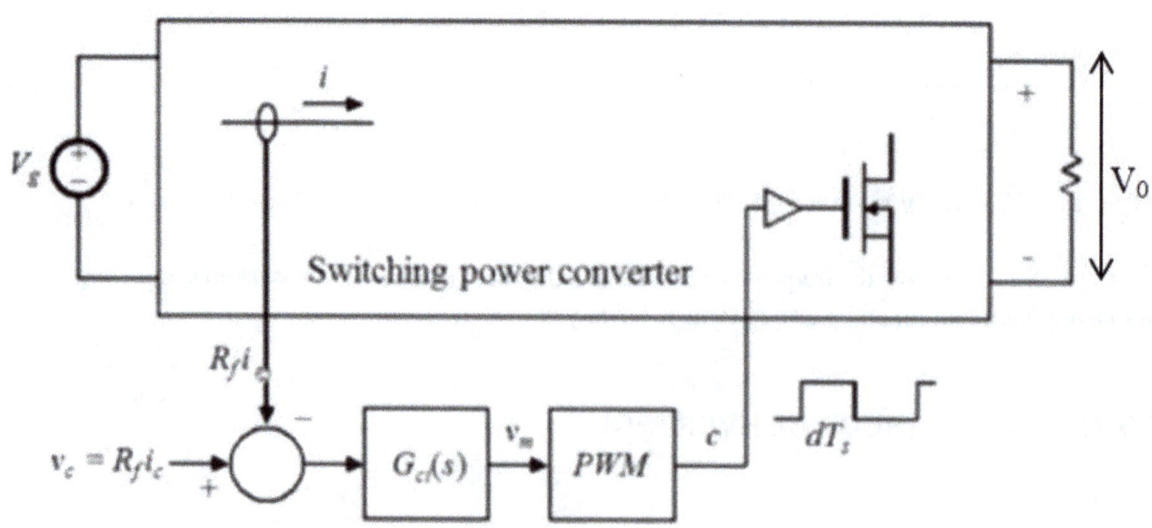

Figure 22. Voltage loop around current control loop

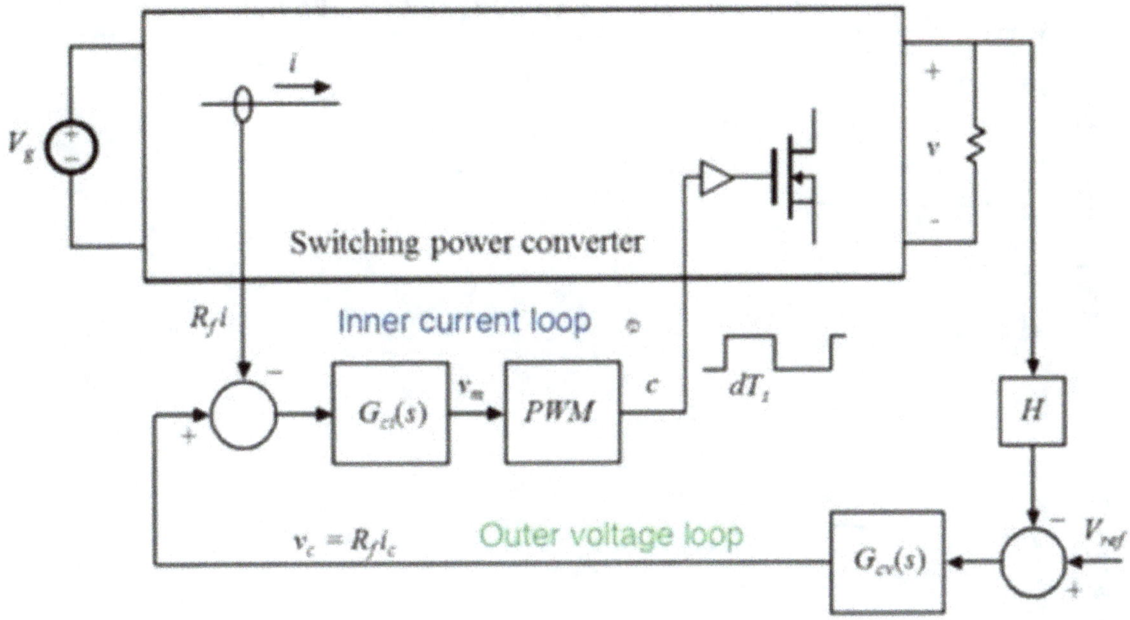


```
% Average current control of a Buck converter%
vg=16
v0=12
r=11
l=1.1e-3
c=84e-6
d=v0/vg
il=v0/r
s=tf('s')
deltai=s^2+(s/(r*c))+(1/(l*c))
sys2=((s+1/(r*c))*(vg/l))/deltai
zpk(sys2)
```

Table 3. Boost converter

Parameter	Value
Input Voltage, V_g	5V
Output Voltage, V_0	12V
Output Resistance, R	12Ω
Inductor, L	4.76μH
Capacitor, C	9.66μF
Duty cycle	0.5833

Figure 23. Bode plot of G_{id}

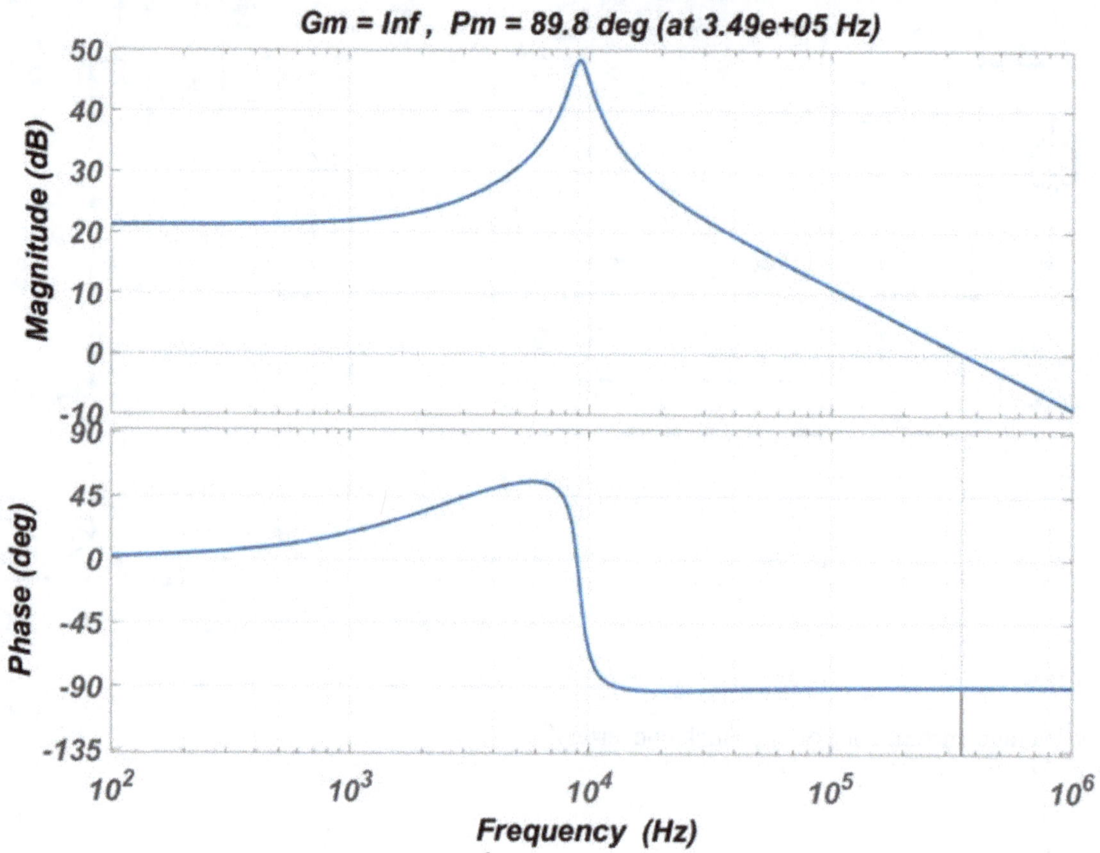


```
rlocus(sys2)
margin(sys2)
```

PEAK CURRENT MODE CONTROL (PCM)

This is a popular technique used in current control technique to protect the switch from overcurrent. Figure 27 shows the control technique involved in PCM.

The control current $i_c(t)$ is used as a reference and compared with the switch or the inductor current $i_s(t)$. When $i_s(t)$ reaches its peak value, the switch (transistor) is switched off. Figure 28 shows a PCM for an ideal buck converter. The analog comparator consists of two inputs viz., (a) Sensed switch current and (2) Control input – which sets the peak current

During the start, the SR latch sets the signal enabling the switch ON condition. The current $i_s(t)$ starts ramping up. When $i_L = i_c = i_s$, reset signal would be enabled, due to the control logic will become zero. This cycle repeats and control is ensured.

One of the major drawbacks of PCM is that all the DC-DC converters are unstable for $D > 0.5$. In detail explanation and the waveforms are presented well in (Erickson & Dragon, 2007).

Figure 24. Root locus of G_{id}

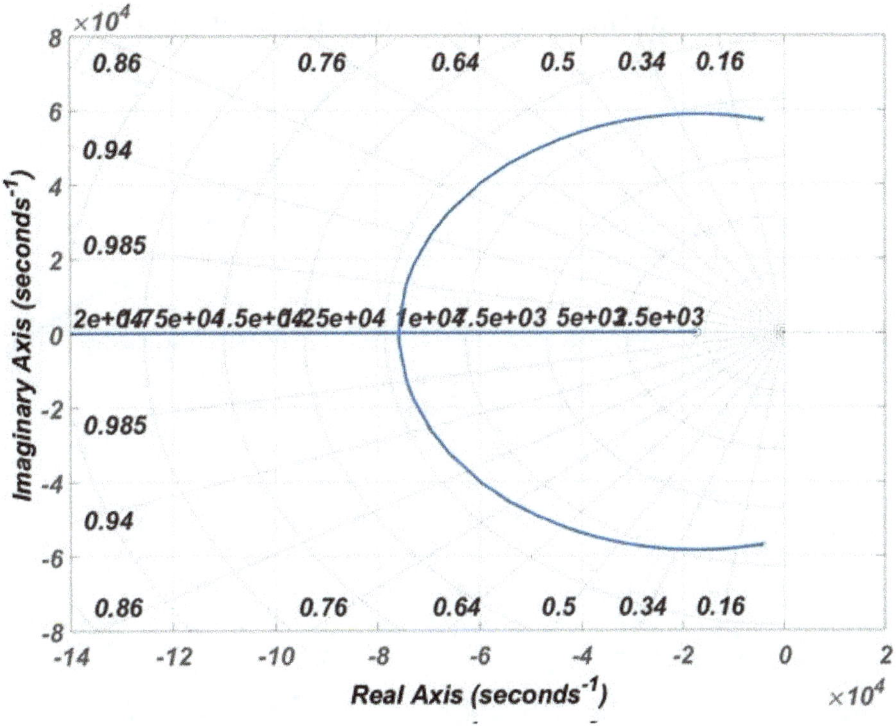

Hence, a new technique for comparing i_s with i_c is proposed. The new control strategy introduced an artificial ramp instead of a DC control signal. The magnitude of the ramp is defined as i_c-i_a and this ramp is compared with i_s. Let m_1 be the increasing slope and $-m_2$ be the falling slopes of i_s. m_a is the slope of the artificial ramp. Figure 29 shows the artificial ramp stabilization. The clock in Figure 28 now not only generates the set and reset signals, but also the artificial ramp signal. This technique ensures that the converter remains stable for all value of D. During the design process, it is suggested in (Erickson & Dragon, 2007) that m_a=m_2 where m_2 is the slope of i_s during (1-D) condition of the switch. Artificial ramp also helps in the reduction of sensitivity towards noise.

Table 4. Buck converter

Parameter	Value
Input Voltage, V_g	16V
Output Voltage, V_0	12V
Output Resistance, R	11Ω
Inductor, L	1.1mH
Capacitor, C	84μF

Figure 25. Bode plot of G_{id}

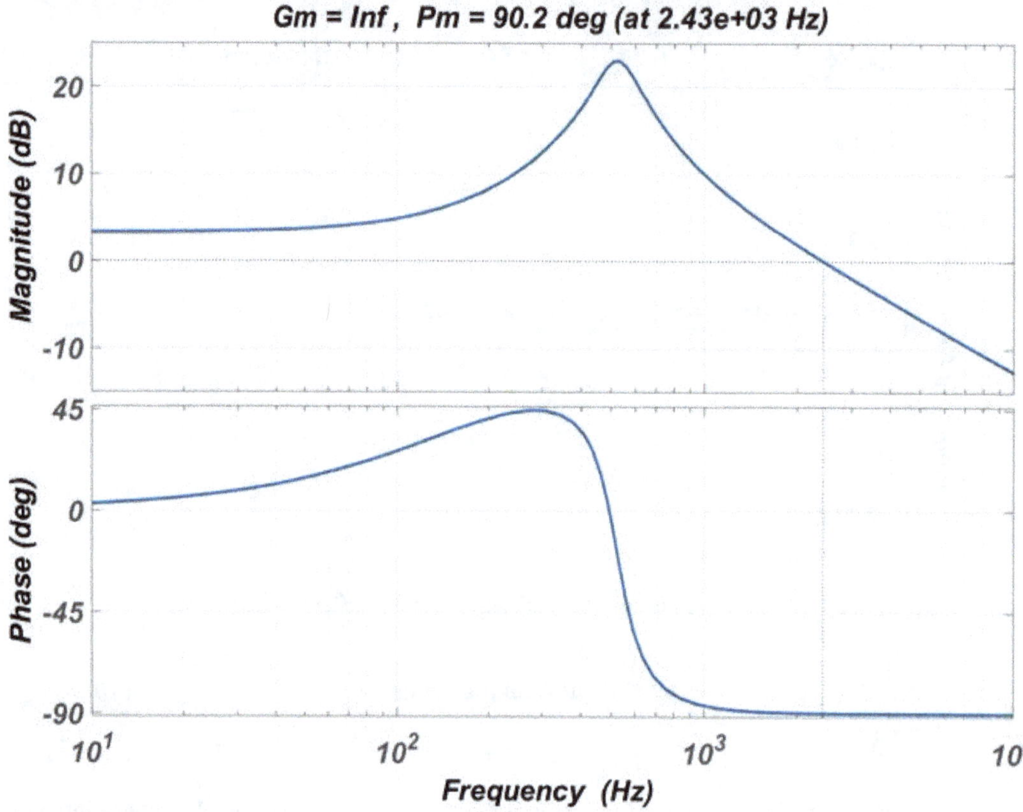

BATTERY MODELING

In the past, several models for batteries have been proposed (Lee & Lai, 2007). Some of the important methods for modeling a battery are shown below

A battery is treated as an ideal voltage source. This method is called poor voltage. Figure 30 shows an illustration of the battery model.

This method does not account for the drop in the terminal voltage $v(t)$ due to the flow of current when a load is connected. However, it provides a starting point to the modeling approach.

In the next method, a resistance is connected end to end with the voltage source to account for the drop in voltage due to flow of current. Under loaded conditions, v(t) rises above the OCV during charging condition. In Figure 31 the phenomenon can be explained in part by placing a resistance in series with the controlled voltage source.

In reality, v(t) slowly decays and reaches its OCV when kept under rest. This is termed as Diffusion Voltage. Figure 33 shows the equivalent battery model considering the diffusion voltage.

The ESC model consists of Five RC pair and is called Enhanced Self-Correcting Model, shown in Figure 34. This model is used to capture the high frequency components in current during pulse charging.

Using library functions provided by MATHWORKS, the battery parameters can be found. The program uses estimation technique and Li ion polymer battery data sheets to estimate the battery parameters.

Figure 26. Root locus of G_{id}

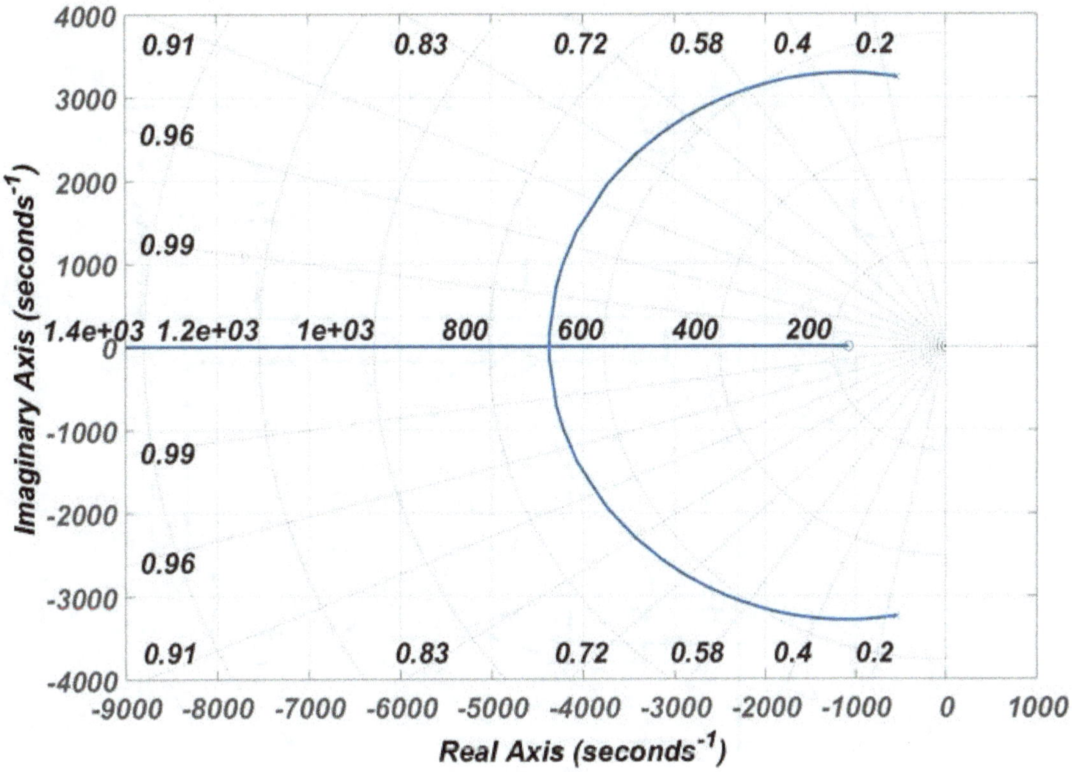

Figure 27. PCM control technique (Erickson & Dragon, 2007)

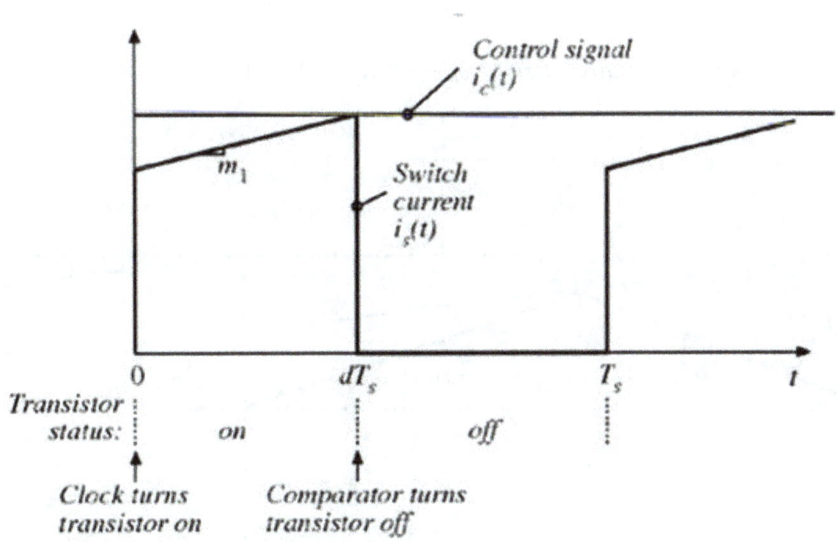

Figure 28. PCM applied to a Buck converter (Erickson & Dragon, 2007)

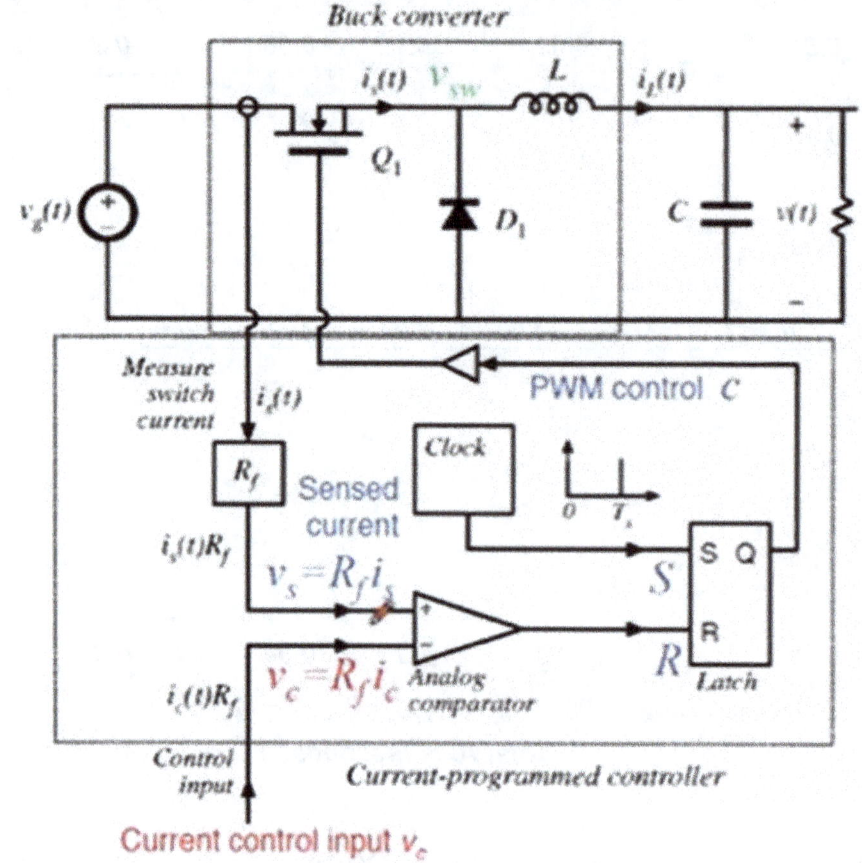

Figure 29. Artificial ramp stabilization (Erickson & Dragon, 2007)

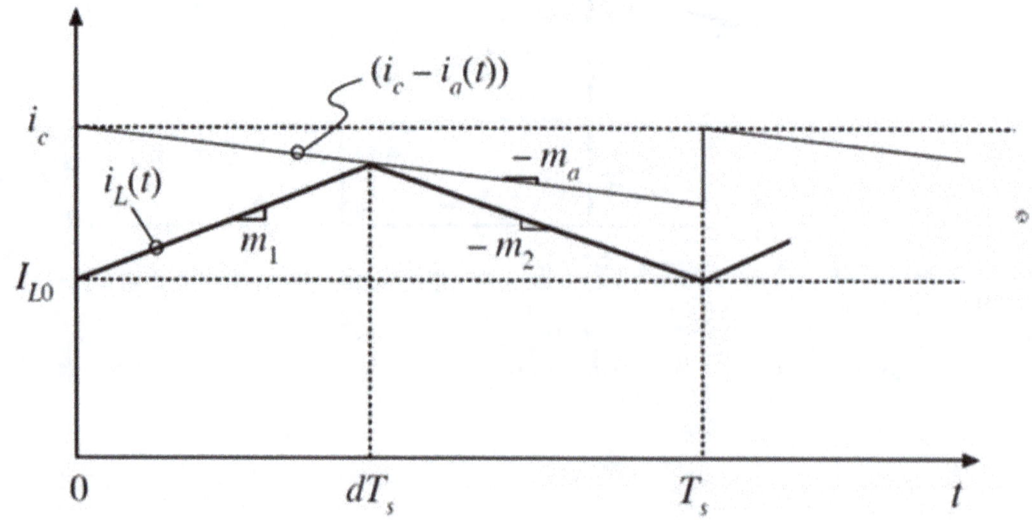

Figure 30. Poor Voltage based method (Williamson, 2013)

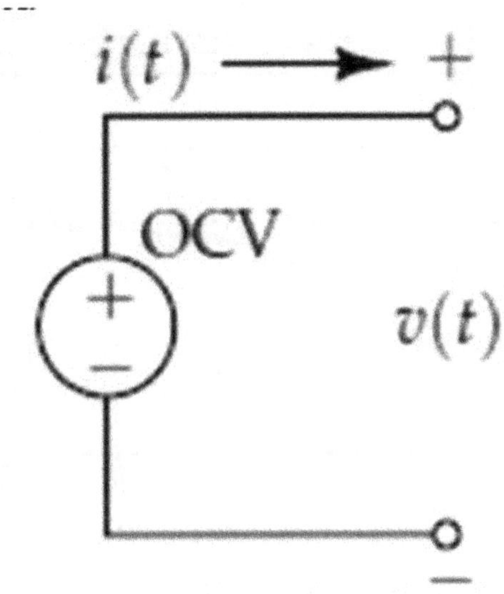

Additional toolboxes have to be installed. The Equivalent Circuit battery block implements a parallel RC component (one or more branches) in series with a voltage source, OCV. In order to replicate the behavior of a battery, five RC pairs are required. However, multiple RC pair modeling leads to complexity and increases the memory requirement.

For the creation of parameters of Equivalent Circuit Battery block, four steps are to be followed. The blocks use optimization methods to find out the number of resistor capacitor parallel combinations. They help in facilitating the initial conditions to the battery model and estimation of parameters are used to force fit the experimental data with the simulation. The results provide the like OCV, terminal voltage, values of R_0, R_1, C_1 and time constants of R_1C_1, R_2C_2, R_3C_3, R_5C_5, ……

Figure 31. ESR method (Williamson, 2013)

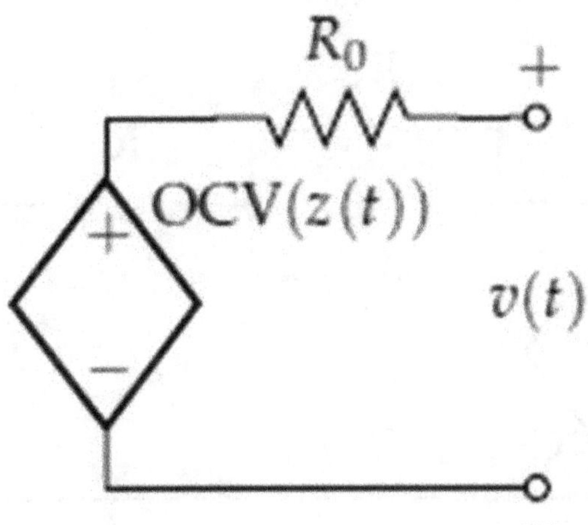

Figure 32. Diffusion voltage in a Li ion battery (Williamson, 2013)

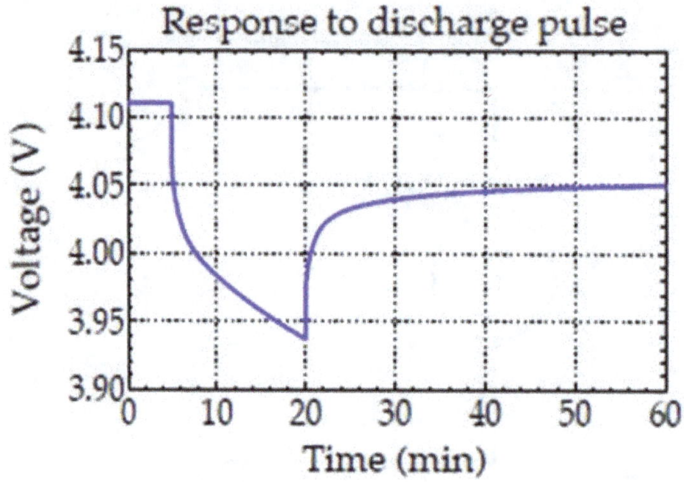

The additional libraries required are Curve Fitting Toolbox, Parallel Computing Toolbox, Optimization Toolbox, and Simulink and Design Optimization.

The generation of battery parameters involves running .m file and two .slx files viz; (a) Estimate Battery Discharge (b) Model BatteryEstim3RC_PTBS (c) Model BatteryEstim3RC_PTBS_EQ

By default, the program gave the values of battery parameters for two RC pairs. The default settings are shown in Table 5.

[15] showed an example of how the parameters in the program can be changed and the importance of each steps. The program has been split into four steps. In order to change the parameters (shown in Table 3), the program had to be run section by section. It was noted that the parameters of the battery were dependent on the SOC and T_{amb}. However, these parameters had no dependency on the discharge current.

Step.1 - provided experimental preloaded data. It consists of series of measured voltage profiles under a constant current discharge followed by long periods of rest. There was a time constant associated

Figure 33. Equivalent battery model (Williamson, 2013)

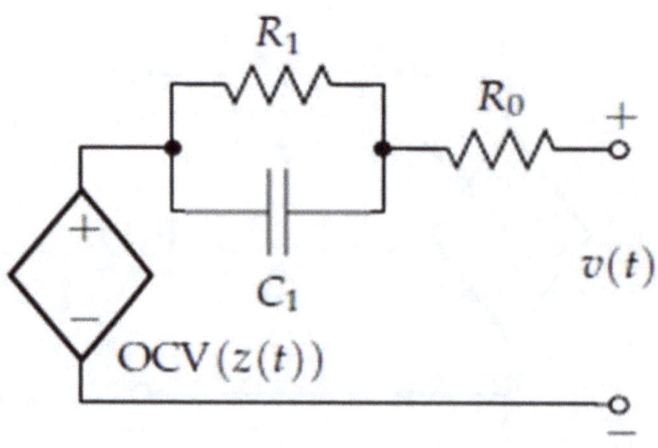

Figure 34. ESC model (Williamson, 2013)

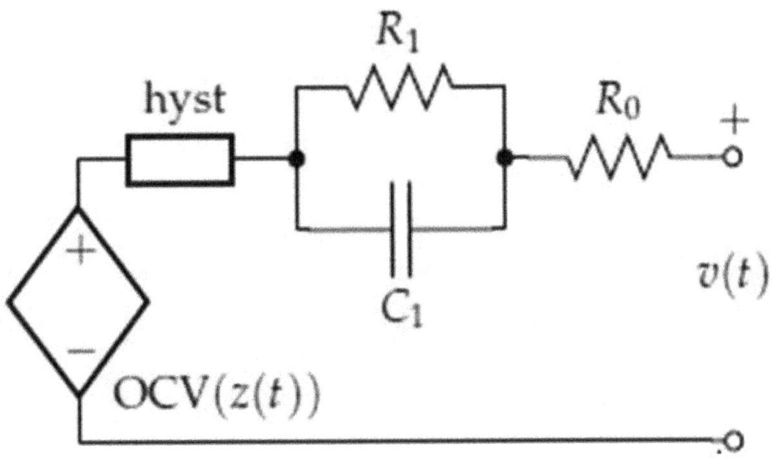

with the response. The voltage relaxes to its OCV after the removal of current. Step.1 created an object required to calculate the parameters of the cell and the data got loaded to the object.

Step. 2 - It is important to break up the pulse discharge profile into individual pulse objects. This step would break up the pulse discharge into smaller fragments. This showed the appropriate places where the pulses started and ended. When the waveforms were zoomed, red and green points were observed. Red points which indicate the load phase and the green points which are the relaxation phase.

Step. 3 – This would approximate the OCV based on the voltage at the relaxation phase. Based on the point at which the load is applied, the value of series ohmic resistance, R_0 would be estimated. After the completion of the estimation of R_0, the plots and values of E_m (OCV), R_0, R_1, C_1 etc.; were populated. These values would be tuned in the next step. It was observed that the pulse sequence had data and simulation points. The points in red indicated the values which were computed during simulation and the blue points were the actual values obtained after performing the experiment. During the dynamic portion of the voltage, the OCV was not estimated to the correct value. Now the initial values for the time constants were entered. The exponential curve was observed at the relaxation phase of each voltage and the curve fitting operation was performed (assuming that there were 3 time constants). It was observed that the estimation of OCV was better than the previous phase.

Step. 4 - This step estimates the values of the resistances R_0, R_1, C_1 etc.; the portion of the code treats the system like a linear system, makes some assumptions and carries out conditioning and processing of data. It uses optimization toolbox for the conditioning and processing of data. We saw that the relaxation plot was modified and the final values of battery parameters were obtained. Later, Simulink optimization tool was used in order to have best possible fit between the simulated and the experimental values.

Table 5. Steps to determine the battery parameters

OCV (V)	Capacity (Ah)	Ambient Temperature (K)	Number of RC pairs
4.1	10	303 to 315.15	2

Figure 35. Experimental data pre-loaded (Xi et al., 2019)

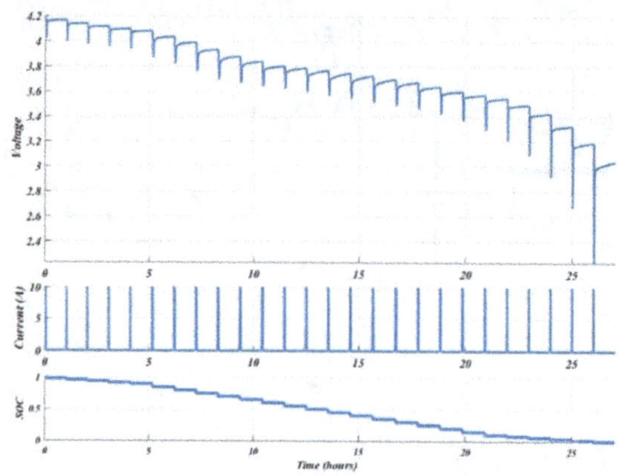

Figure 36 shows the pre-loaded experimental data for a Li ion Polymer 10Ah battery. This was available as a .mat file in Step 1. Figure 36 shows the split pulse discharge profiles into individual pulse objects. Figure 37 shows a comparison of estimated OCV versus the experimental data. The corresponding error was in the order mV.

The values of E_m and battery parameters for 5 RC pairs can be either provided by the user or defaulted to the values shown in Table 6.

It was found that the battery parameters were dependent on SOC and independent of T_{amb}. The mean residual compared to the experimental data was in the order of mV. With the correct initial guess for battery parameters, the response in the parameter estimation was faster. The value of R_1 for 1RC was around 9mΩ, around 2mΩ for 2RC and 3RC, around 1mΩ for 4RC and 0.25mΩ for SOC=100%. Hence, as the order of the RC increases, R_1 estimation provides better results.

Figure 36. Pulse identification

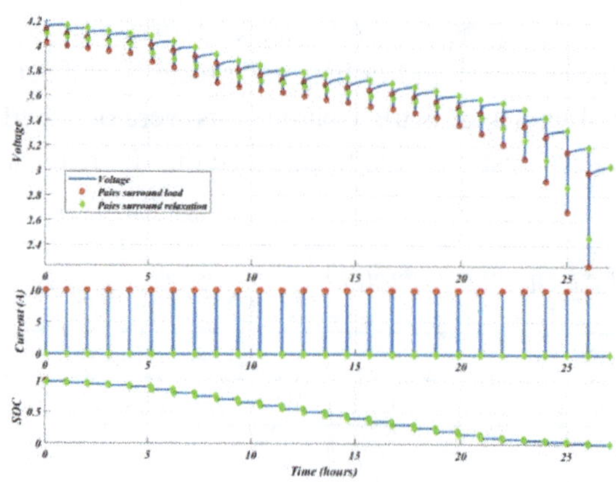

Figure 37. Comparison of estimated OCV with experimental data

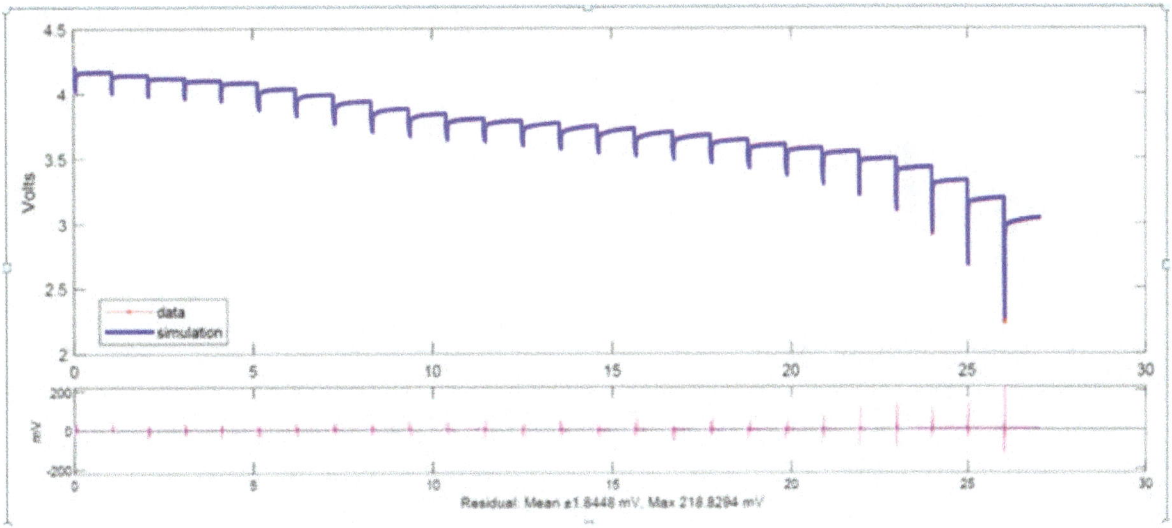

ESTIMATING CORE BATTERY TEMPERATURE

During high acceleration period, high current is drawn from battery due to which T_c reaches to a higher value than T_s. The difference can vary from 6°C to 10°C. Hence, T_c reaches a larger value than T_s. T_c can be used to operate the cooling system by designing a thermal management system. SOC or State of Health (SOH) algorithm uses T_c estimation for higher accuracy.

The temperature inside the battery or the electrolyte temperature is referred to as core temperature (T_c), T_s as surface temperature and T_{amb} as ambient temperature. Thermo couples are used for this purpose.

Thermistor (NTC/PTC) can be used to measure different temperatures in EV's. Once the resistances are calibrated, thermistor can be used in Wheatstone's bridge. A calibration chart would be provided wherein the change in temperature would be proportional to the voltage in mV. The software converts

Figure 38. 5RC modeling Vs. SOC showing E_m, R_1..R_5, τ_1..τ_5 (Xi et al., 2019)

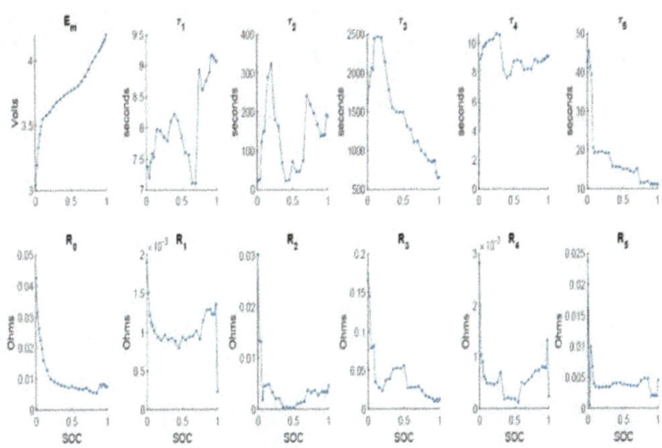

Figure 39. 4RC modeling Vs. SOC showing E_m, $R_1..R_4$, τ_1, $..\tau_4$ (Xi et al., 2019)

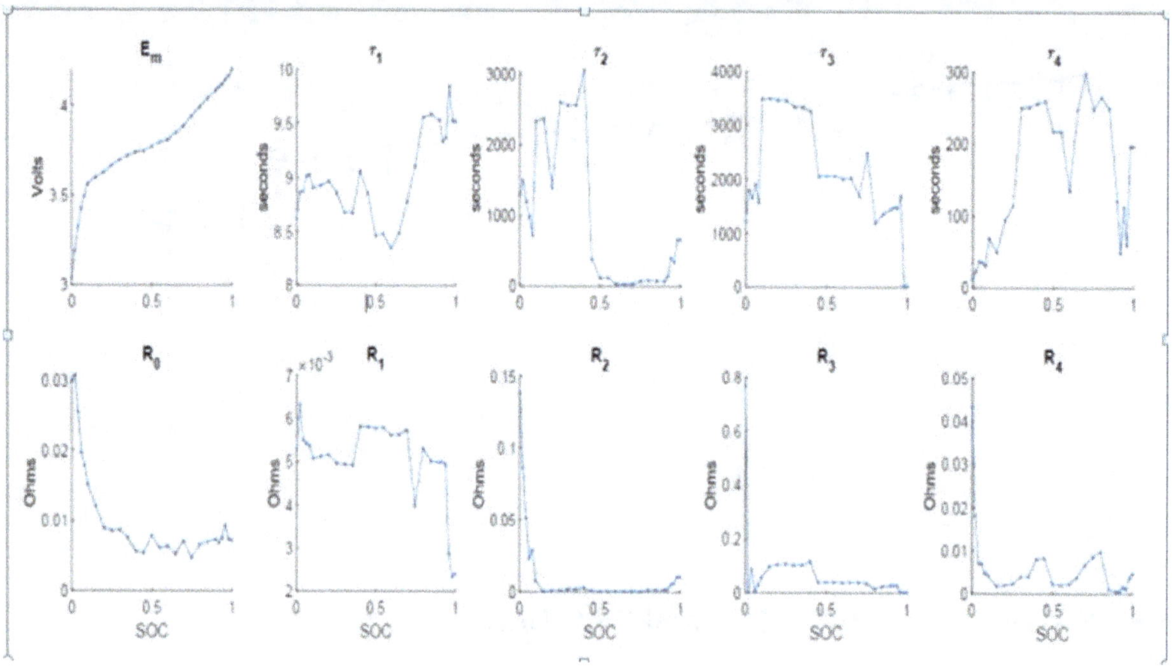

the measured voltage into the thermistor temperature. A thermal control for a Lithium-ion battery is necessary as Li ion batteries are very sensitive to temperature. Figure 43 shows the cell voltage of a Li ion battery (V) versus T_s. From the plot, it was observed that for the efficient working of the battery, temperature has to be < 60°C (Williamson, 2013).

Figure 40. 3RC modeling Vs. SOC showing E_m, $R_1..R_3$, τ_1, $..\tau_3$ (Xi et al., 2019)

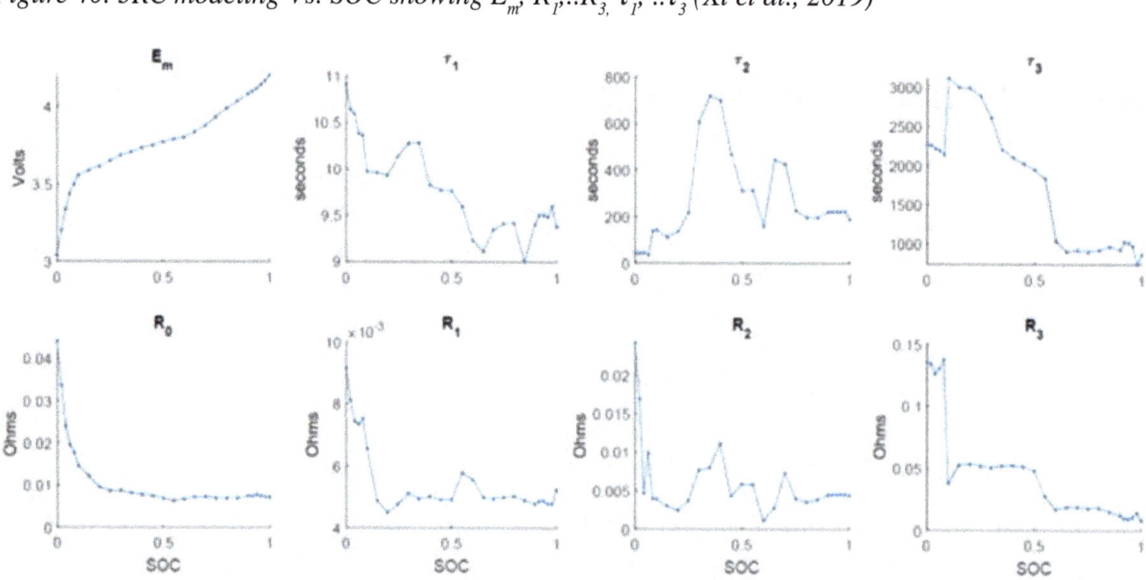

Figure 41. 2RC modeling Vs. SOC showing E_m, R_1, R_2, τ_1, τ_2 (Xi et al., 2019)

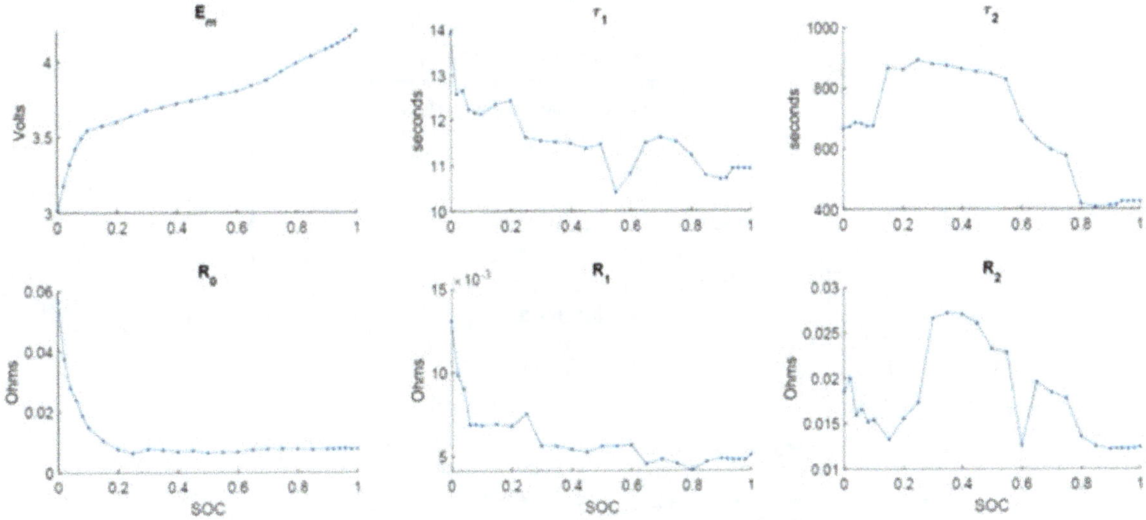

Figure 42. 1RC modeling Vs. SOC showing E_m, R_1 and τ_1 (Xi et al., 2019)

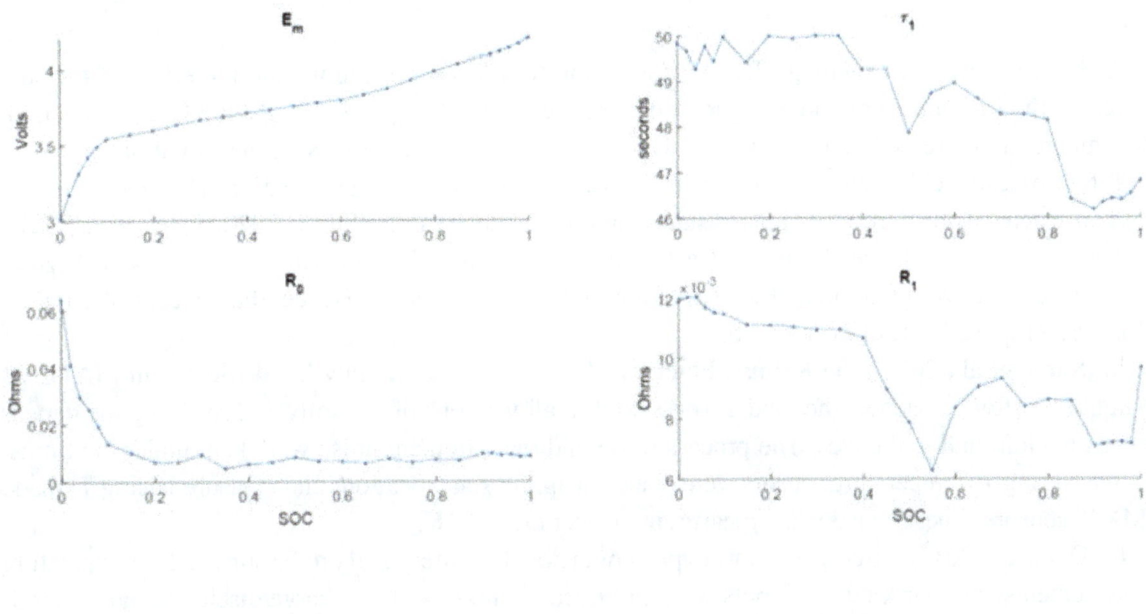

Table 6. Default values in the estimation of ECM parameters (Xi et al., 2019)

E_m (V)	R_0 (Ω)	τ
Nominal – 4.1	Nominal – 0.001	5
Min – 3.4	Min – 0.0001	50
Max – 4.2	Max – 0.1	200

Figure 43. Cell voltage Vs Ambient Temperature(Niculescu & Iancu, 2000)

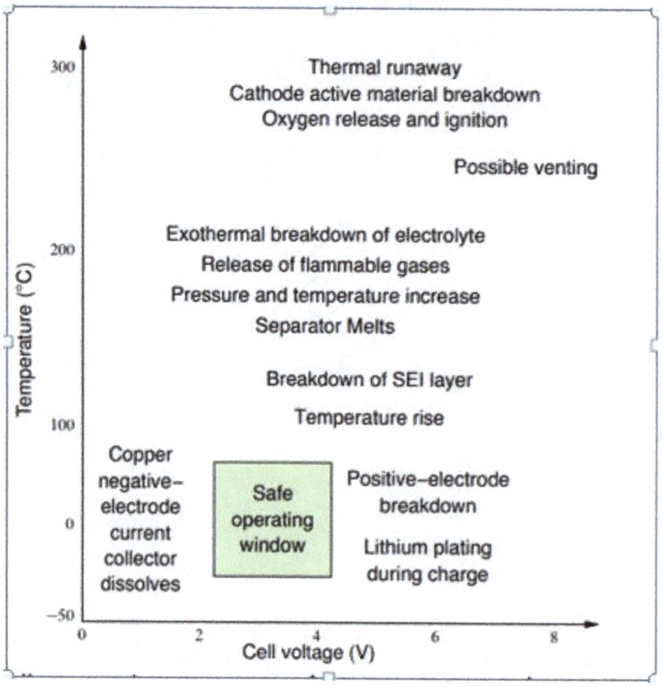

A fast charger was mathematically modeled using MATLAB / Simulink and the effect of fast discharge on the Li ion battery core temperature was investigated (Surya, 2020). As the C rate increased, the temperature increased to higher values which causes detrimental effects on the Li ion battery.

T_c for Pb (lead) acid battery was predicted and used a Kalman Filter for estimation. The thermal model was constructed using the convection resistance and capacitance values (Surya, 2020; Surya et al., 2021).

The thermal capacitance C_s had no effect on the estimation of T_c as the change in T_s was made zero. However, C_s role was predominant in estimation of T_s from known T_c. Hence, the concept of wireless BMS was proposed in (Surya et al., 2021).

In (Surya et al., 2021), the Kalman Filter was built using 'Commonly Used Blocks' in MATLAB / Simulink software. Hence, the model works well in all versions of the software, reducing the dependence on additional tool boxes. The process noise and measurement noise were kept under the limits. For measuring $T_{s,}$ temperature sensors were placed at appropriate locations and measured using TI made LMX35 sensors. These sensors had measurement accuracy of 1K.

In (Dai et al., 2015), three sensors were present inside the battery and on the surface for temperature measurements. Similar kinds of inputs were provided to the model. A charge/discharge pulse at 2C-rate was applied with ambient temperature of 24°C. It was assumed during the simulation that that the thermal parameters of battery were constants and did not vary with temperatures as their variation is not significant. The simulation run time was 6000s.

Figs. 44 and 45 show the current and the quantity of heat supplied to the thermal model.

The measured T_c in (Dai et al., 2015) is shown in Figure 46. The estimated T_c from the current algorithm is shown in Figure 47.

Figure 44. Current at 2C rate

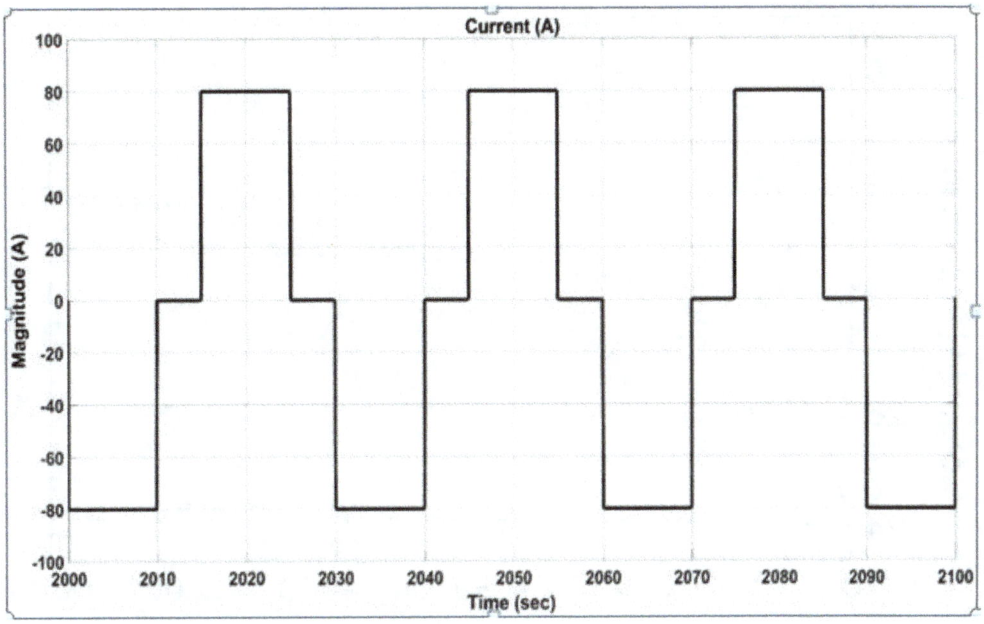

Figure 45. Heat Generated at 2C

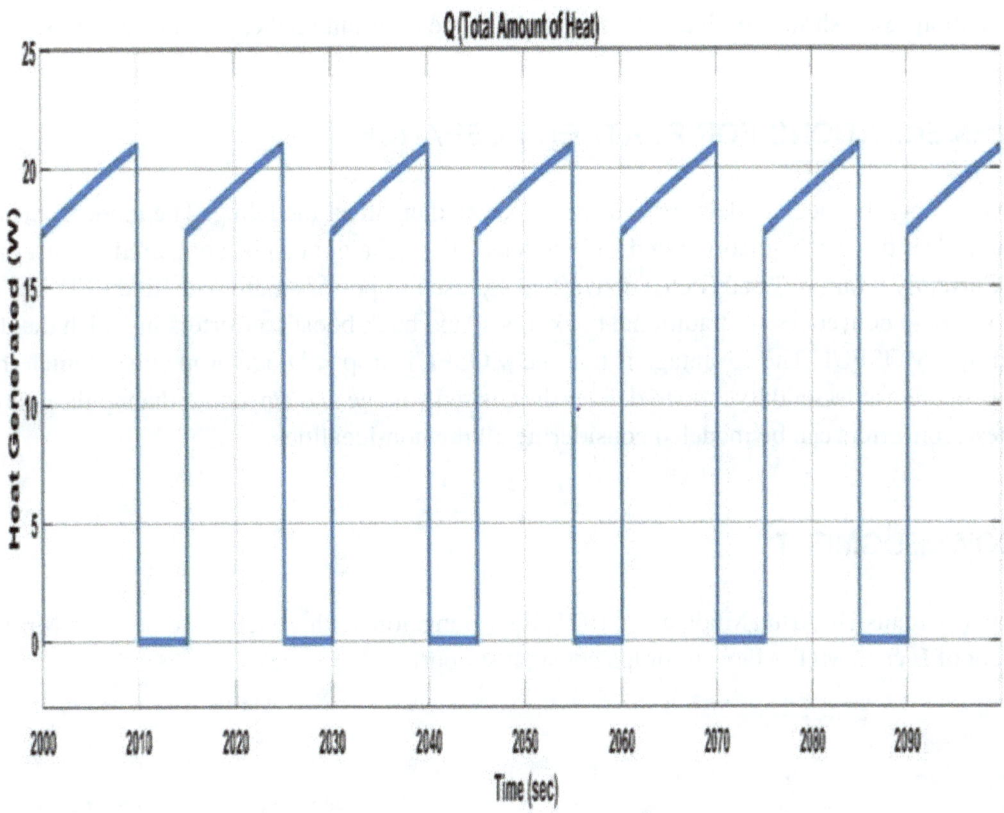

Figure 46. Measured Tc (Dai et al., 2015)

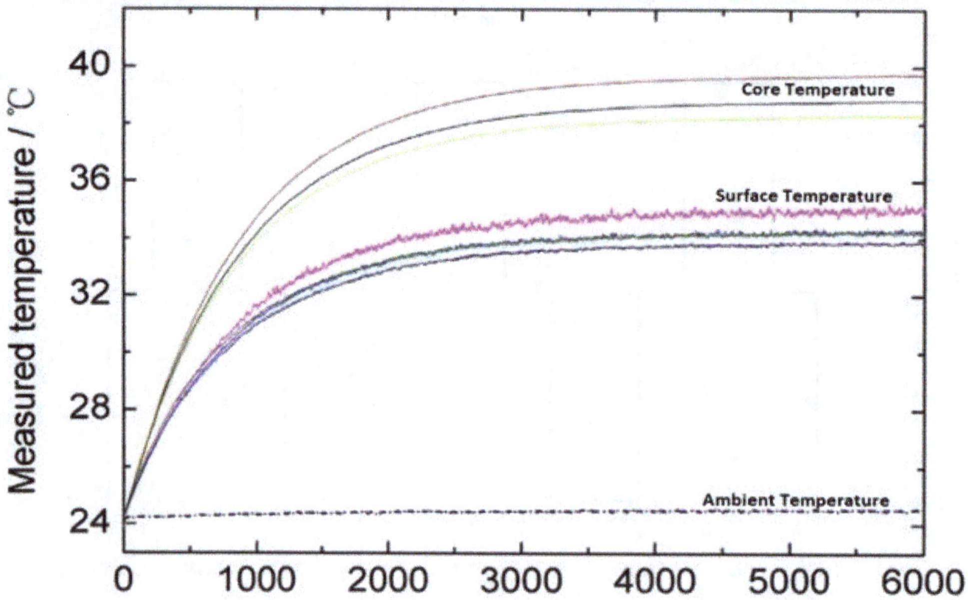

T_c was measured by inserting three temperature sensors inside a battery (Dai et al., 2015). Hence, with one temperature sensor, T_c will be the average of three temperature values which is around 38.5°C. The simulation result shown in Figure 46 depicted a value of around 38°C with an error 0.5°

RECOMMENDATIONS FOR FURTHER RESEARCH

In this work, only the ideal models are considered for mathematical modeling. The models considering all the non-idealities can be incorporated and modeled. The behavior can be compared against the ideal model. Currently, Point of Load (POL) converters are used to power small loads like LED and microprocessor. These converters are traditional buck, boost and buck-boost converters in which the diode is replaced by a MOSFET. The advantage is that the MOSFET drop < Diode drop due to which the converter can operate at lesser duty ratio to deliver the desired voltage and current, reducing the conduction loss. These converters can be modeled considering all the non-idealities.

ACKNOWLEDGMENT

I would like to thank Dr. Arjun Mudlpaur, CEDT, IISc, Bangalore for his encouragement and Mr. Vineeth Patil, Dept of E & E, MIT, Manipal for his constant support

Figure 47. Estimated T_c from the algorithm

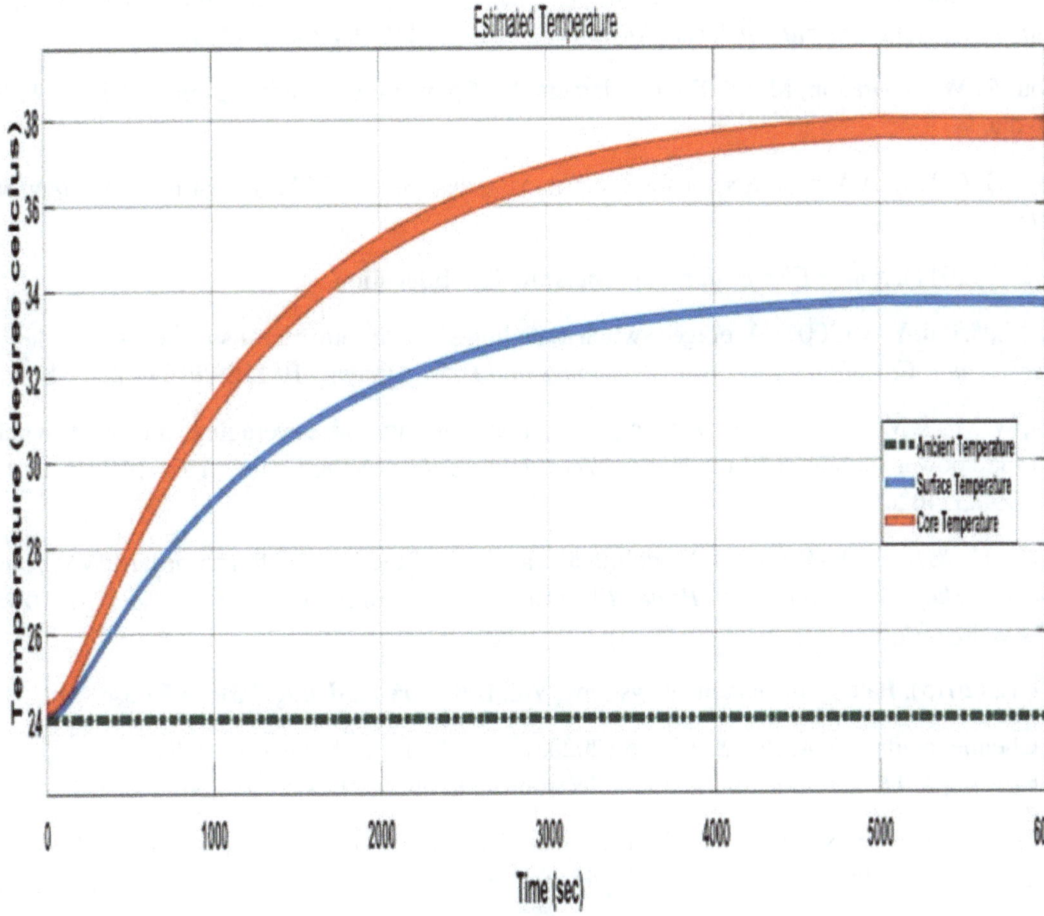

REFERENCES

Bertoldi, B. (2018). A Non-Ideal SEPIC DCM Modeling for LED Lighting Applications. *IEEE 4th Southern Power Electronics Conference (SPEC)*. 10.1109/SPEC.2018.8635896

Canalli, V. M. (1996). Behavioral large signal averaged model for dc/dc switching power converters. *PESC Record. 27th Annual IEEE Power Electronics Specialists Conference*, 2. 10.1109/PESC.1996.548806

Cuk, S., & Middlebrook, R. D. (1979). A general uniðed approach to modeling switching DC- to-DC converters in discontinuous conduction mode. *Proc. IEEE PESC*, 36–57.

Czarkowski, D., & Kazimierczuk, M. K. (1993). Energy-conservation approach to modeling PWM DC-DC converters. *IEEE Transactions on Aerospace and Electronic Systems*, *29*(3), 1059–1063. doi:10.1109/7.220955

Dai, H., Zhu, L., Zhu, J., Wei, X., & Sun, Z. (2015). Adaptive Kalman filtering based internal temperature estimation with an equivalent electrical network thermal model for hard-cased batteries. *Journal of Power Sources*, *293*, 351–365. doi:10.1016/j.jpowsour.2015.05.087

Eng, V., & Bunlaksananusorn, C. (2009). Modeling of a SEPIC converter operating in discontinuous conduction mode. *6th International Conference on Electrical Engineering/Electronics, Computer, Telecommunications and Information Technology*, 1. 10.1109/ECTICON.2009.5136983

Erickson, R. W., & Dragon, M. (2007). Fundamentals of power electronics. Springer Science & Business Media.

Gruosso, G. (2020). A Virtual Sensor for Electric Vehicles' State of Charge Estimation. *Electronics*, *9*(2), 278.

Hart, D. W. (2011). *Power Electronics*. Tata McGraw-Hill Education.

Lee, C.-M., & Lai, Y.-S. (2007). Averaged switch modeling of dc/dc converters using new switch network. *7th International Conference on Power Electronics and Drive Systems*. 10.1109/PEDS.2007.4487891

Moussa, W. M., & Morris, J. E. (1990). Comparison between state space averaging and PWM switch for switch mode power supply analysis. In *IEEE Technical Conference on Southern Tier*. IEEE. 10.1109/STIER.1990.324626

Niculescu, E., & Iancu, E.-P. (2000). Modeling and analysis of the DC/DC fourth-order PWM converters. *7th Workshop on Computers in Power Electronics. Proceedings (Cat. No. 00TH8535)*. 10.1109/CIPE.2000.904696

Plett, G. L. (2015). Battery management systems: Vol. I. *Battery modeling*. Artech House.

Surya, Chennegowda, Datar, Jha, & Victor. (2020, Dec.). *Accurate Battery Modeling Based on Pulse Charging using MATLAB / Simulink*. Power Electronics Drives and Energy System, PEDES.

Surya, Chennegowda, & Naraharisetti. (2020). Generalised Circuit Averaging Technique for Two Switch DC –DC Converters. In *8th International Conference on Power Electronics Systems and Applications*. PESA.

Surya, S. (2020). Development of thermal model for estimation of core temperature of batteries. *International Journal of Emerging Electric Power Systems*.

Surya, S., & Arjun, M. N. (2020). Effect of fast discharge of a battery on its core temperature. In *International Conference on Futuristic Technologies in Control Systems & Renewable Energy (ICFCR)*. IEEE. 10.1109/ICFCR50903.2020.9249999

Surya, S., Marcis, V., & Williamson, S. S. (2021). Core Temperature Estimation for a Lithium ion 18650 Cell. *Energies*, *14*(1), 87. doi:10.3390/en14010087

Surya, S., & Williamson, S. (2021). Generalized Circuit Averaging Technique for Two-Switch PWM DC-DC Converters in CCM. *Electronics (Basel)*, *10*(4), 392.

VandenBos, G., Knapp, S., & Doe, J. (2001). *Primary Inductor Selection by IET Labs*. https://www.ietlabs.com/pdf/Datasheets/1482.pdf

Williamson, S. S. (2013). *Energy management strategies for electric and plug-in hybrid electric vehicles*. Springer. doi:10.1007/978-1-4614-7711-2

Xi, Z., Dahmardeh, M., Xia, B., Fu, Y., & Mi, C. (2019). Learning of battery model bias for effective state of charge estimation of lithium-ion batteries. *IEEE Transactions on Vehicular Technology*, *68*(9), 8613–8628. doi:10.1109/TVT.2019.2929197

Chapter 6
Microgrid-Based Sustainable E-Bike Charging Station

Ghanishtha Bhatti
Department of Electrical and Electronics Engineering, Vellore Institute of Technology, Vellore, India

Raja Singh R.
Department of Energy and Power Electronics, Vellore Institute of Technology, Vellore, India

ABSTRACT

This chapter focuses on developing a sustainable architecture for public electric motorbike charging stations. Electric motorbikes or electric bicycles (both referred to as e-bikes) are compact electric vehicles which are primarily battery-powered and driven solely by electric motors. This work conceptualizes a microgrid architecture which utilizes the integration of distributed generation energy resources providing the charging station nodes with sustainable power and increased fault tolerance. The charging stations proposed in the study increase the long-time energy savings of the infrastructure maintenance authorities while also reducing reliance on the public grid during peak hours. The photovoltaic-based DC microgrid is integrated with e-bike charging infrastructure, moving towards a future of eco-friendly and power-efficient technology.

INTRODUCTION

The rapid depletion of polluting resources such as petroleum is a threat to existing technology and the civilization that it supports, and this has proliferated to a scale which cannot be ignored. As a result, sustainable technology and innovation has become the central focus of industry and also policymakers worldwide. The increased support for sustainability has caused the emergence of various scientific studies into employing technologies with greater renewable penetration like distributed generation. Even so, energy harvesting technology and DG resources are still in their nascent stage. Therefore, shifting entirely to DG supplied infrastructure is not yet cost-effective or resource-friendly, and the multitudes of protocols, standards, and infrastructure that is predominantly compatible only with conventional power supply is an additional hurdle. To overcome this impediment, the concept of Microgrids has been emergent, combining

DOI: 10.4018/978-1-7998-7626-7.ch006

conventional sources along with simultaneous injection from more sustainable DG resources (Locment and Sechilariu, 2015). This grid configuration is a feasible alternative to boost DG penetration during local utilization, while also truncating the cost of supply (Manna and Goswami, 2017; Patterson, 2012). The low-carbon automobile revolution is said to be spearheaded by plug-in electric vehicles, which are predicted to dominate the automotive market in the coming years. Presently, a noteworthy amount of car manufacturers have plug–in electric vehicle (PEV) options in their fleet, either battery-based or hybrid EVs. Another statistical forecast suggests that in the near future PEV models will be part of every major automotive company's fleet (Locment and Sechilariu, 2015). Although there is considerable scepticism about shifting to electric vehicles within the mean consumer demographic, this shift attributes to various customer perks such as a reduction in fuel costs and minimizing maintenance as the electric vehicle have approximately a tenth of the moving parts that mainstream vehicles have. Government promotion is key in the public perspective towards EVs. Additionally, many state-owned communities and municipal authorities provide added incentives like free parking services and insurance waivers to encourage purchasing greener alternatives (Bhatti and Singh, 2020). So far we have discussed the numerous merits of EVs, however, concomitantly they also increase the loading of grid supply sizeably during charging. Studies show the substantial impact of PEVs on the supply, noting that distribution transformers saw up to 68% increase in loading during peak hours (Aljanad and Mohamed, 2015). The same study notes that the peak power requirement due to PEVs increased by 30% at a penetration level of 17%. A detrimental effect on the public grid has been observed due to the sizeable current that is drawn during EV charging. A study on constant voltage charging shows that PEV current demands can vary from 64A to 400A, in fast charging mechanisms (Dharmakeerthi et al., 2011). This substantial impact indicates the need for renewing the conventional charging architecture of PEVs. Furthermore, a solution to incentivize the purchase and use of electric vehicles may stem from innovating its commercial model and infrastructure.

To tackle the aforementioned problems, this study is focused on realizing an energy-efficient charging station and a viable commercial model to bring electric vehicles into the mainstream transportation, even in developing countries. While similar studies have developed models for charging stations, this work uniquely conceptualizes a comprehensive utilization scheme, whose merits are expounded through the usage of Electric Bikes. The novelty of this study stems from its multifaceted savings approach in every aspect including:

- The development of Microgrid for charging over conventional grid-based charging.
- The usage of a rental transport mechanism over personal ownership.
- The usage of Electric Bikes over fuel-driven motorcycles.
- The focus on two-wheelers for commuting over four-wheelers.

The impacts of these design choices are illustrated in the results section of this paper. To the knowledge of the authors, this is the first study of its kind that provides an exhaustive scheme for the sustainable charging of electric bikes, while also promulgating a viable commercial model for incentivizing EVs. This paper focuses on conditions and development in suburban India, to ensure that the model is viable for use in developing nations.

The remainder of this paper is organized as follows. A practical consideration and implementation strategy for the Microgrid is presented in Section II. Section III discusses the foundational mathematical relations describing the systems involved in constructing a microgrid from an electrical perspective. It also expounds the use of photovoltaic arrays as the renewable energy resource used as supply to the

Microgrid. The input constraints and data used for simulation purposes is expounded in Section IV. This section also evaluates the criterion observed when modelling a Microgrid for use in E-bike charging. The results of the simulation performed on MATLAB Simulink and analysis of these results are discussed in Section V. Section VI provides a conclusion which encapsulates the contributions made by the proposed framework and examines its novelty.

Background

Microgrid architectures present prospects of innovation in power supply, and as a result, a plethora of research has integrated this configuration for PEV infrastructure optimization. Kuperman et al. (2013) discuss a novel charging methodology allowing an increased charging rate of 25 kWh per hour (peak rate) by utilizing a bidirectional AC-DC converter. A topology favoured by academics is integrating photovoltaic arrays within electric vehicle charging stations (Corchero et al., 2012; Guo et al., 2016; Mouli et al., 2016). However, these models often fail to consider practical constraints and market deployable implementation, which is crucial for commercial success. Selling energy back to the grid is another viable strategy to wax long-term energy savings and efficiency. It can be implemented through models such as vehicle to grid (V2G) (Armstrong et al., 2012). Incorporating this framework with smart energy management such as predictive scheduling can warrant long term profits. Van Roy et al. (2014) depicts an analogous approach that profits through utilization of excess energy while vehicles are charging. Several integrated schemes have also been covered such as Mohamed et al. (2014) which conceptualizes an unprecedented architecture for PV-integrated grid-focused charging stations. The model maximizes benefits through real-time energy management algorithms that propagate through feedback loops to allow seamless operation. In the case of composite loads like building complexes, multiple agents are requisite and a layered approach must be employed as demonstrated by Zhu et al. (2012). The DC Microgrid is incorporated in building architecture, uses various diverse renewable sources, and showcases a paradigm for smart construction in the future. Other academia concentrates on the control of parameters concerning PEV charging during peak hours in order to reduce the burden on public grids (Goli and Shireen, 2014). However, predominantly it can be observed that such works lack systems description during grid power disconnected state. Conventionally PEVs rely majorly on Lithium-Ion batteries, though recent research demonstrates the prospects of utilizing distinctive storage media such as magnetic superconductors (Liu et al., 2015; Strunz et al., 2014). Once again, power electronic converters act as bridges for the interface of these media to a DC bus. Apart from the reduction of reliance on the grid, a microgrid also presents benefits from an economic standpoint. It has been studied that, in case of geographically small projects, Microgrid offers a 16% increase in overall profitability as compared with standard grid supply (Uski et al., 2018). Literature shows that the diversity of supply provides increased reliability to the system in the face of grid power outages, and Microgrids can run at an average of 66% efficiency, which is substantially larger than the 50% peak efficiency of conventional power delivery (Dirkman, 2015).

Majorly, the research and production of PEVs are inclined towards electric cars. However, another promising area of development is electric two-wheelers, consisting of electric bicycles and electric motorbikes. An additional factor that impacts the popularity of two-wheelers is the advent of the urban traffic menace coupled with narrow roads and a lack of large parking spaces. The result of this convenience is seen in automotive demographic studies which showed that 80% of the 26.27 million vehicles purchased in India were two-wheelers. Furthermore, it was observed that a majority of the 1.56 lakh EV sales (excluding E-rickshaws) in India were attributed to electric two-wheelers (India Brand Equity

Foundation, 2020). With growing government interest towards sustainability, a large amount of research has been going into developing infrastructure to support the movement towards energy efficiency and renewable resources. Researchers have explored various frameworks for two-wheeled EV charging such as Widjanarka and Sutiyatno (2020), which devises a feasible strategy for remote charging using wireless energy transfer using FETs, proposed by, is shown to be a viable charging strategy for motorcycles in remote locations. Further analysis of literature shows that two-wheeled EVs can be categorized as a unique hybrid EV class which can support flexible supply inclusive of petroleum and renewable resources (Wu and Xu, 2019). Inácio et al. (2020) propose a proof-of-concept logic controller for authenticated use of public charging stations, although, the RFID system used in the preliminary study cannot fulfil the required reliability and security constraints. However, it may be observed from a review of literature that these studies that the study of microgrids for usage with Electric Bikes are generally absent. Furthermore, the studies focus on the charging of individual Electric Bikes, as opposed to the creation of a market deployable model for charging fleets of Electric Bikes.

To develop a commercially popular model, as well as an energy-efficient one, it is necessary to study the approach followed by businesses in the transportation industry. In India, the new generation of office workers and city dwellers are constantly subject to long commutes, with an average of 90 minutes spent in commute (Pant et al., 2017). Purchasing personal vehicles, thereby putting more vehicles on the road, is not sustainable and not economically viable for most of the target demographic. A study on the impact of rising fuel prices and commute time showed that the percentage of the population in urban Indian sample who preferred public transportation to personal vehicle ownership increased from 19% in 2010 to 49% in 2018 (Sivarajan et al., 2018). One rental bike startup Bounce™ has notably procured funding in the order of 105 million dollars, boasting of a customer base of 2.1 million (Wicked Ride Adventure Services, 2020). While such services are immensely popular, they invest substantial amounts into refuelling bikes exacerbated by the increasing prices of crude oil by 2.5% from 2018 to 2019 (PPAC, 2020). This setup can be optimized by adopting E-bikes as well as introducing sustainable charging.

MICROGRID ARCHITECTURE OF E-BIKE CHARGING STATION

A DC Microgrid forms the foundation of this charging station infrastructure. As shown in Figure 1, the DC Microgrid receives supply from two disparate types of sources: two photovoltaic arrays (10kW and 50kW) as well as a 220V 60Hz AC supply from the public grid. The grid supply along with power electronic converters form the backbone of the Microgrid. The supply is rectified and regulated at a level of 48 V DC to facilitate charging of the E-bike batteries, which are docked at the station. A purely grid reliant charging station would present several aspects of convenience being the major commercially utilized model. However, the Microgrid opted in this model paves the path for introducing resilient systems in small-scale applications to minimize costs and reduce peak-hour public grid consumption, which an emerging concern in metropolitan areas (Koh and Lim, 2015).

Using diverse power source puts into place a system contingency, which is an effective countermeasure for power outages and the presence of faults or disturbances in the public grid supply. Operational resilience is increased through this contingency, which reduces instances of maintenance for the owners along with uninterrupted and reliable service to the end-user. Additionally, secondary source status has been given to the public grid, which invokes the prospect of selling excess energy back to the grid. This Vehicle to Grid model (V2G) is used in various studies, though it is not deemed necessary for the purpose of this study. By demoting the utility grid, this system not only presents flexibility, but also the potential of expansion either by the means of adding other DG resources or by later liberating this system from the public grid altogether.

The operation of microgrids takes place in various modes, which make them a flexible alternative to conventional rigid grid configurations. The two prominent modes of operation are delineated as follows (Prakash and Paul, 2016).

- **Hybrid Mode:** The breaking mechanism (B1) is closed to connect to the utility grid. The utility grid equips all the feeders and the distributed generation based Microgrid provides some surplus power to the main grid.
- **Islanding Mode:** The breaker (B1) is opened as per the microgrid control strategy during peak grid hours or grid failures. Therefore, this mode constitutes the independent operation through DG resources preventing the station from going offline when the utility grid cannot be availed.

Both these modes are simulated as sections of this study and their operation is individually validated to ensure that the model is feasible. After deliberation on the merits and challenges of various renewable sources, solar energy has been adopted as the distributed generation source for the following reasons:

1. From a human perspective, the energy that can be tapped from the Sun is limitless.
2. Solar power has comparably high power density amongst other energy resources.
3. The resource does not pollute directly and substantially alleviates carbon footprint.
4. It demonstrates compatibility with low and high power device categories.
5. Apart from the array capital and maintenance costs, solar energy is free.
6. Solar cells are known for their low maintenance requirements and operational durability.
7. Through effective placement strategies (pole-mounting) it will not use up valuable ground space.
8. Are suitable for placement near residential regions as it does not contribute to noise pollution.
9. Most open areas are suitable for high harvesting rates through solar panels.

Overall, the model integrates aspects of automation which enables operation with minimal intervention by humans, which reduces labour costs. Furthermore, the aspect of fuelling the vehicles is eliminated, therefore, great benefits to the operating agency are incurred. The primary source in this setup up consists of the two solar arrays. These arrays shall be ideally shed-mounted over the charging docking station of the E-bikes, which provides the additional perks of shading the vehicles and low transmission loss while harvesting. While other DG sources may provide larger volumes of power, the proposed PV system causes the least disruption to surrounding construction and is the most economical. The systems design has been visualized as a schematic diagram in Figure 1.

The commercial utilization scheme of this model is derived from the existing model of rental bike services in India. However, the scheme has been modified to increase savings and sustainability, while reducing the environmental impact of the system. The exhaustive nature of this architecture and its usage as a product is illustrated in Figure 2 and is the major salient feature of this work. By targeting an average commuter demographic (21-26years), who statistically prefer not to own personal vehicles, the product can easily garner a customer base (Sivarajan et al., 2018). This scheme incentivizes the wide-scale use of electric vehicles, and it does not require customers to make a permanent commitment, unlike purchasing a personal EV. The reduction of dependence on the grid allows the business to reduce electricity

Figure 1. Microgrid architecture of the E-bike charging station

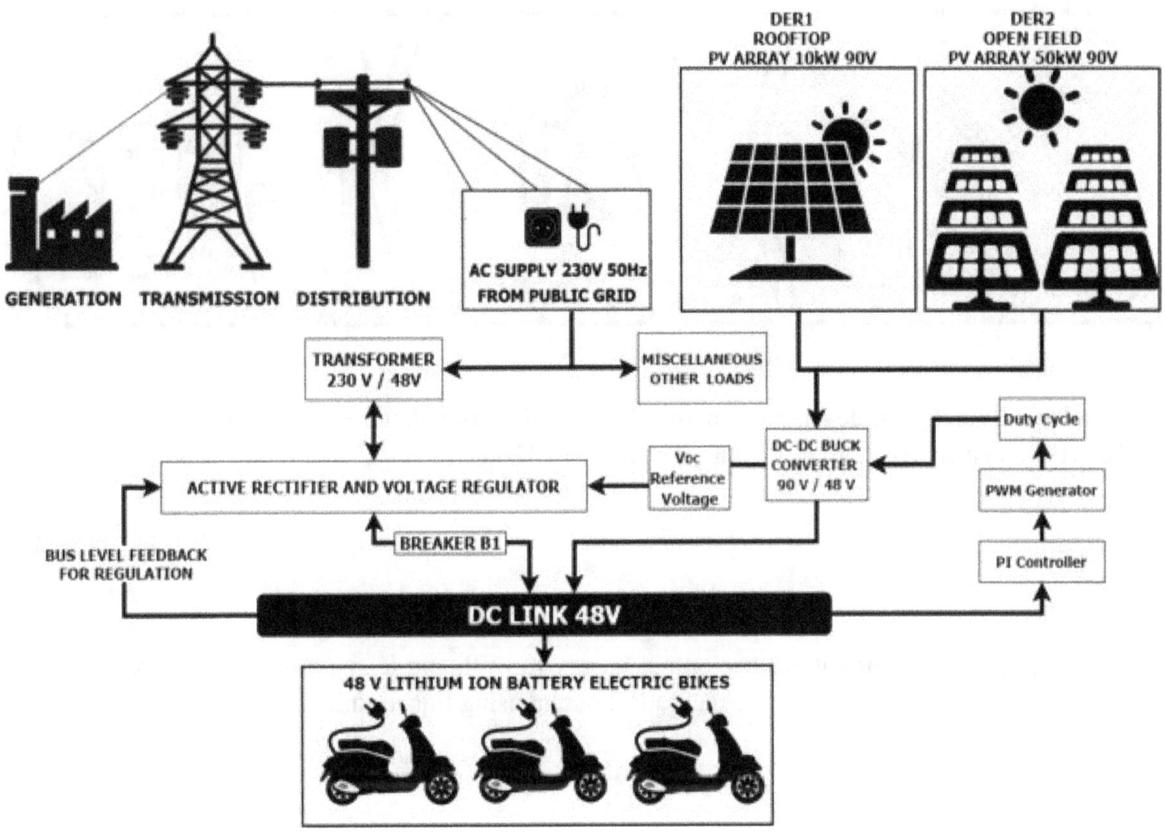

expenses and possibly allowing them to offer the rental service at a lower price than their competitors. They are also able to provide greater consumer satisfaction as the probability that the charging station has to be taken offline during power outages is minimized by the resilience of the microgrid architecture.

MATHEMATICAL MODELLING OF PV POWERED E-BIKE CHARGING STATION

Computation of the mathematical relations that govern the physical and electrical components is critical for successfully simulating the DC microgrid. It is also necessary to priory determine the battery modelling parameters as they have a profound impact on the charging behaviour and output analysis. Additionally, state of charge (SOC) and voltage behaviours must also be established to effectively evaluate the framework performance.

Photovoltaic Array

The two solar systems that act as distributed energy resources for the microgrid must be sized appropriately to ensure efficient usage. In order to create a sustainable power supply, it must be ensured that energy supply is always greater than energy demand, accounting for uncertainty in the incident radiation

Figure 2. Commercial utilization of the e-bike charging station

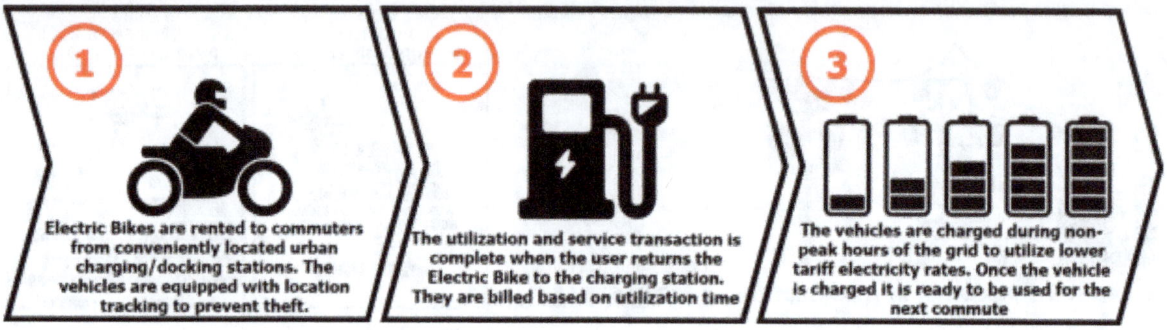

to the solar cells and possible aberrant energy demands. To this end, we must establish the quality factor of the array as given in Equation 1 (Skunpong and Planklang, 2011).

$$Q = \frac{E_{Load}}{E_{th}} \tag{1}$$

Here, E_{Load} is the empirical electric output energy in kWh and E_{th} is the theoretical output of the system in kWh. We can determine the theoretical output using Equation 2.

$$E_{th} = \frac{P_{peak} \times E_{glob}}{I_{STC}} \tag{2}$$

The P_{peak} value refers to the peak power of the PV array, whereas E_{glob} represents the global radiation in kWh/m^2. This value is generally calculated under standard testing conditions (STC), where temperature is maintained at 25°C and irradiance (I_{STC}) is 1000 W/m^2 (Skunpong and Planklang, 2011). It now becomes essential to determine the efficiency (η_s) of the PV cell, shown in Equation 3.

$$\eta_s = \frac{I_{max} \times V_{max}}{P_{peak}} \tag{3}$$

Here, I_{max} and V_{max} denote maximum current and maximum voltage, respectively. Once this computation is completed, the calculation of PV array sizing can be undertaken, as per Equation 4. A_{array} is the size of the solar system in m^2.

$$A_{array} = \frac{E_{th}}{E_{glob} \times \eta_s} \tag{4}$$

Power Electronic Converters for the Microgrid

DC-DC converters provide the necessary interface for the photovoltaic supply to the charging nodes at the DC bus. The objective of these converters is to regulate the output voltage at a preset level corresponding to the specifications of loads. Specifically for this model, a buck converter in tandem to the individual PV arrays is required to step down the voltage to levels which are acceptable for the E-bike load. The variables encompassing a description of the DC-DC buck converter, concerning its electrical characteristics, is given below (Gowda et al., 2014):

1. F_s: minimum switching frequency of the converter.
2. $V_{IN(min-max)}$: the input voltage range.
3. V_{OUT}: nominal output voltage.
4. $I_{OUT(max)}$: maximum output current.

In order to determine the value of inductance, duty cycle (D) of a converter must be calculated using Equation 5 where η_c represents efficiency of the converter. For this work η_c is assumed to be 0.9.

$$D = \frac{V_{OUT}}{V_{IN(\min)} \times \eta_c} \tag{5}$$

The constituent passive element values are the next step of the converter design process. These calculations are detailed in Equation 6 and Equation 7. L and C represent inductance and capacitance, respectively.

$$C = \frac{\Delta I_R}{8 \times Fs \times \Delta V} \tag{6}$$

$$L = \frac{D \times \left(V_{IN} - V_{OUT}\right)}{\Delta I_R \times Fs} \tag{7}$$

Here, ΔI_L represents ripple current, which is 30% of the maximum output current. ΔV is the output voltage ripple which as per is computed as below. Here, ΔV_{OUT} is the desired voltage output ripple, which is determined with consideration for ripple added by the output capacitor equivalent series resistance (ESR) shown in Equation 8. The value of ESR is taken to be 0.03Ω as per Gowda et al. (2014).

$$\Delta V_{OUT} = ESR \times \Delta I_L \tag{8}$$

Through a similar process, the determination of capacitance on the input side is taken from datasheets. Ideally, to provide voltage stability for the erratic supply, this value is kept as low as possible.

Modelling Electrical Vehicles

The validation of this charging station architecture relies solely on the charging profile shown by the E-bikes. Therefore, it is justified to focus the EV modelling primarily on the charging subsystem and the battery circuit equivalent parameters. With this consideration, a pivotal aspect for the simulation is ascertaining the battery chemistry. In accordance with typical EV battery chemistry, Lithium-Ion battery specifications have been employed throughout this chapter.

EV manufacturers generally favour Lithium-Ion batteries as they possess Coulombic Efficiency (CE) above 99%, which surpasses the value for most other conventional batteries. CE is defined as a charge efficiency of electrons transportation within batteries. Specifically, it refers to the total charge obtained from the battery per total charge given to the battery during a full cycle (Miao et al., 2019). The usable battery capacity is characterized by SOC of battery defined as the ratio of remaining capacity to fully charged capacity (Equation 9).

$$SOC = \frac{Q - Q_{min}}{Q_{max} - Q_{min}} \tag{9}$$

Here, Q denotes the charge that is presently stored, Q_{min} represents the residual charge that persists in the battery when during open circuit or minimal voltage state, whereas Q_{max} is the charge at maximum voltage state. To visualize the charging of a battery, SOC must be defined, computed in Equation 10.

$$\mathrm{SOC} = \mathrm{SOC}_0 - \frac{1}{\mathrm{SOC}_0} \int \mathrm{i}\, \mathrm{dt}. \tag{10}$$

The initial SOC level of the battery is given by variable SOC_0 and i represents battery current. For an exhaustive approximation, additional parameters must be subsequently computed. Principally, the voltage-based SOC (SOC_V) must be determined through Equation 11.

$$\mathrm{SOC}_v = \frac{1}{\mathrm{a1}} \left(V_{OC} - a_0 \right) \tag{11}$$

Here, V_{OC} denotes the open-circuit voltage (dependant on the battery temperature), a_0 is terminal voltage at 0% SOC, and a1 represents the terminal voltage at 100% SOC. Current-based SOC (SOC_i) can be found through the Coulomb Counting method (Equation 12). Here, C_p constitutes battery capacity.

$$\mathrm{SOC}_i = \frac{1}{C_p} \left(C_p - \int \mathrm{i}\, \mathrm{dt} \right) \tag{12}$$

Accurate estimation is made possible by including the systematic correction factor (η), which is derived from initial SOC level. The discharging profile may then be corrected through Equation 13.

$$\eta = \left(1 - \frac{SOC_0}{100}\right) \quad (13)$$

Using the defined constraints an optimized equation for SOC is derived, as shown in Equation 14. An additional weighting factor (w) is added to ensure consistency.

$$SOC = w \times SOC_v + (1 - w)(SOC_i - \eta) \quad (14)$$

The Lithium-ion batteries used in E-bikes are designed in such a way that even with large discharge currents (commonly occurring in electric vehicles) they maintain their capacity. The required battery capacity (C_B) for traversal depends mainly on its riding characteristics and can be calculated using Equation 15 (Thirugnanam et al., 2012). In this equation, P_D refers to the power demand per distance travelled and D represents the average distance travelled by the vehicle.

$$C_B = P_D \times D \quad (15)$$

In order to gain perspective on the behaviour of the E-bike battery, it is important to systematically categorize its discharging profile based on riding characteristics. For this purpose on important constraint is weight of the entire vehicle inclusive of luggage and passengers. Vehicle weight has the greatest impact on the range in the speed range from 0 to about 25-30km/h and the uphill climbs. The air resistance (R_{Air}) and rolling resistance (R_{Roll}) can then be defined by Equation 16 and 17 (Electric Bike Simulator, 2013). For this purpose, the assumption of suitably flat terrain is made.

$$R_{Air} = 0.5 \times 1.226 \times K_{Aero} \times S_{face} \times N_t \quad (16)$$

$$R_{Roll} = M \times K_{Roll} \times A \quad (17)$$

Here, K_{Aero} and K_{Roll} are the coefficients for aerodynamics and rolling, respectively. S_{face} is the face surface, N_t is the speed of the vehicle, M is its mass, and A is the acceleration of Earth.

IMPLEMENTATION OF THE E-BIKE CHARGING STATION

The proposed DC microgrid was constructed using electrical systems design blocksets in MATLAB Simulink. The simulation diagram with annotations delineating the various sources is shown in Figure 3. PV arrays, converters, and E-bikes loads are all cohesively modelled within this framework.

Photovoltaic Array Parameters

Temperature and Irradiance variation encompasses the physical considerations while sizing the solar system. Two separate solar systems have been realized as DERs in order to validate the resilience of the model as a Microgrid. The first PV array will be roof-mounted locally at the charging system, whereas

Figure 3. Microgrid simulation diagram

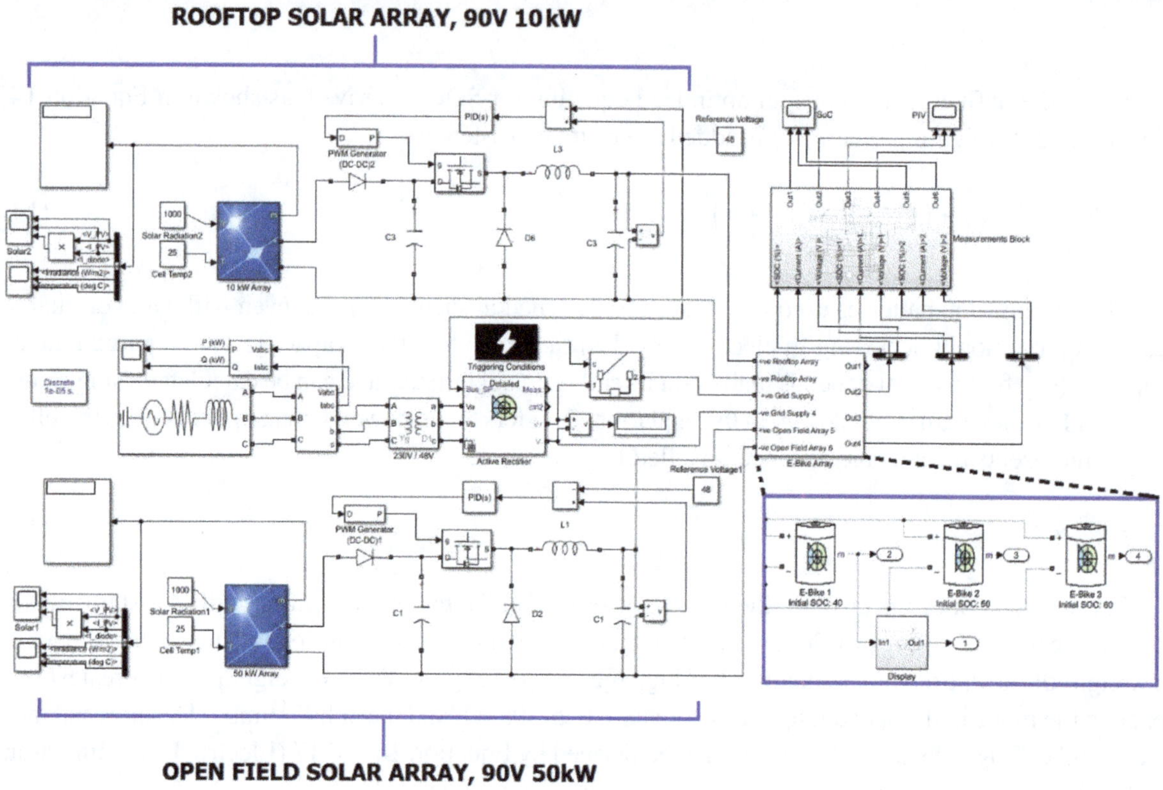

the second array will be situated in an open field. Indian market models for solar cells are utilized in this chapter, maintaining homogeneity with the target geographic location. As mentioned in Section III, standard testing conditions have been utilized for the purpose of simulation. The parameters of the two solar systems are detailed in Table 1 and 2. As the system is grid-connected, no storage mechanism (battery storage) has been modelled as this is the configuration used for 98% of installed on-grid solar systems (Pillai et al., 2020). The use of two geographically separated arrays provides greater reliability.

Public Grid Parameters

A standard three-phase source models the public grid at 230V phase voltage and 230 MVA (Chowdhury, 2015; Rodney, 2020) operating at 50Hz. A 230V/48V transformer is used to streamline the voltage to bus levels. The grid supply and its connection to the DC link are visualized in Figure 4.

Power Electronic Converter Parameters

Power electronic conversion is majorly facilitated in the microgrid by these two circuits.

- **Three-phase AC-DC converter:** The converter in a microgrid bridges the energy generation and consumption points. A standard 3-phase PWM active rectifier is used for this purpose. The recti-

Table 1. Simulation parameters of the rooftop PV array

Sl. No.	Parameter	Value
1	Quality Factor	0.6
2	Peak Power of Array	10 kWp
3	Efficiency of Solar Cell	0.187
4	Array Area	112m^2

Table 2. Simulation parameters of the open-field PV array

Sl. No.	Parameter	Value
1	Quality Factor	0.75
2	Peak Power of Array	50 kWp
3	Efficiency of Solar Cell	0.187
4	Array Area	560m^2

fier controller is a derivative model of a PI regulator, which receives feedback (reference level) from the grid and PV array electrical parameters including voltage and current.

- **DC-DC Buck Converter:** The converter's function is to step down the voltage level from the solar array (90V) to the stable voltage of the E-bike battery (48V). A MOSFET is used to provide the switching action needed for stepping down the voltage. This has been chosen over IGBTs due to its superior efficiency during low voltage operation. Converter specifications are delineated in Table 3.

In order to maintain the DC link voltage and facilitate CV charging without significant distortion, separate strategies are utilized for the AC and DC supplies. For the PV array supply, a PI controller is fed the output of the buck converter (as seen in Figure 3). The output generated by the controller after tuning supplies a PWM generator. This dynamically varies the switching of the MOSFET to achieve regulated output voltage levels. In the case of the public grid supply, the PWM rectifier achieves the same result by receiving feedback from the DC link. This allows the voltage level to be maintained with minimal distortion. The converter circuit, constructed on MATLAB Simulink, is shown in Figure 5.

Figure 4. Simulink diagram of grid supply and connections

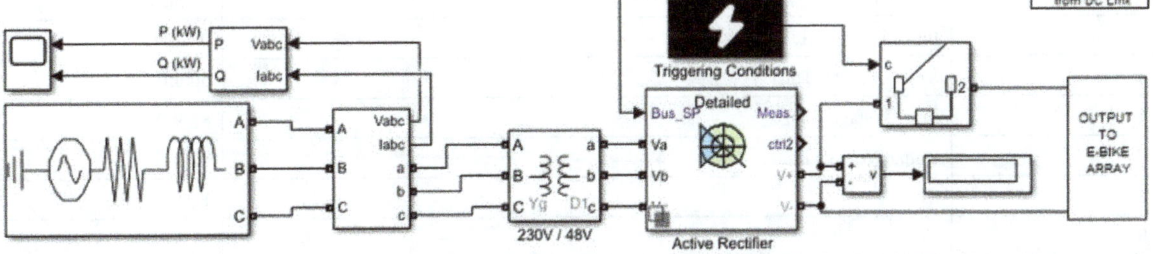

Figure 5. DC-DC buck converter circuit

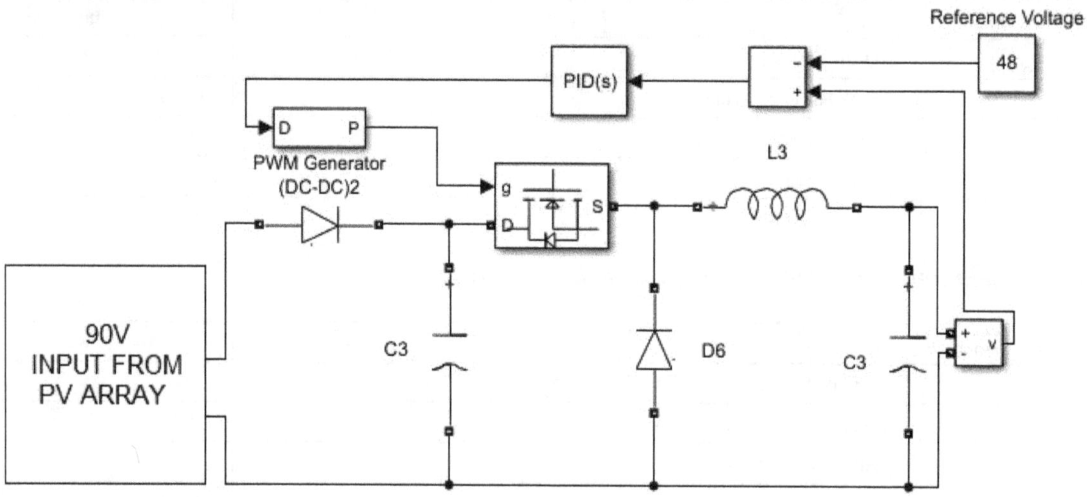

Model Modes and Microgrid Control Action

In order to evaluate the feasibility of microgrid architecture, the model runs in different modes during simulation, which entail various energy source combinations. The modes are as follows:

1. **Hybrid Mode:** The ideal functioning mode for a microgrid, which utilizes power from both AC public grid as well as two separate DG sources (in this case, PV arrays).
2. **Islanding Mode:** In case a fault or failure to the public grid occurs, the microgrid islands itself and supplies power independently from DG sources to the load.

The model is built with circuit breakers for switching between the different modes of the microgrid. This breaker may be triggered either by detection of a fault within the grid. Generally, faults in grid-connected systems can be caused by sudden disturbances, fault conditions, single or multiple contingen-

Table 3. Parameters of the DC-DC buck converter

Sl. No.	Parameter	Value
1	Input Capacitance	183μF
2	Output Capacitance	10μF
3	Inductance	1.23μH
4	Initial Duty Cycle of Switching	0.5269
5	Switching Frequency	600kHz
6	Input Voltage Range	88-92V
7	Output Voltage Range	47-49V
8	Kp	1
9	Ki	12

cies, line overloading or load increases. In such cases, if the charging station must continue to function it must island itself. The circuit breaker is operated with accordance to the standard IEEE 1159–2019 and EN 50160, for fault-based tripping. According to these protocols a 10% deviation from the nominal system voltage for greater than 1 minute, will be categorized by the system as an overvoltage or under-voltage fault depending on whether the change is a net increase or decrease, respectively (Singh et al., 2020). If such a long-term voltage instability occurs the switch B1 is programmed to open and protect the system. From the system stability point of view, +/-10% voltage fluctuations of the rated voltage have been allowed. However, the voltage reduces drastically during the high X/R ratio in the distribution network. The X/R ratio is calculated using Equation 18 and 19. In which, X is the reactive part and R is the resistive part of impedance (Z_N) (Sarkar et al., 2016).

$$X = \frac{Z_n}{\sqrt{1 + \frac{1}{XRR^2}}} \tag{18}$$

$$R = \frac{Z_n}{\sqrt{1 + XRR^2}} \tag{19}$$

The breaker is also triggered during normal operation to reduce the load on the public grid during peak hours, by islanding. This is triggering is facilitated by the microgrid control algorithm. The microgrid has been designed to render the maximum benefits to the users while placing the least load on the public grid especially during peak grid hours. By distributing the utilization of the public grid to hours when commercial load is not at its peak, the charging station can utilize the maximum benefit of low electricity tariff rates, therefore, increasing savings. As per the Indian standard daily load curve, the peak morning hours are taken between 9:00 AM and 1:30 PM (Buckley et al., 2019). Therefore, the microgrid is operated in islanding mode during these hours, as long as the PV generation can meet the energy demand of the E-bike load. The control algorithm is expounded in Figure 6.

Systems Design of E-Bike Battery

The load is modelled based on Vespa Elettrica™, a representative economic class E-bike which is readily available in the Indian market (Piaggio and C, 2018). To fulfil simulation requirements without tampering with model accuracy, commercial customizations have been made to the Simulink battery blockset making them compatible with the E-bike batteries. Other tertiary variables that affect battery performance such as ageing and temperature effects were not expounded during this study. To ensure the practicality of the setup, the various E-bikes simulated have different levels of initial charge within them as shown in Table 4. This enables validation of charging system against simultaneous charging of varying loads. The discharge characteristics highlighting the calculated riding behaviour of the E-bike is given in Table 5.

The microgrid control algorithm shown in Figure 6 is centred around the commercial demand curve in India, as well as the time-of-day tariff system. The algorithm begins execution when a load is placed on the charging station. The system evaluates the condition of the grid, according to the overvoltage, undervoltage, and power outage categories (IEEE 1159–2019 and EN 50160). If the grid is online and no faults are detected by the system, the algorithm proceeds to the time of the day criterion. The daytime peak hours of the load on the public grid are from 9:00 AM to 1:30 PM and this is when the electricity tariff is highest. Therefore, provided that sufficient irradiance is present to fulfil the output demand using the DERs, the algorithm islands the microgrid from the public grid supply. Once the peak hours have passed the microgrid switches to its ideal operating state in the hybrid mode, where higher performance charging is facilitated.

RESULTS AND DISCUSSION

The MATLAB Simulink platform is used to construct the systems simulation and validate the DC Microgrid for PEV charging. The developed model conceptualizes the charging station as a standalone Microgrid. This framework is then executed and analyzed during islanding mode and hybrid mode operation. To optimize the simulation execution without creating processor-hungry instances, it has been executed on Simulink Accelerator mode using Just-in-time algorithm for a run-time of 10 seconds. As part of the accelerator mode, the platform excludes referenced models and produces data only for the top-level model in memory. The C compiler is then free during simulation, which has a positive impact on the speed of execution and reduces processor load. The charging profile obtained for the PEVs in various operational modes is shown in Figure 7.

Observed Characteristics

To ensure commercial compatibility, this model attempts to implement a Constant Voltage (CV) charging profile for every EV. As seen in Figure 8b and 9b all the nodes are uniformly charging at same voltage level despite their varying initial SOC levels (varying load). The CV method gives the model robustness when being charged by a DC link and also makes the manufacturing relatively uncomplicated. However, it may be observed that the initial spike in charging current within each mode which is a commonly attributed characteristic of CV charging and can cause the battery temperature to rise that may cause damage to the battery if high voltage is used.

On studying the charging profiles for each of the three E-bikes, it is evident that when simultaneously connected to the charger in parallel configuration, the vehicle with the lowest initial SOC utilizes the largest portion of available energy. This results in a comparatively faster-charging rate for the vehicles, which have lower initial SOC, introducing dynamic adaptability to this charging station and boosting its efficiency when subject to real demand in the form of E-bike loads.

Hybrid Mode

This mode introduces the microgrid architecture in place of a common public grid supply. Predictably, Figure 7 shows that the rate of SOC growth is 0.02544, 0.01818, and 0.01299 per second for the three E-Bikes at 40%, 50%, and 60% initial SOC, respectively. The charging characteristics, shown in Figure 8, show consistent charging pattern with respect to current and power. However, a noticeable ripple is present with the waveforms for both current and power that is attributed to incomplete rectification of

Figure 6. Microgrid control algorithm for the operation

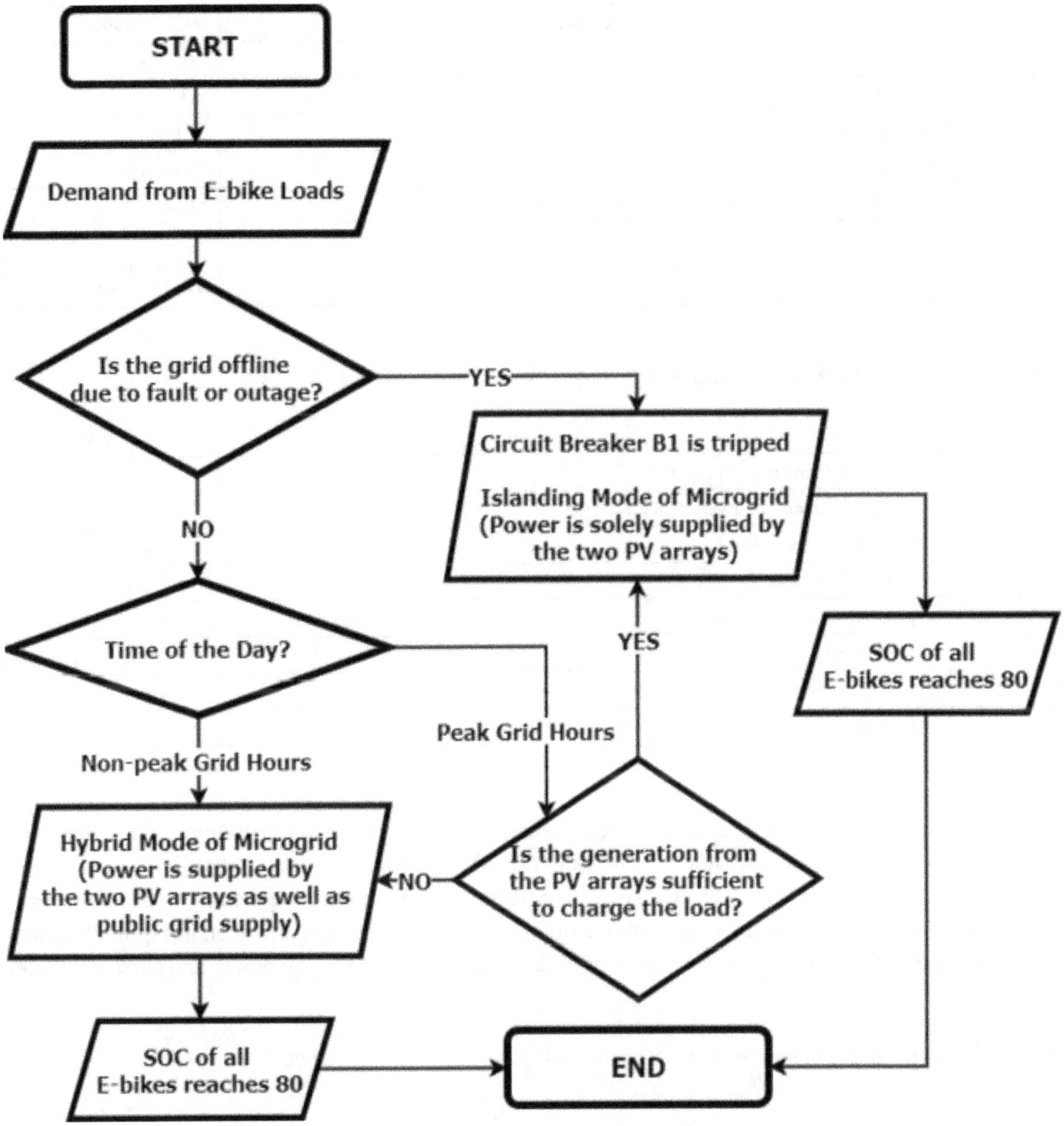

the AC supply of the public grid. It may be further optimized by using digital filtration or capacitor-based regulation.

Islanding Mode

Islanding, a salient feature of microgrids when exposed to electric faults in a public grid, is presented in this mode. The rate of charging, seen in Figure 7, is observed to be 0.02169, 0.01484 and 0.01008

Table 4. Parameters of the electric bike model

Sl. No.	Parameter	Value
1	Nominal Battery Voltage	48V
2	Nominal Battery Capacity	86Ah
3	Battery Energy	4.2 kW/h
4	Forecasted Battery Life	1,000 cycles
5	Optimal Charging Time (charging from 10% initial SOC)	4 hours
6	Initial Battery SOC (Varying Loads)	40% - E-bike 1 50% - E-bike 2 60% - E-bike 3

Table 5. E-bike riding and discharge characteristics

Sl. No.	Parameter	Value
1	Required Capacity for Traversal	4132Wh
2	Calculated Battery Capacity during Traversal	86.1Ah
3	Assumed Average Speed	60km/h
4	Rolling Resistance Coefficient	0.013
5	Aerodynamic Coefficient	0.9
6	Face Surface	0.46m^2
7	Rolling Resistance	7.65N
8	Air Resistance	33.10N

per second for E-bike 1, 2, and 3, respectively. While comparing to the hybrid mode, this represents an average drop of 18.5% in the SOC growth rate, it is observed that renewable distributed generation framework is able to supply enough energy to support the system without requiring it to be taken completely offline for maintenance. Furthermore, from the charging characteristics seen in Figure 9, it may be discerned that the utilization of this mode provides a more stable charging power and current profile,

Figure 7. Charging profile for e-bikes at initial SOC of a) 40% b) 50% c) 60%

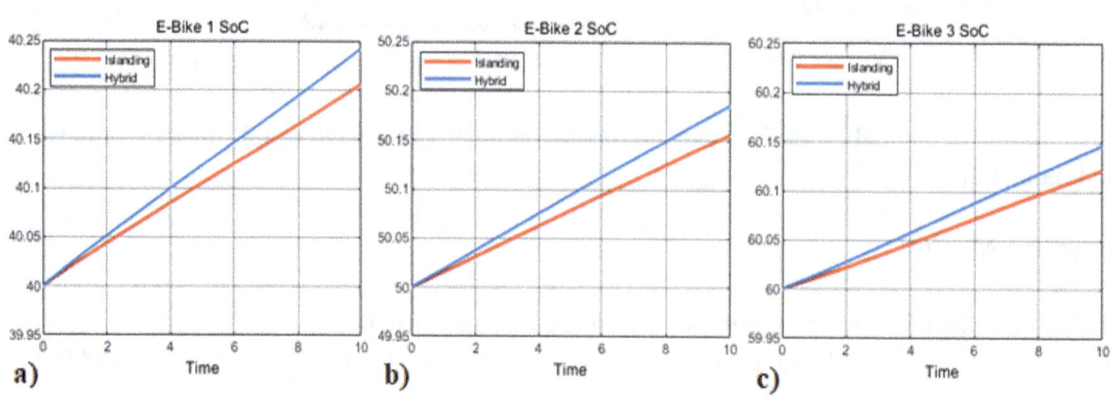

Figure 8. Hybrid mode charging characteristics a) Current b) Voltage c) Power

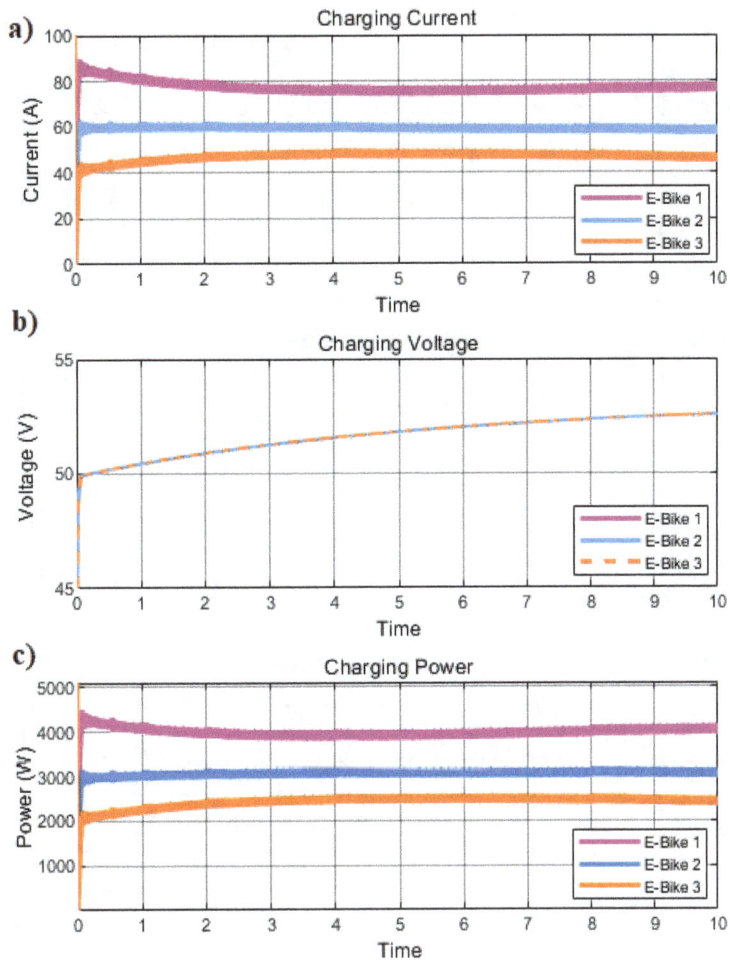

which should be considered as an incentive for efforts to switch to more sustainable DC sources, such as PV systems. A smaller current ripple constitutes safer charging and a smaller likelihood of deterioration of the battery due to increased temperature.

Techno-Economic Analysis

As this study focuses on developing a market deployable model, the economic constraints must be studied and the environmental impact of the proposed architecture. The economic and environmental comparison has been streamlined for the case study area of India. These findings have been summarized in Table 6. This analysis has been performed on Homer Grid™. The key takeaways from this analysis are that due to the initial capital cost the initial net present cost is higher for the model proposed by this paper. However, on recovery of the initial capital, the microgrid architecture presents considerable savings (an operating cost reduction of 53%), as well as superior renewable penetration and lower carbon emissions (reduction by 41.2%). Furthermore, there is potential for the architecture to sell back excess

Figure 9. Islanding mode charging characteristics a) Current b) Voltage c) Power

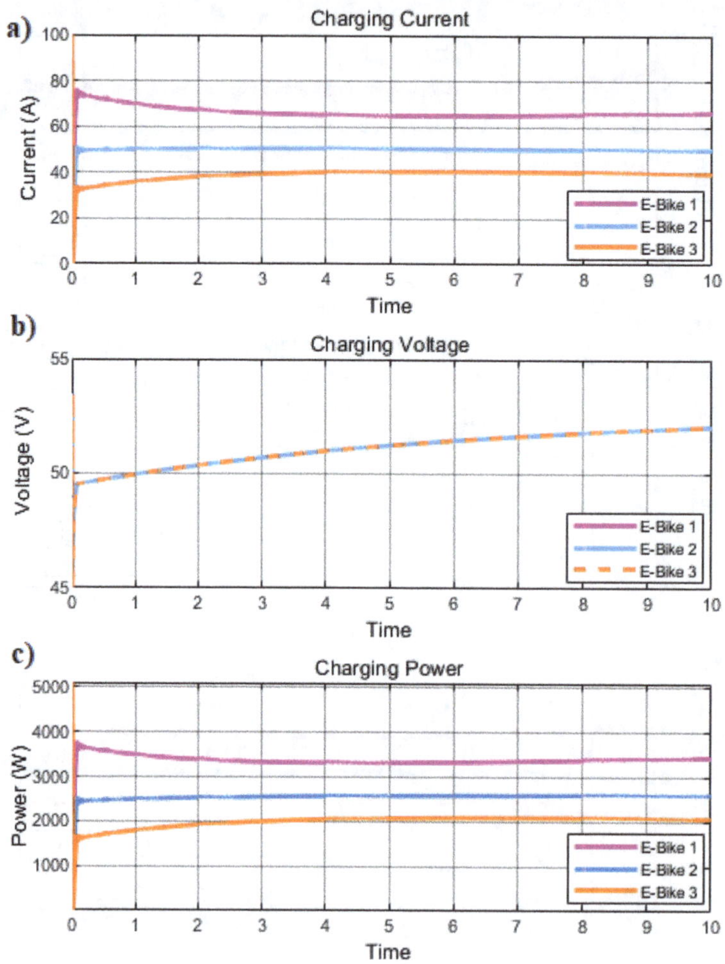

energy, produced during peak sunlight hours, back to the grid. This substantiates the novelty and merits of this model from the standpoint of rental businesses.

FUTURE RESEARCH DIRECTIONS

The model developed in this study was analysed against the metric of ideal EV battery load, in which it was found to sufficiently charge at a comparable rate. The objective of this study was to establish a proof-of-concept for the usage of microgrids in commercial EV charging stations. Therefore, there is a limited venture into the complexities of battery capacities and real-time vehicle dynamics that may impact the usage of this architecture. The future scope of this chapter would be to build on these current limitations and incorporate intelligent systems that would aid in power flow optimization.

Table 6. An economic comparative study.

Sl. No.	Parameter	Conventional E-Bike Charging	Proposed Microgrid E-bike Charging
1	Total Net Present Cost	$123,166.50	$215,003.20
2	Annual Operating Cost	$9,527.47	$4,477.17
3	Annual Excess Electricity Potential	0 kWh	2,438 kWh
4	Renewable Fraction	0%	51.6%
5	Annual Carbon Dioxide Emissions	85,787 kg	50,459 kg

CONCLUSION

Slowly but surely, the market inches towards adopting electric vehicles especially those that are more cost-effective such as E-bikes, as they offer a myriad of incentives. To keep pace with this growing EV market, it is crucial to begin developing sustainable infrastructure to maintain EVs. A prime example of such infrastructure is EV charging stations and rental services. Especially in developing nations, these commodities are a rarity and continue to negatively impact the public outlook towards electric motorcycle and bicycles. Therefore, this study develops a sustainable model for a charging station, based on Microgrid architecture. The model is targeted at rental bike businesses, which have become increasingly popular in urban cities. By utilizing a photovoltaic setup with two separate DERs, the charging station model achieves a renewable penetration of 51%, as well as a sizable reduction in running costs and carbon emissions. Using the islanding and hybrid modes of the Microgrid, the charging station deters overdraw from the public grid during the peak hours of operation and can increase savings through utilization of lower time-of-day tariff rates. The charging station simulated in this study has been validated for stable simultaneous charging of varying E-bike loads, which fulfils the objective of implementing a viable commercial architecture.

ACKNOWLEDGMENT

The authors would like to express our gratitude to the faculty and staff members at Vellore Institute of Technology who supported us in the procurement of the necessary software and development of the proposed model. This research received no specific grant from any funding agency in the public, commercial, or not-for-profit sectors.

REFERENCES

Aljanad, A., & Mohamed, A. (2015). Impact of Plug-in Hybrid Electric Vehicle on Power Distribution System Considering Vehicle to Grid Technology: A Review. *Research Journal of Applied Sciences, Engineering and Technology*, *10*, 1404–1413. doi:10.19026/rjaset.10.1841

Armstrong, M., Moussa, C., Adnot, J., Galli, A., & Rivière, P. (2012). Optimal Recharging Strategy for Battery-Switch Stations for Electric Vehicles in France. *Energy Policy, 60*. Advance online publication. doi:10.2139srn.2169221

Bhatti, G., & Raja Singh, R. (2020). Design of a Microgrid Architecture for Rental E-Bike Charging Stations. *IOP Conference Series: Materials Science and Engineering*, Volume 906, *Virtual Conference on Advances in Electric Drives, Process Control and Automation.*

Buckley, T., Gupta, A., Garg, V., & Shah, K. (2019). *Flexing India's Energy System Making the Case for the Right Price Signals through Time-Of-Day Pricing*. Institute for Energy Economics and Financial Analysis Press Release.

Chowdhury, T. (2015). Design of a Micro-Grid System in Matlab/Simulink. *International Journal of Innovative Research in Science. Engineering and Technology., 04*, 5262–5269. doi:10.15680/IJIRSET.2015.0407030

Corchero, C., Cruz-Zambrano, M., Heredia, F., Cairo, J., Igualada-Gonzalez, L., & Romero-Ortega, A. (2012). Optimal sizing of microgrids: A fast charging station case. *Proceedings of 9th International Conference on the European Energy Market*, 1-6. doi: 10.1109/EEM.2012.6254677

Dharmakeerthi, C.H., Nadarajah, M., & Saha, T. (2011). Overview of the impacts of plug-in electric vehicles on the power grid. *Innovative Smart Grid Technologies Asia (ISGT), 2011 IEEE PES.* doi:10.1109/ISGT-Asia.2011.6167115

Dirkman, J. (2015). *Microgrids: Changing the Face of Energy Output and Consumption.* Retrieved from https://electricenergyonline.com/energy/magazine/884/article/Microgrids-Changing-the-Face-of-Energy-Output-and-Consumption.htm

Electric Bike Simulator. (2013). *Power Calculator.* Retrieved from http://www.electricbikesimulator.com/index,en.html

Goli, P., & Shireen, W. (2014, May). PV Integrated Smart Charging of PHEVs Based on DC Link Voltage Sensing. *IEEE Transactions on Smart Grid, 5*(3), 1421–1428. doi:10.1109/TSG.2013.2286745

Gowda, M., Kiran, Y., & Parthasarthy, S. (2014). Modelling of Buck DC-DC Converter Using Simulink. *International Journal of Innovative Research in Science, Engineering and Technology, 3.*

Guo, Y., Xiong, J., Xu, S., & Su, W. (2016, May). Two-Stage Economic Operation of Microgrid-Like Electric Vehicle Parking Deck. *IEEE Transactions on Smart Grid, 7*(3), 1703–1712. doi:10.1109/TSG.2015.2424912

Inácio, C., Silva, I., Faria, J., Pombo, J., Calado, M., & Mariano, S. (2020). *Controller Design and Experimental Validation of a Solar Powered E-bike Charging Station.* KnE Engineering. doi:10.18502/keg.v5i6.7104

Indian Automobile Industry Report. (2020). India Brand Equity Foundation.

Koh, S. L., & Lim, Y. (2015). *Evaluating the economic benefits of peak load shifting for building owners and grid operator.* doi:10.1109/ICSGCE.2015.7454265

Kuperman, A., Levy, U., Goren, J., Zafransky, A., & Savernin, A. (2013). Battery Charger for Electric Vehicle Traction Battery Switch Station. *Industrial Electronics. IEEE Transactions on.*, *60*, 5391–5399. doi:10.1109/TIE.2012.2233695

Liu, Y., Tang, Y., Shi, J., Shi, X., Deng, J., & Gong, K. (2015). Application of Small-Sized SMES in an EV Charging Station with DC Bus and PV System. *IEEE Transactions on Applied Superconductivity, 25*(3), 1-6. doi:10.1109/TASC.2014.2374174

Locment, F., & Sechilariu, M. (2015). Modeling and Simulation of DC Microgrids for Electric Vehicle Charging Stations. *Energies*, *8*(5), 4335–4356. doi:10.3390/en8054335

Manna, D., & Goswami, S. (2017). Micro-Grid Operation and Control. *Proceedings of National Conference on Recent Trends in Engineering and Technology (NCRTET-2017).*

Miao, Y., Hynan, P., von Jouanne, A., & Yokochi, A. (2019). Current Li-Ion Battery Technologies in Electric Vehicles and Opportunities for Advancements. *Energies*, *12*, 1074–1094. doi:10.3390/en12061074

Mohamed, A., Salehi, V., Ma, T., & Mohammed, O. (2014, April). Real-Time Energy Management Algorithm for Plug-In Hybrid Electric Vehicle Charging Parks Involving Sustainable Energy. *IEEE Transactions on Sustainable Energy*, *5*(2), 577–586. doi:10.1109/TSTE.2013.2278544

Mouli, C., Ram, G., Bauer, P., & Miro, Z. (2016). System design for a solar powered electric vehicle charging station for workplaces. *Applied Energy*, *168*, 434–443. doi:10.1016/j.apenergy.2016.01.110

Pant, P., Habib, G., Marshall, J., & Peltier, R. (2017). PM 2.5 exposure in highly polluted cities: A case study from New Delhi, India. *Environmental Research*, *156*, 167–174. doi:10.1016/j.envres.2017.03.024

Patterson, B. T. (2012, November-December). DC, Come Home: DC Microgrids and the Birth of the "Enernet". *IEEE Power & Energy Magazine*, *10*(6), 60–69. doi:10.1109/MPE.2012.2212610

Petrol Planning and Analysis Cell. (2020). *Crude Oil Price (Indian Basket).* Retrieved from https://www.ppac.gov.in/content/149_1_PricesPetroleum.aspx

Piaggio & C. (2018). *Vespa Elettrica.* Retrieved from https://elettrica.vespa.com/_common/pdf/EN-DIGITAL-VespaElettrica-10-2018.pdf

Pillai, A. M., Kanna, R. R., & Raja Singh, R. (2020). Advancement of Inventive Solar Power Based Frameworks for Rural India. *IOP Conference Series: Materials Science and Engineering*, Volume 906, *Virtual Conference on Advances in Electric Drives, Process Control and Automation.*

Prakash, M. M., & Paul, J. (2016, May). Control of Microgrid for Different Modes of Operation. *International Journal of Engineering Research & Technology (Ahmedabad)*, *5*(5). https://dx.doi.org/10.17577/IJERTV5IS051001

Raja Singh, R., Yash, S. M., Chavan, S. S., Indragandhi, V., Vijayakumar, V., Saravanan, P., & Subramaniyaswamy, V. (2020, November). IoT embedded cloud-based intelligent power quality monitoring system for industrial drive application. *Future Generation Computer Systems*, *112*, 884–898.

Rodney, T. (2020). *MPPT Solar Charge Controller Model.* MATLAB Central File Exchange. Retrieved June 26, 2020, https://www.mathworks.com/matlabcentral/fileexchange/73115-mppt-solar-charge-controller-model

Roy, J. V., Leemput, N., Geth, F., Büscher, J., Salenbien, R., & Driesen, J. (2014, October). Electric Vehicle Charging in an Office Building Microgrid with Distributed Energy Resources. *IEEE Transactions on Sustainable Energy*, *5*(4), 1389–1396. doi:10.1109/TSTE.2014.2314754

Sarkar, T., Dan, A. K., & Ghosh, S. (2016). Effect of X/R ratio on low voltage distribution system connected with constant speed wind turbine. *2nd International Conference on Control, Instrumentation, Energy and Communication, 417-421.*

Sivarajan, N., Mathew, N.P., Gowda, N.D., & Thomas, N. (2018). Rising Fuel Prices in Bangalore - Causes and Impact. *International Journal of Research and Analytical Reviews, 5*(4).

Skunpong, R., & Plangklang, B. (2011). A Practical Method for Quickly PV Sizing. *Procedia Engineering*, *8*, 120–127. doi:10.1016/j.proeng.2011.03.022

Strunz, K., Abbasi, E., & Huu, D. N. (2014, March). DC Microgrid for Wind and Solar Power Integration. *IEEE Journal of Emerging and Selected Topics in Power Electronics*, *2*(1), 115–126. doi:10.1109/JESTPE.2013.2294738

Thirugnanam, K., Saini, H., & Kumar, P. (2012). Mathematical modeling of Li-ion battery for charge/discharge rate and capacity fading characteristics using genetic algorithm approach. *2012 IEEE Transportation Electrification Conference and Expo. ITEC*, *2012*, 1–6. doi:10.1109/ITEC.2012.6243431

Uski, S., Rinne, E., & Sarsama, J. (2018). Microgrid as a Cost-Effective Alternative to Rural Network Underground Cabling for Adequate Reliability. *Energies*, *11*, 1978. doi:10.3390/en11081978

Wang, Z., Wang, L., Dounis, A., & Yang, R. (2012). Integration of plug-in hybrid electric vehicles into energy and comfort management for smart building. *Energy and Building*, *47*, 260–266. doi:10.1016/j.enbuild.2011.11.048

Wicked Ride Adventure Services Pvt Ltd. (2020). *Bounce™*. Retrieved from https://bounceshare.com/

Widjanarka, W., & Sutiyatno, S. (2020). Experiment electrical motorcycle used solar energy with wireless transfer energy method for transportation in isolated regions. *Journal of Physics: Conference Series*, *1517*, 012098. doi:10.1088/1742-6596/1517/1/012098

Wu, C. H., & Xu, Y. X. (2019). The Optimal Control of Fuel Consumption for a Heavy-duty Motorcycle with Three Power Sources Using Hardware-in-the-Loop Simulation. *Energies*, *13*, 22. doi:10.3390/en13010022

ADDITIONAL READING

Aluisio, B., Dicorato, M., Ferrini, I., Forte, G., Sbrizzai, R., & Trovato, M. (2019). Optimal Sizing Procedure for Electric Vehicle Supply Infrastructure Based on DC Microgrid with Station Commitment. *Energies*, *12*(10), 1901. doi:10.3390/en12101901

Beer, S., Gomez, T., Dallinger, D., Momber, I., Marnay, C., Stadler, M., & Lai, J. (2012). An Economic Analysis of Used Electric Vehicle Batteries Integrated Into Commercial Building Microgrids. *IEEE Transactions on Smart Grid*, *3*(1), 517–525. doi:10.1109/TSG.2011.2163091

Salehpour, M., & Moghaddas-Tafreshi, S. M. (2020). Contract-based utilization of plug-in electric vehicle batteries for day-ahead optimal operation of a smart micro-grid. *Journal of Energy Storage*, *27*, 101157. Advance online publication. doi:10.1016/j.est.2019.101157

KEY TERMS AND DEFINITIONS

Distributed Generation: An approach that uses small-scale sources and converters to produce electricity close to the utilization point of that power. This form of generation is commonly associated with modular and renewable-energy generators.

Electric Vehicle: This is an emerging class of vehicles that utilize only electric motors or traction motors to propel themselves.

Hybrid Electric Vehicle: This class of electric vehicles combine the conventional internal combustion engine with electric energy for propulsion.

Insulated Gate Bipolar Transistor (IGBT): A semiconductor device with three-terminal power majorly used as a controlled switch in applications requiring fast switching.

Metal-Oxide-Semiconductor Field-Effect-Transistor (MOSFET): A type of voltage-controlled field-effect transistor that is manufactured by the controlled oxidation of a semiconductor such as silicon. It is commonly used in automotive electronics for controlled switching applications.

Microgrid: It is referred to as a cluster of interconnected loads and various energy resources within electrical boundaries that are predefined, and can function as a sole controllable structure with respect to the grid.

Photovoltaic: This refers to any technology or media that harness energy from solar irradiation and convert it to electrical energy.

Plug-In Electric Vehicle: A type of electric vehicle which can be recharged by external energy sources, commonly electric sockets which gives rise to the term "plug in."

Chapter 7
Power Converters for Hybrid Electric Vehicles

M. Nandhini Gayathri
SASTRA University (Deemed), India

ABSTRACT

Hybrid electric vehicles have great potential to reduce fuel consumption and palliate harmful emissions for better population health and climate change. These power trains are essential to meet current emissions legislation, contribute to more environmentally friendly mobility and improve air quality in big cities. In the past two decades, due to the advent growth of modern electric drives, battery technology, and intelligent charging methodology, the overall performance of the electric vehicles have improved. Many research works have been conducted to improve the efficiency and reliability. It increases the applications of electric vehicles. Apart from grid power, the use of renewable sources makes zero-emission rate and zero fuel cost. This chapter deals with the role of power converters in hybrid electric vehicles (PHEVs). It deals with power converters used in hybrid electric vehicles, types of battery chargers, emerging power electronic devices, and thermal management of HEV power electronics, which are presented with the simulation using MATLAB software.

INTRODUCTION

Power electronics replaces the conventional internal combustion engine-powered vehicles to electric motor-driven vehicles step by step. In this chapter, the application of power electronics in hybrid electric vehicles and plug-in hybrid vehicles are explained. Figure 1 shows the configuration of a series HEV. The typical electric vehicle made up of engine-driven three-phase permanent magnet synchronous generator, induction motor-driven wheel mechanism, and HV battery backup. Also, the usual components such as headlights, wipers, music system, auxiliary battery and so on. So there is always a need for energy conversion and it is accomplished by power electronic converters. The power electronic converters used in automotive applications need unique features like wider operating temperature (-20 to 50°C), EMI, thermal management and vibration & shock. The following issues are addressed (Bose, 2000; Jahns et al., 2001; Miller, 2003; Ehsani et al., 1997; Rahman, 2004; Rajashekara, 2003; Emadi et al., 2006;

DOI: 10.4018/978-1-7998-7626-7.ch007

Hamilton, 1996; Vezzini et al., 1996; Emadi ed al., 1999; Amrhein et al., 2005; Gargies et al., 2006; Onoda et al., 2004; Williamson et al., 2006; Johnson et al., 2004; Traci et al., 1999; Alaoui et al., 2005 and White et al., 2004) in the design of HEV power electronic converters. They are electrical design, control algorithm design, magnetic design, EMC design and mechanical & thermal design.

Power Electronics

Figure 1. The basic structure of series HEV

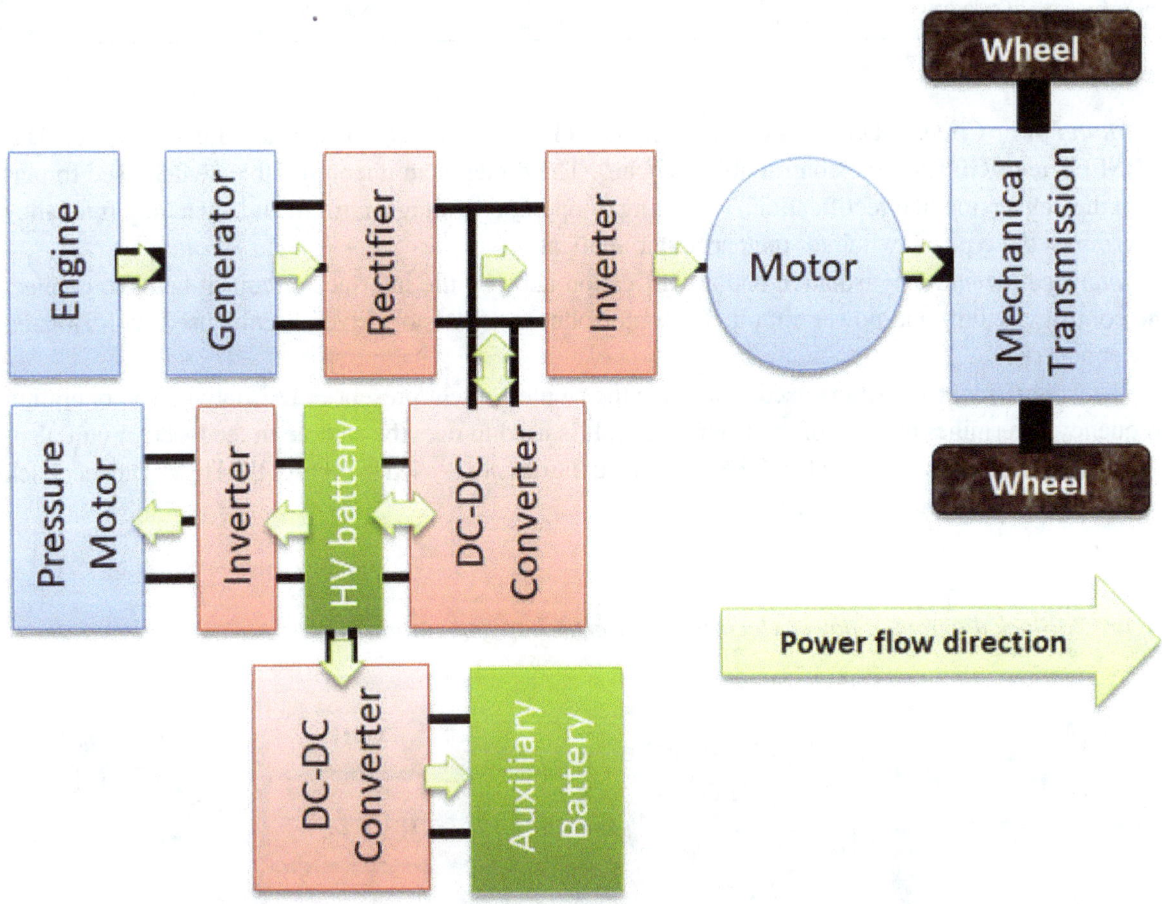

Power electronics is defined as the control and conversion of electric power using semiconductor devices. Based on the type of conversion, the converter topologies are classified into four. The type of conversion, converters and control parameters are shown in Table. 1

The multistage power converters are used in HEV to improve the overall performance. The bidirectional power flow also enabled to utilize the energy more effectively. Multi-input and multi-output (MIMO) topologies support to integrate multiple sources and multiple loads simultaneously. Any power converter implementation is made up of the following parts as shown in the figure. 2. They are,

Table 1. Type of conversions and its typical usage in HEV

Conversion	Converter	Control parameter	Further classifications	Used in the HVE
AC-DC	Rectifier	Voltage	Controlled rectifier, Uncontrolled rectifier	To charge the battery either from the grid or engine-driven PMSG
DC-AC	Inverter	Voltage and frequency	Voltage source inverter, Current source inverter	To run the induction motor-driven wheel mechanism
DC-DC	Chopper	Voltage	Isolated converter, Non- isolated converter	To run dc loads including charging an auxiliary battery from the battery
AC-AC	Cycloconverter	Voltage and frequency	AC voltage controller, Matric converter	Not preferred

Main Power Circuit: It consists of power semiconductor devices (switches and diodes). Power BJT, MOSFET and IGBT are the controllable switches. The diode is an uncontrollable switch used to perform the conversion like rectification. The rating of operating voltage, current and switching frequency determines the type of switch for the particular application.

Interface Circuit: The isolation and amplification done by the interface circuit. It helps to connect the control circuitry and power circuit. The optoisolators, transistor driver circuits used with isolated power sources

Filtering Circuit: LC filters used to remove the ripple content present in DC output. The operating frequency determines the size of the filter circuit. It is used to take the sample in feedback circuit also.

Sensor circuit: The electrical and mechanical quantities are sensed to control the HEV. The feedback and feed-forward controls are used.

Figure 2. Block diagram of power electronic converter implementation

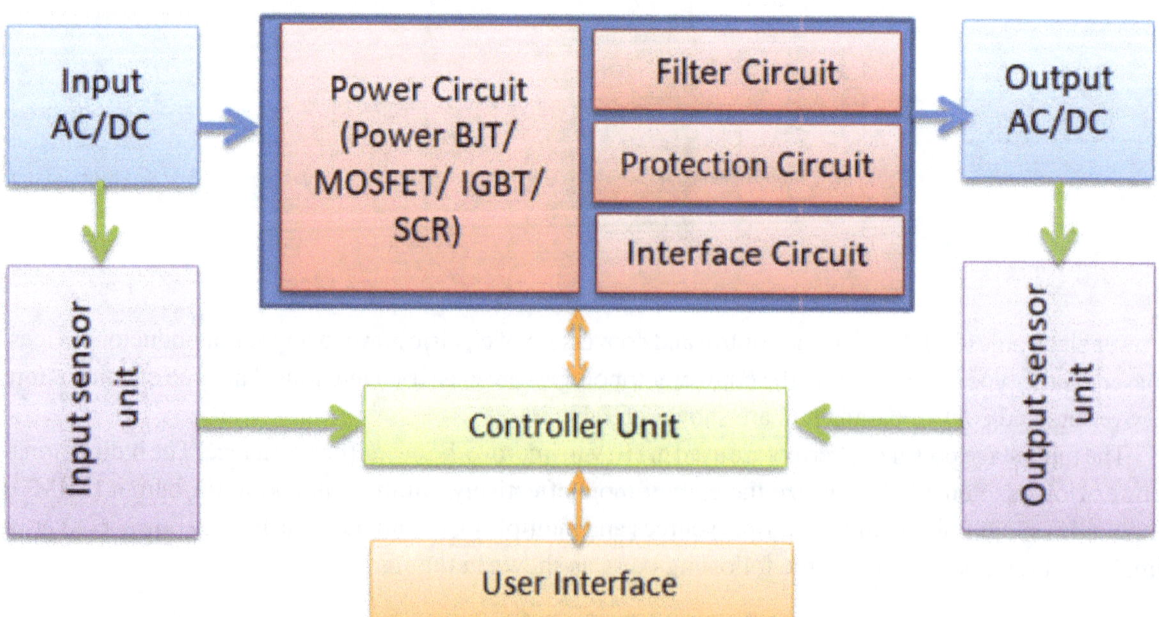

Table 2. International Standard for AC and DC Charging

	Charging Power Levels	Location for Charger	Expected Power Level
SAE Standard	Basic: Level 1 Charging • Vac= 230 (EU)/120 (US)	Single Phase • On-board	1.4kW with (12A) and 1.9kW with (20A)
	Main: Level 2 Charging • Vac=400 (EU)/ 240 (US)	Single Phase/ Three Phase • On-board	4kW with (17A), 8kW with (32A) and 19.2kW with (80A)
	Fast: Level 3 Charging • Vac=208-600	Three Phase • Located Off-Board	50kW and 100kW
	Level 1: DC Charging • Vdc=200-450	• Located Off-Board	40kW with (80A)
	Level 2: DC Charging • Vdc=200-450	• Located Off-Board	90kW with (200A)
	Level 3: DC Charging • Vdc=200-600	• Located Off-Board	240kW with (400A)
IEC STANDARDS	AC Power Level 1	Single Phase • On-board	4-7.5kW with (16A)
	AC Power Level 2	Single Phase/ Three Phase • On-board	8-15kW with (32A)
	AC Power Level 3	Three Phase • On-board	60-120kW with (250A)
	DC Rapid Charging	• Off-Board	1000-2000kW with (400A)

Control circuit: High-end microcontrollers and automotive dedicative control chips are used with sensor feedback and user interface. AI algorithms are implemented to optimize overall performance.

The major power electronic converter topologies used in HEV are diode rectifier, motor drive and battery charger. Table 2 lists the AC and DC charging power levels and the location with respect SAE and IEC standards(Habib et al., 2020). Apart from the main function served, the power converter topologies should obey the safety standards. The safety standards(Habib et al., 2020) describe the safety measures required for HEV listed in table 3.

RECTIFIERS USED IN HEVS

Due to a lack of controllability, the controlled rectifier made up of thyristors is not preferred in automotive applications. So diode rectifiers only discussed. The pure sinusoidal voltage shown in figure 3 is taken into consideration as a supply voltage of the rectifier circuit. It can be expressed as,

$$V_{an} = V_m \sin(\omega t) \tag{1}$$

$$V_{bn} = V_m \sin\left(\omega t - 2\pi/3\right) \tag{2}$$

Table 3. Safety Standards for EVs

Technical Code	Details
SAE J-2464:	Safety rules for recharge energy storage systems(RESS).
SAE-J2929:	Safety for EV propulsion battery system.
ISO 6469-1: 2009 (IEC):	Electrically propelled road vehicles, inside and outside the protection of a person, on-board RESS.
UL 2202:	Protection of charging system.
IEC 61851-21:	Requirements for conductive connection of EVs.
ECE R100:	Electric shock protection.
NFPA 70/70E:	Workplace safety, branch circuit protection, and charging system safety
IEC TC 69/64	Electric shock protection, EV infrastructure safety, and Electrical installation.

$$V_{cn} = V_m \sin\left(\omega t + {}^{2\pi}\!/_{3}\right) \tag{3}$$

Where, V_m is the maximum value of phase to neutral voltage.

Ideal Rectifier

In the ideal rectifier, the forward voltage drop and reverse leakage current are zero. Apart from that the power handling capacity and device switching frequency are limitless. In practical rectifier, all are having certain limitations and the effect of source impedance also causes a reduction in output voltage due to diode commutation time. The output voltage of a single-phase rectifier and three-phase rectifier with the presence of ideal diode characteristics can be expressed as,

$$\text{For single-phase, } V_o = \frac{1}{T/2}\int_0^{T/2} V_m \sin\left(\omega t\right) dt = 0.637 V_m \tag{4}$$

$$\text{For three phase, } V_o = \frac{1}{\pi/3}\int_{-\pi/6}^{\pi/6} 3\sqrt{3} V_m sin\left(\omega t\right) dt = 1.654 V_m \tag{5}$$

Practical Rectifier

The single-phase and three-phase diode rectifier circuits are shown in figure 4 and figure 5 respectively. Three-phase diode rectifier produces more output power at reduced ripple content when compared with single-phase diode rectifier. Figure 6 shows the output voltage of a practical three-phase rectifier circuit. But at the same time, it injects the third-order harmonics into the three-phase source. In HEV applica-

Figure 3. 230V, 50Hz three-phase AC voltage (Phase to neutral)
(X axis: 0.02s/ div, Y axis: 100V/div)

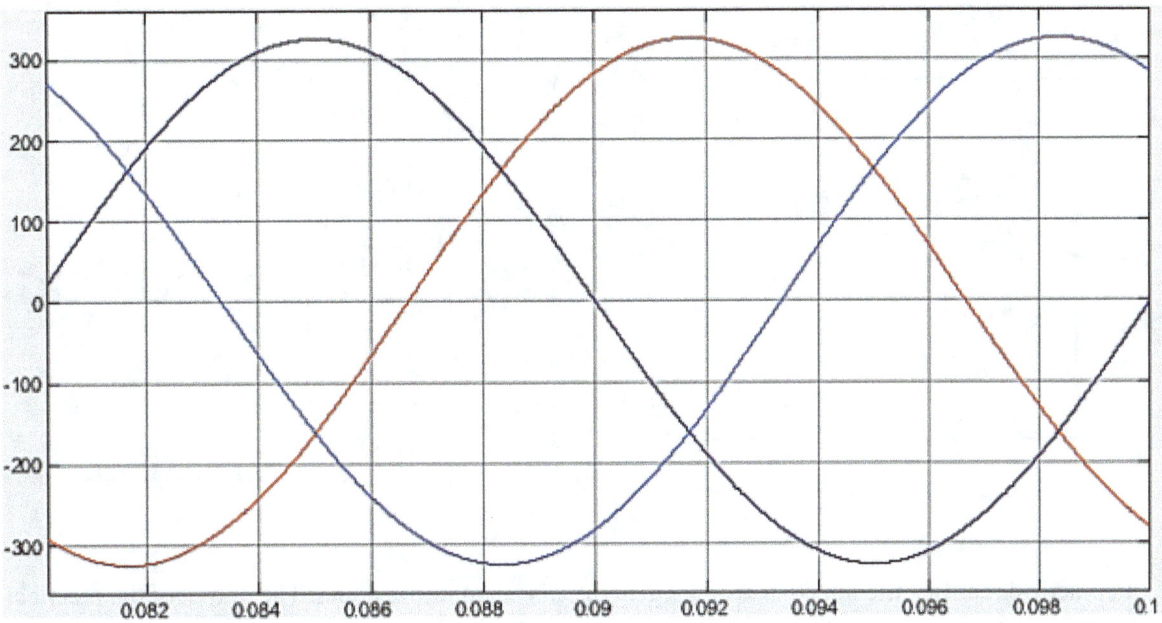

tions, it is connected to the synchronous generator to avail DC voltage output. Due to the effect of source impedance, i.e. the generator impedance, the output voltage reduction occurred and it is increased with an increase in load current. It is measured in terms of voltage regulation. The output voltage of the single-phase and three-phase diode rectifier at no load is equal to the maximum voltage of line to neutral and line to line respectively. Because the filter capacitor charged to the maximum value of the supply voltage.

$$V_o = V_m \tag{6}$$

Figure 4 Single-phase diode rectifier

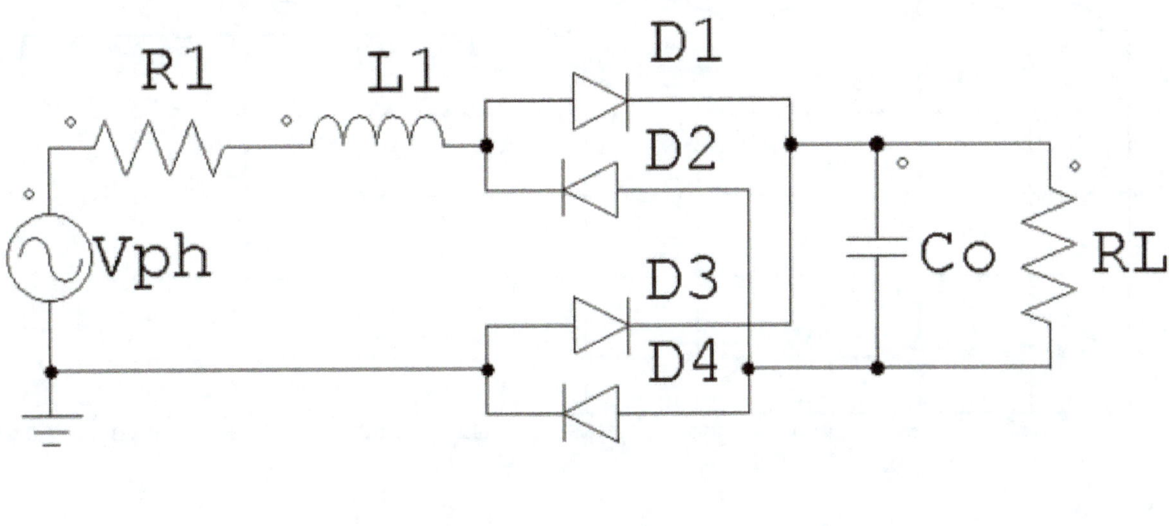

Figure 5. Three-phase diode rectifier

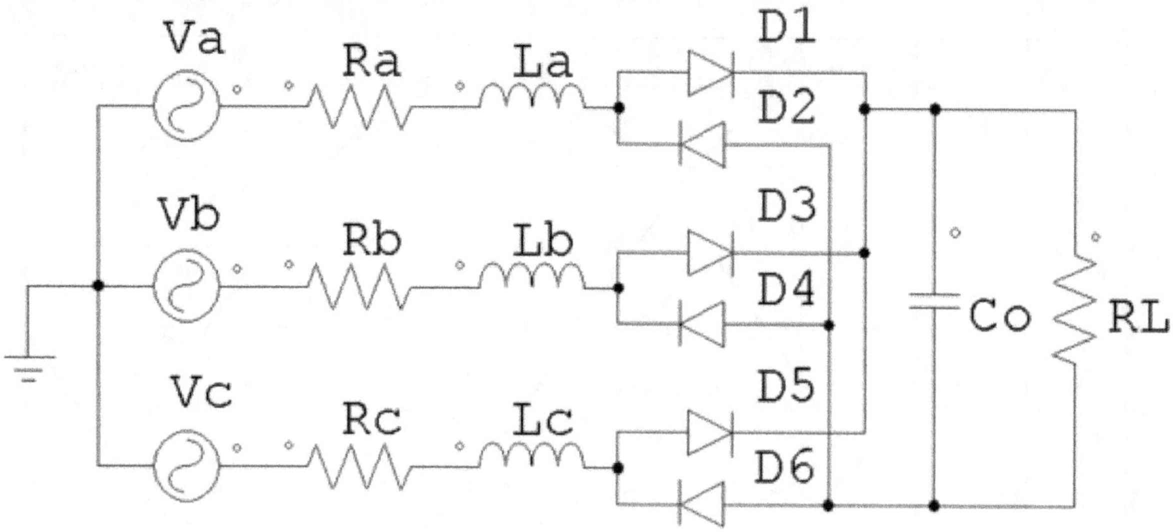

In practical rectifier, the impedance of the generator is taken into account. The output of the practical rectifier in continuous conduction mode can be expressed as,

$$V_o = \sqrt{\frac{R}{\pi}\int_{\theta_1}^{\theta_2} V_m sin\omega t^{*} i(t)\, d(\omega t)} \tag{7}$$

Figure 6. The output voltage of a three-phase diode rectifier
(X axis: 0.02s/ div, Y axis: 100V/div)

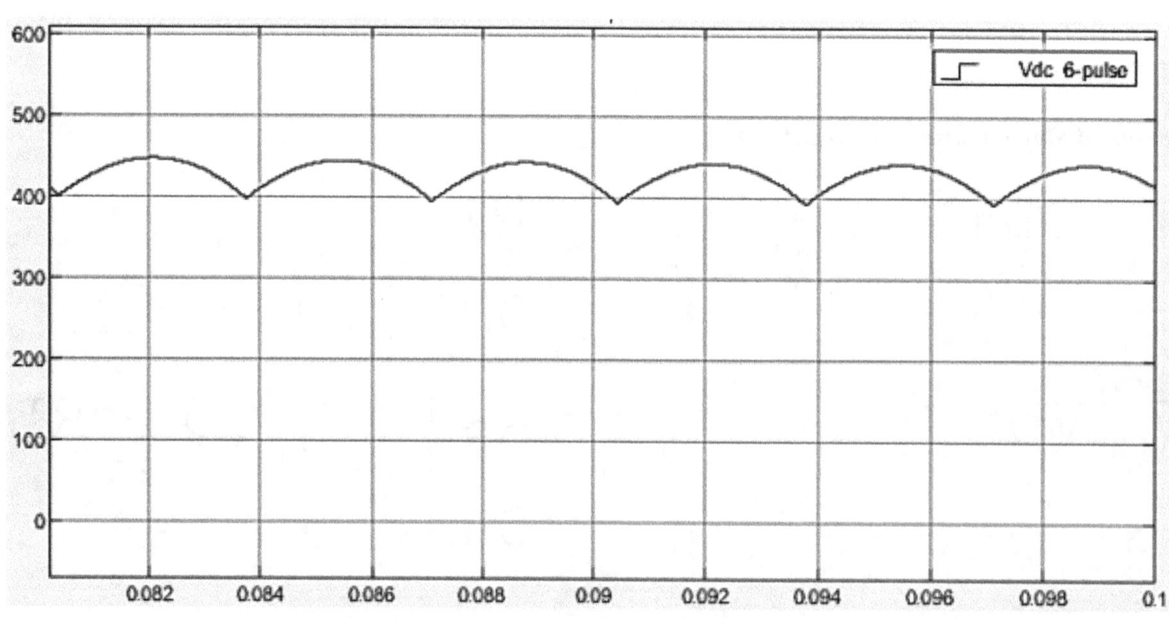

Where R is the load resistance and$\theta 1$, $\theta 2$ are the conduction period and the blocking period of the thyristor pair

The percentage voltage regulation may be expressed as,

$$\Delta V_o = \frac{V_1 - V_0}{V_o} \times 100\% \tag{8}$$

Where V_1 is the capacitor voltage at no load.

Voltage Ripple

The voltage ripple is expressed under the assumption of constant load current with 50% of capacitor contribution,

$$\Delta V_r = \frac{1}{2}\frac{1}{C}\frac{V_0}{R}\frac{\pi}{\omega} \tag{9}$$

The AC input is discontinuous when the conduction period of the diode is less than 180°. During that period, the load current is supplied by the capacitor.

CONVERTER TOPOLOGIES FOR EV BATTERY CHARGERS

Many converter topologies are available for EV battery charging. Each one has its advantages and limitations. The source and power rating are the major factor that determines the desired converter topology for EV charging. The various EV chargers (Habib et al., 2020) shown in figure 7.

The dominant types of power converter topologies are discussed below.

BUCK CONVERTER USED IN HEVs

Operating Principle

A buck converter shown in figure 8 is step down the input DC voltage. In HEV application, the buck converter steps down the high voltage battery (400V) to auxiliary battery voltage (14V). The large step-down ratio (0.035) becomes more difficult to realize. There is a need for careful design considering voltage and current ripple.

The buck converter is made up of a series-connected high switching frequency switch and freewheeling diode with LC filter.

The output voltage, ripple current and ripple voltage can be expressed as,

Output voltage, $V_o = DV_d$ (10)

Figure 7. EV charging solutions using various power electronic converters

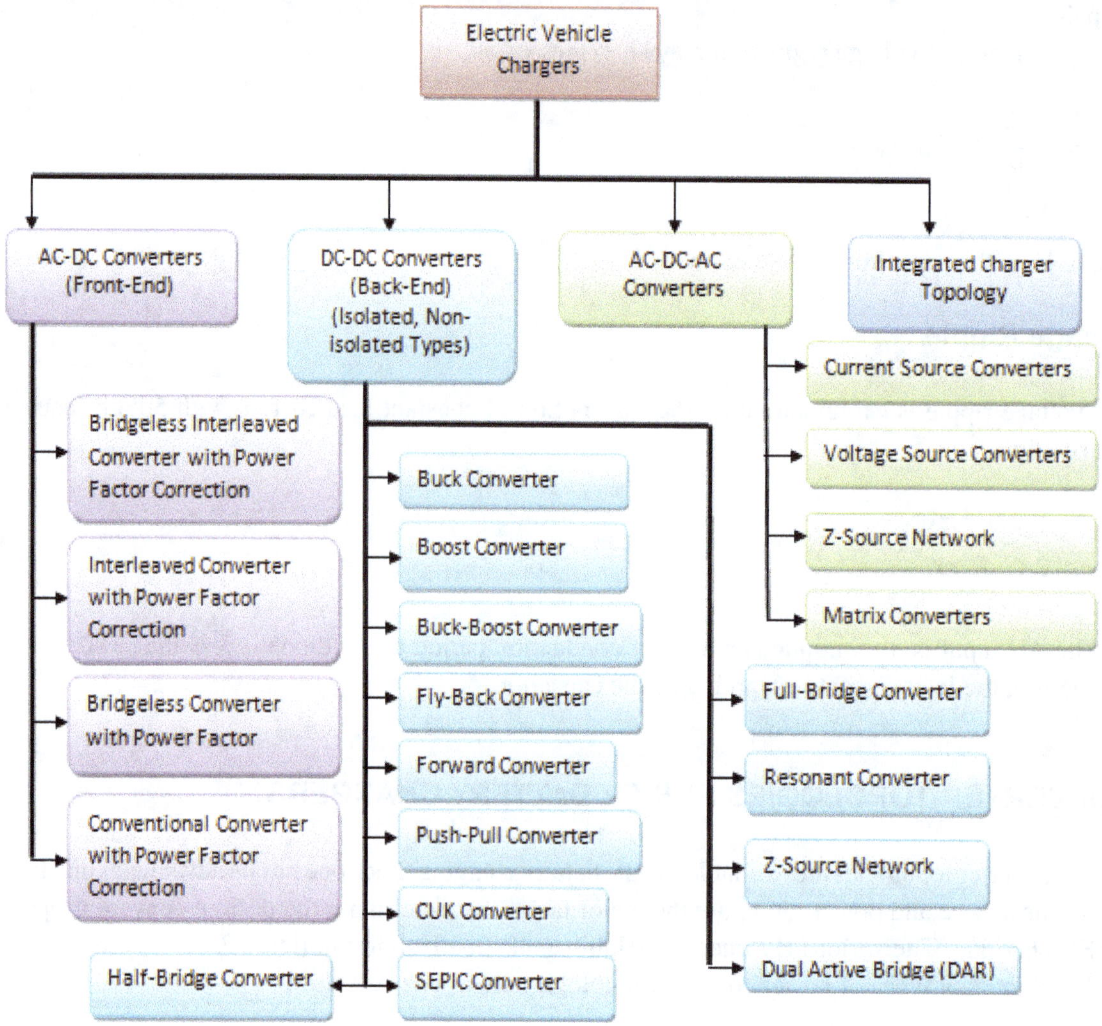

Figure 8. Buck converter

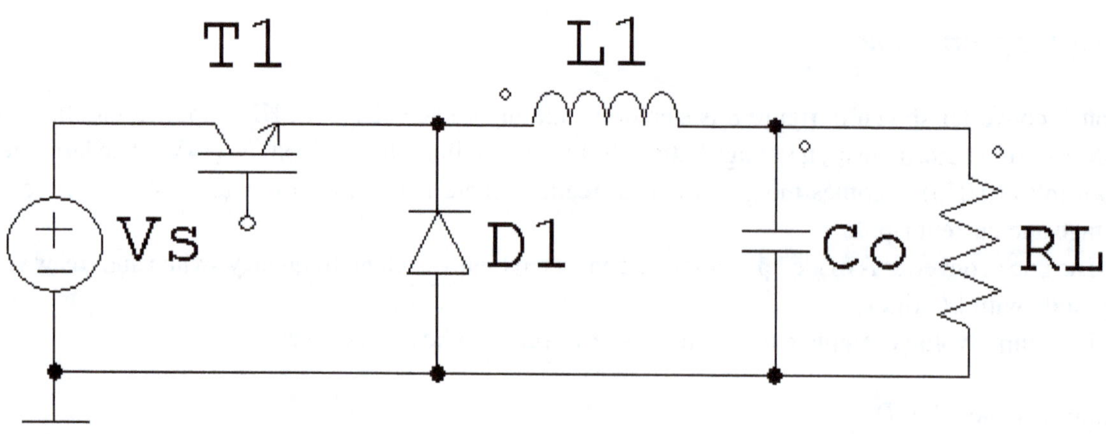

Figure 9. Voltage and current waveforms: Output voltage and source current.
(X axis: 0.2ms/ div, Y axis(top to bottom): 50V/di and 5A/div)

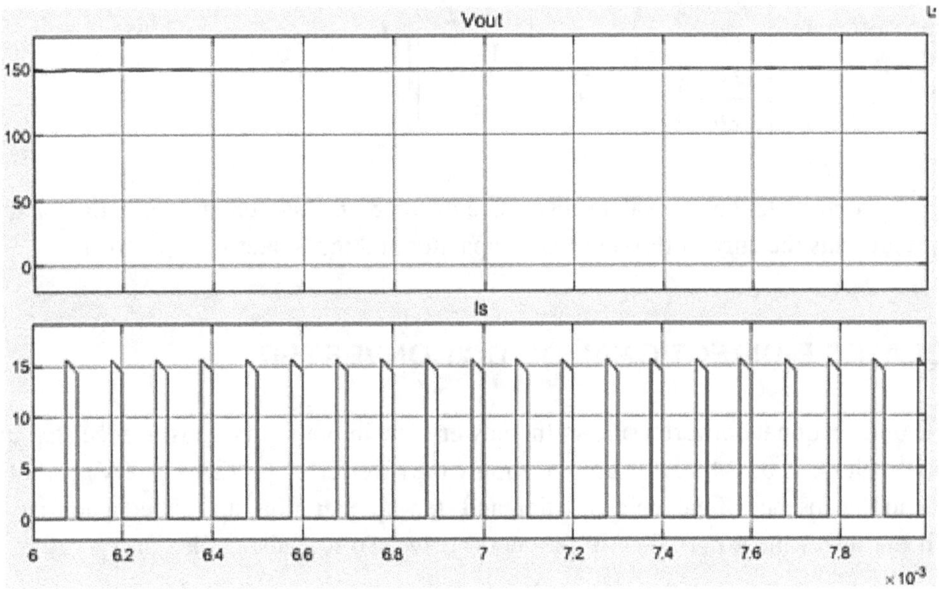

Current ripple, $\Delta I_L = \frac{1}{L} V_o (1-D) T_s$ (11)

Voltage ripple, $\Delta V_o = \frac{1}{C} \frac{1}{2} \frac{\Delta I_L}{2} \frac{T_s}{2} = \frac{T_s^2}{8LC} V_o (1-D)$ (12)

Figure 9 shows the MatLab simulation output voltage and source current waveform of a typical buck converter circuit. The supply voltage of 400V converted into 150V at a switching frequency of 10 kHz.

Nonlinear Model

In a real-time implementation, the closed-loop control involves and the desired output voltage is obtained. But the steady-state characteristics are not linear and belong to the boundary. So the following state-space model used to analyses the circuit with more accuracy. The following equations can be solved using a computer simulation program.

During switch-off state, $\begin{bmatrix} \frac{di_L}{dt} \\ \frac{dV_o}{dt} \end{bmatrix} = \begin{bmatrix} -\frac{r_d + r_s + r_L}{L} & -\frac{1}{L} \\ \frac{1}{C} & -\frac{1}{CR} \end{bmatrix} \begin{bmatrix} i_L \\ V_o \end{bmatrix} + \begin{bmatrix} \frac{1}{L} \\ 0 \end{bmatrix} V_d$ (13)

During switch-off state, $$\begin{bmatrix} \frac{di_L}{dt} \\ \frac{dV_o}{dt} \end{bmatrix} = \begin{bmatrix} -\frac{r_d + r_L}{L} & -\frac{1}{L} \\ \frac{1}{C} & -\frac{1}{CR} \end{bmatrix} \begin{bmatrix} i_L \\ V_o \end{bmatrix} \quad (14)$$

Where, r_d, r_s, r_L represents the equivalent resistance of the diode, switch and the inductor respectively and i_o, i_c, i_L represents the current through load, capacitor and the inductor respectively

NON-ISOLATED BIDIRECTIONAL DC-DC CONVERTER

The DC-DC bi-directional converter allows the power flow in both directions. Hence step up and step down of input voltage is possible. It is used to supply the inverter motor drive (600V) from HV battery (400V). Due to the absence of galvanic isolation and step-up operation, it is called a non-isolated boost converter. It optimizes the power flow to the motor drive. So it is also called an energy management converter.

Operating Principle

The proposed topology as shown in figure 10 can be operated in buck and boost mode either in stand-alone operating mode or simultaneous operating mode. If T1 enables with PWM and T2 disabled, then it behaves like a buck converter. The power transfer takes place from the high voltage side to the low voltage side. Similarly, if T2 enabled with PWM and T1 disabled, then it behaves like a boost converter. The inductor used to boost the low voltage output and power transfer takes place from the low voltage side to the high voltage side. In simultaneous operation, both the switches are enabled by PWM with the following duty ratio relation. In the above said all conditions, the amount of power transfer is determined by duty ratio.

Figure 11 shows the output voltage and inductor current of the non-isolated DC-DC converter. A 100V dc input is step up to around 170V at a switching frequency of 10 kHz.

Maintaining Constant Torque Range and Power Capability

The maximum power delivery of the battery is limited and it is represented by its internal resistance. During normal analysis, the effect of internal resistance is omitted. But it is not a negligible parameter, it causes a voltage drop at battery terminals when large power is drawn. It results in battery terminal voltage variation and internal loss. Apart from that, the battery terminal voltage varies with the battery state of charge (SOC). It affects the motor constant torque region. To overcome this issue a DC-DC converter is connected in between the battery source and inverter. It acts as a front end converter and provides the constant DC output voltage to the inverter input

Figure 10. Non-isolated DC-DC converter

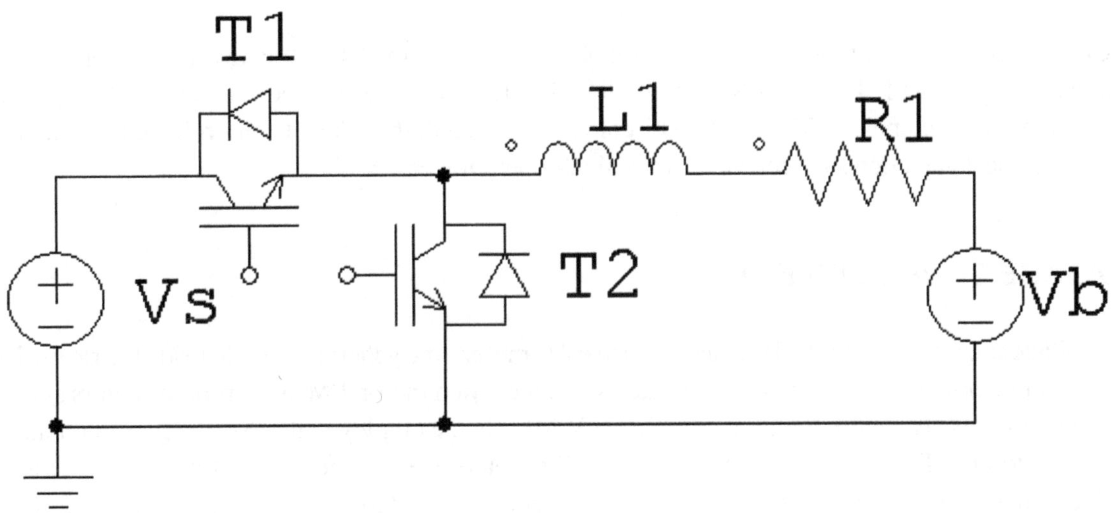

Reducing Current Ripple in the Battery

The DC side source harmonics introduced by high device switching frequency inverter and causes current ripples. The amount of ripple current on a battery depends on switching scheme, switching frequency and DC bus capacitance. In DC-DC front end converter, the presence of the ripple current is decided by DC bus capacitance and the ratio between capacitor impedance and battery impedance. The higher value capacitance reduces the ripple current to lower value. But at the same time, the cost, size and weight of the filter circuit increased. So there is a need for optimum capacitor design.

Figure 11. Output voltage and inductor current waveform
(X axis: 0.05ms/ div, Y axis (left to right): 20V/div, 2A/div)

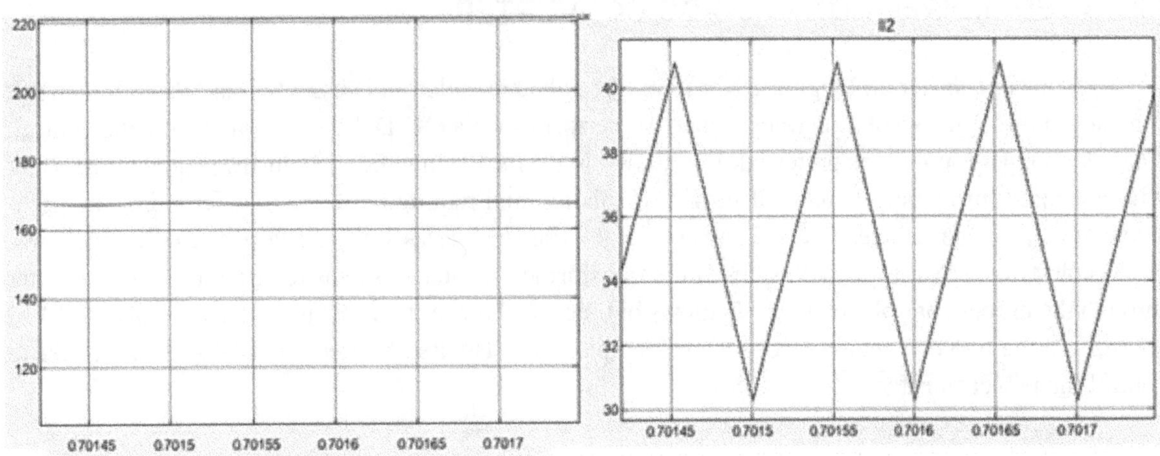

Regenerative Braking

The regenerative braking is achieved on both topologies. Without the DC-DC converter, the regenerative braking is very difficult. During braking operation, the DC bus voltage reduced and makes the regenerative braking more complex. When a DC-DC converter is added, then the DC bus voltage is maintained constant almost and allows the regenerative braking operation easier.

VOLTAGE SOURCE INVERTER

The voltage source inverters (VSI) are used to drive AC motors and generators in a hybrid electric vehicle. The most preferable ac motors in hybrid vehicles are, induction motor, PM synchronous motors and PM brushless motors. The pulse width modulation (PWM) scheme employed in VSI's to control the motor speed and torque. The power switch used in the VSI is determined by the rating of voltage, current and device switching frequency. High device switching frequency used to eliminate the harmonics from the fundamental frequency. So, power BJT and thyristors are not preferred. IGBT preferred for high voltage applications. MOSFET used for low power applications. The feedback diodes are mandatory to protect the power switches from the reversal of stored energy on motor load. Figure 12 shows a three-phase voltage source inverter with an induction motor load.

CURRENT SOURCE INVERTER

The current source inverter (CSI) employed to provide constant torque from the constant current source. The working of CSI's similar to VSI's, the PWM scheme used to control the output voltage. The shoot-through problem of VSI utilized as the advantage. That helps to boost the input voltage level. In CSI, thyristors are used. In hybrid EV applications, for high device switching frequency MOSFET and IGBT are preferred in series with power diode or thyristor as shown in figure 13 In CSI, feedback diodes are not required like VSI.

ISOLATED BIDIRECTIONAL DC-DC CONVERTER

For high power transfer and high transfer gain, the galvanic isolation is essential and desirable one. It is achieved through an isolation transformer. To perform such a DC-DC power conversion and control, AC link chopper topology is preferred. It is shown in figure 14. In which a high-frequency transformer with a desired turn's ratio is used. It provides isolation and required voltage transfer gain. The high-frequency operation reduces the size and cost of the transformer as well as filters circuit components. PWM converters are used on both sides of the transformer. It can be used as rectifier as well as inverter through the proper control technique. It allows bidirectional power flow. Phase angle control and PWM control are the two different control schemes employed in Isolated bidirectional DC-DC converter to control the power transfer.

Figure 12. Three-phase voltage source inverter

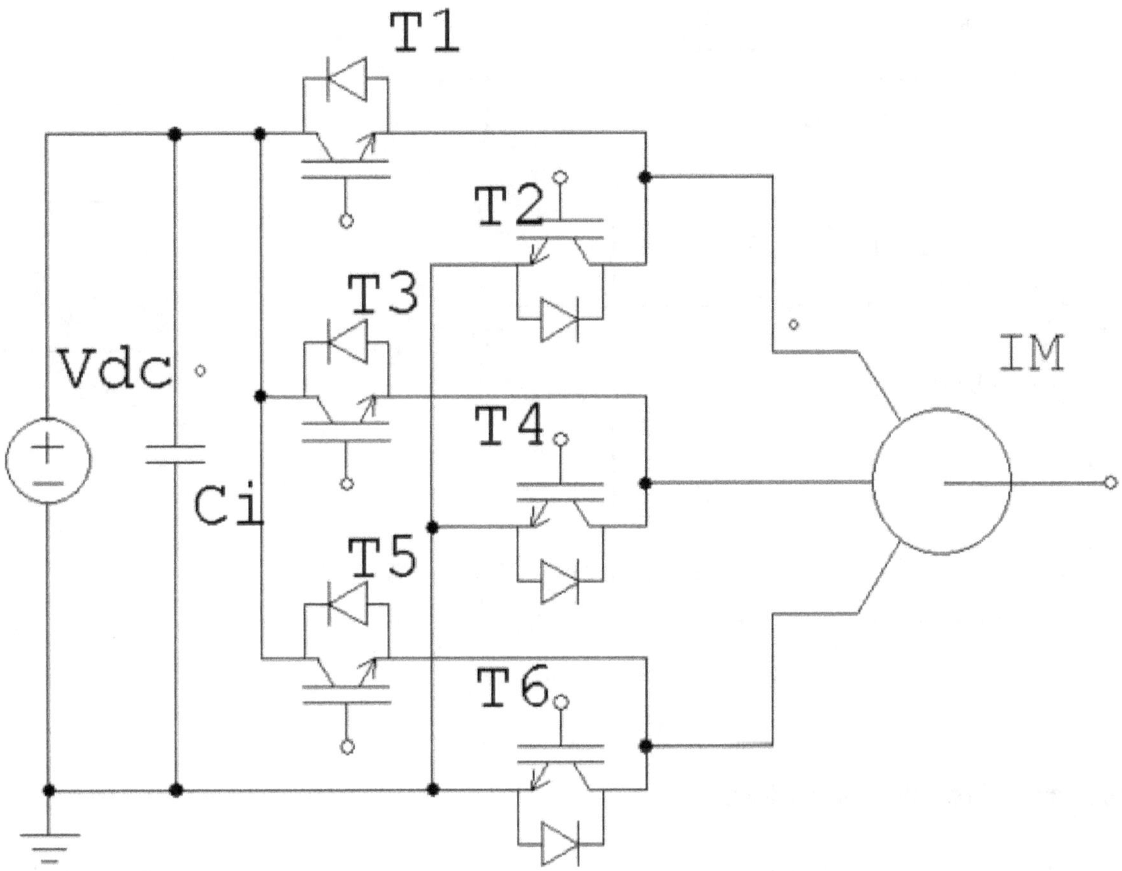

PWM RECTIFIERS IN HEVS

Rectifier Operation of Inverter

The topology of PWM rectifier is same as that of VSI. PWM rectifier is shown in figure 15, the power flow from AC to DC can be controlled through PWM scheme. During regenerative braking, the motor acts as a generator and stored energy returned to the battery. Based on the motor terminal voltage and speed, either boost operation or field weakening method used to perform the regenerative braking operation.

EV AND PHEV BATTERY CHARGERS

In EVs and PHEVs, the battery charging is the challenging part needs optimum design to compromise many factors like efficiency, cost, charging time etc. Also, the safety and economic operation of charger obtained through isolation and bidirectional power transfer capability respectively. Many charging schemes are used in practice. They are listed in table 4.

Figure 13. Three-phase current source inverter

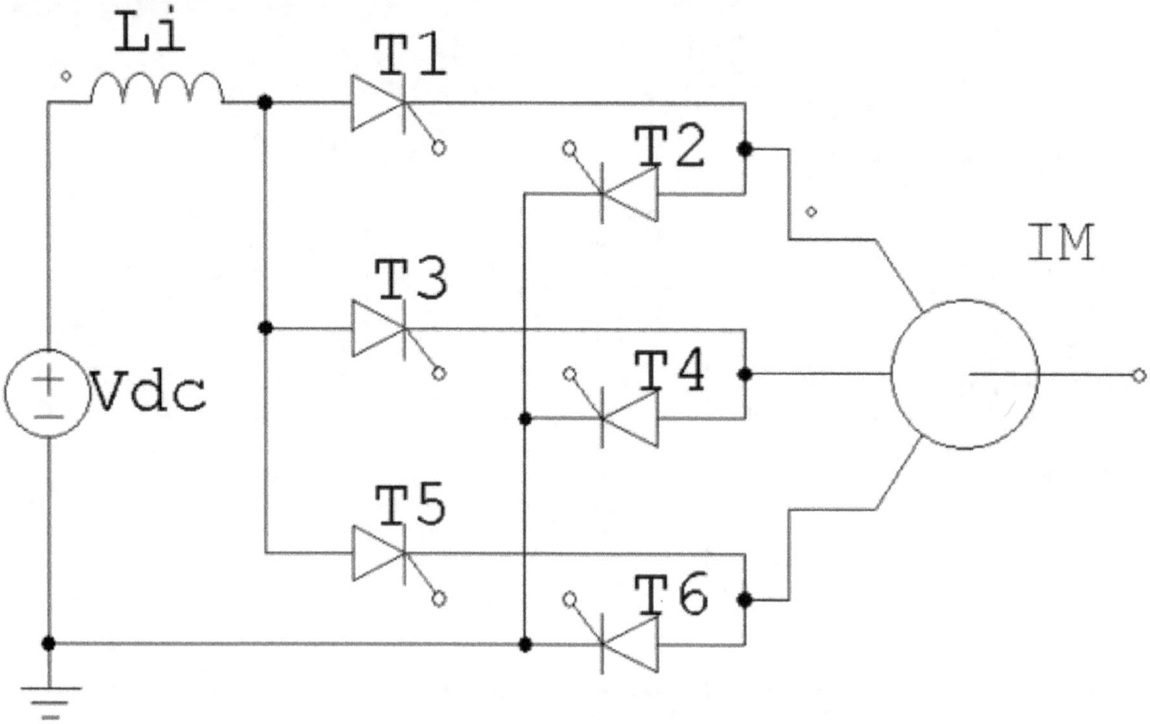

Forward/ Flyback Converters

Figure 16 and 17 represent the construction and operating waveforms of a forward converter topology. The working principle of the forward converter is the same as that of the buck converter. The transformer used to obtain high voltage transfer gain. At the same time, the energy released from the transformer leakage inductance causes spikes. So snubber circuit is required in the forward converter. Similarly, the flyback converter also needs a snubber circuit. Figure 18 and 19 show the construction and operating waveforms of flyback converter topology.

Figure 14 Bidirectional DC-DC converter with galvanic isolation

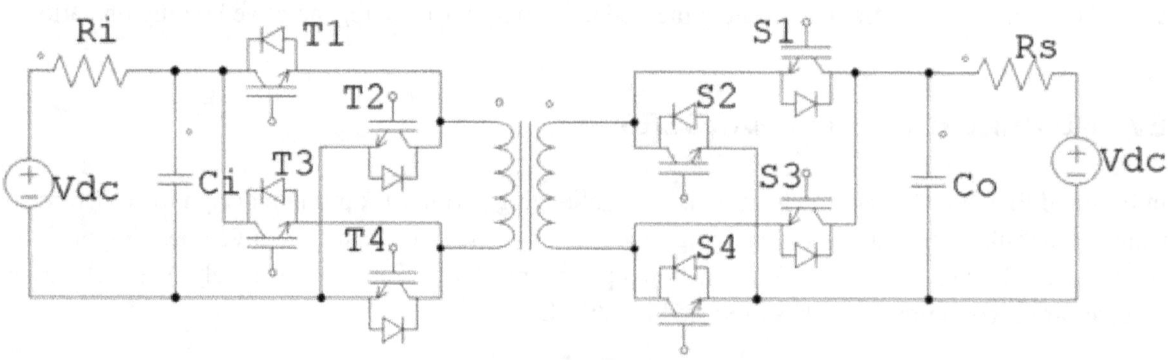

Figure 15. PWM Rectifier

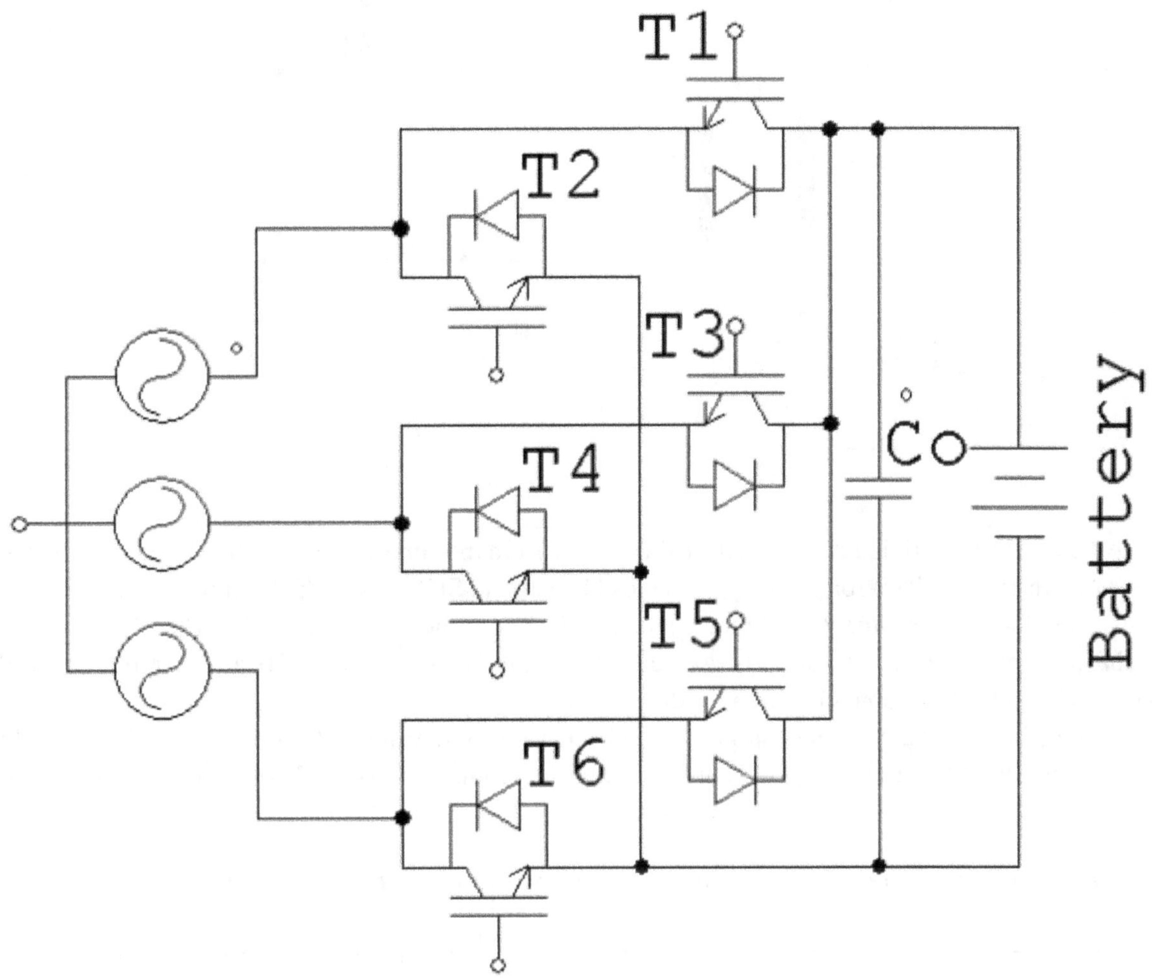

Table 4. Technological Trends in EV Charging

Technological Trends	Descriptions
Fast Charging	Fast charging solutions through universal fast chargers, bidirectional fast chargers and ultra-fast chargers.
Wireless Charging	Several wireless charging solutions such as inductive power transfer, capacitive power transfer are being provided by considering the reduced recharging time of EV such as dynamic wireless charging including charger alignment and 2) foreign object identification etc.
Battery Swapping	Battery swapping network is a fast and efficient source of recharging. Battery degradation, expensive infrastructure and incompatible available batteries are the drawbacks of this type.
Renewable Based Charging	Use of renewable sources like solar, wind and make the self-charger as independent of grid power.
Integrated Solutions	Integrating the above methods to improve the vehicle to vehicle, vehicle to grid and vehicle to any technologies.

Figure 16. DC-DC Forward Converter topology

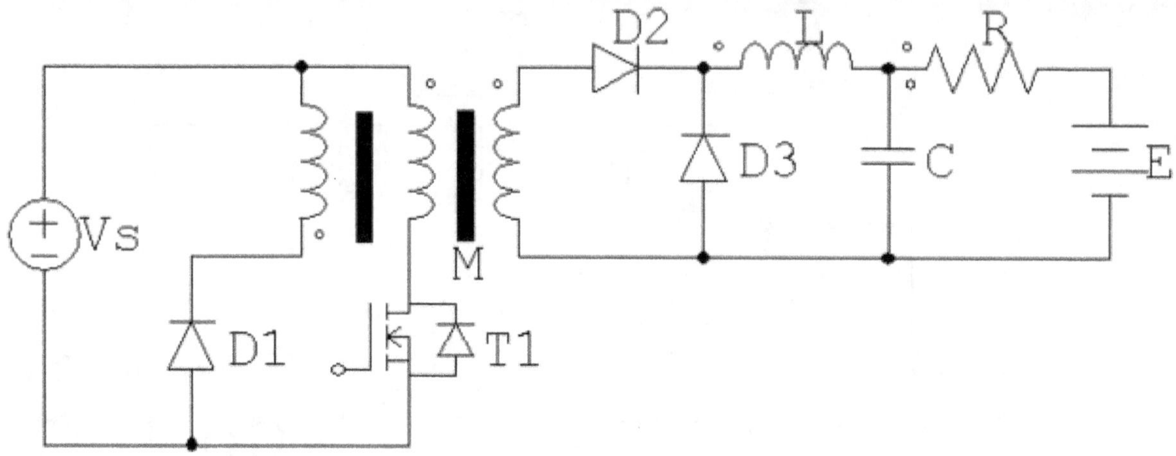

A 60km electric driving range is selected typically from the survey taken by the US Electric Power Research Institute (EPRI) for passenger vehicles like car and SUV. A 6-12kWh Lithium-ion battery is considered for the above-mentioned range. It offers 60-70% of the nominal capacity. The PHEV battery is charged from the grid (110 or 220V AC) either through the onboard or off-board charger. A typical battery charging of 2-8 hours required for the next day use of EV.

Various battery chargers and charging schemes have been developed for different applications. In EV and PHEV battery charging, the high power transfer requires safety mode of operation. Hence gal-

Figure 17. Operating voltage and current waveforms: Source current and inductor current

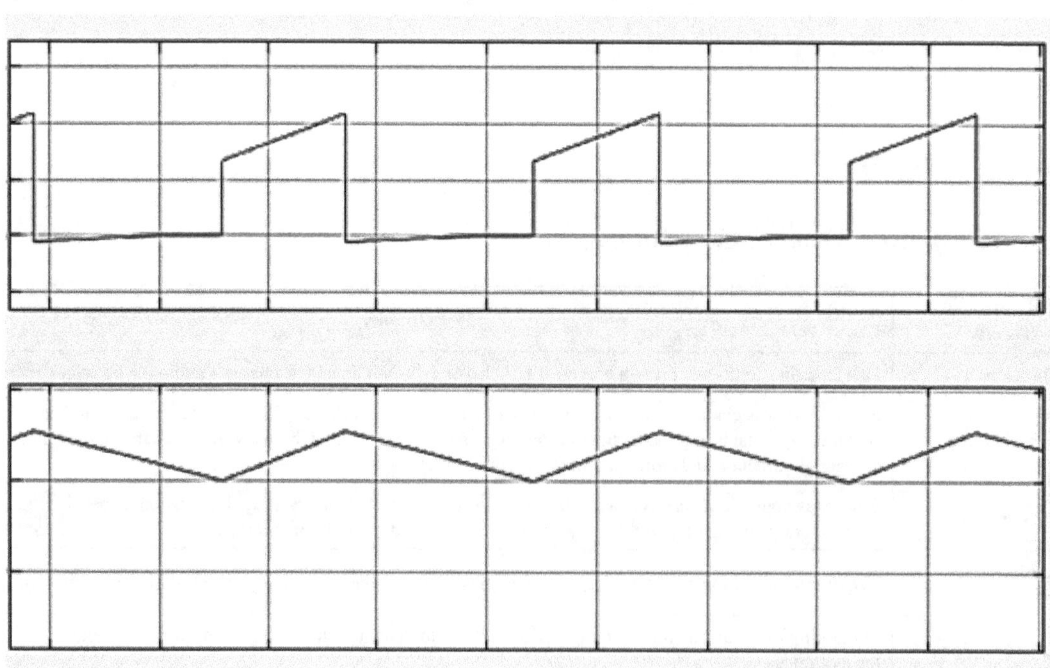

Table 5. Conventional resonant topologies and its benefits

Type of Resonant	Benefits
Series resonant converter (SRC)	• EMI and switching losses are reduced by ZVS results in an increase in efficiency • Size of magnetic components reduced. • ZVS and ZCS can be achievable above and below the resonant frequency respectively.
Parallel resonant converter (PRC)	• No issue on output voltage regulation at no-load condition. • Continuous current output • Suitable for higher output current and low output voltage application. • Wide frequency adjustment not required to maintain the regulation of output voltage.
Series parallel resonant converter(SPRC). Also known as LLC resonant converter	• Circulating energy is small when compared with PRC. • Regulation of output voltage under no-load condition. • Less sensitivity to load variations. • High resonant frequency causes high efficiency.

vanic isolation is provided to assure the safety. To improve the reliability and performance, feedback control used including the critical battery parameters. The major power converters topologies used in battery chargers are Flyback converter, forward converter, half-bridge converter, full-bridge converter and resonant converters. Different types of resonant converters and their advantages are listed in table 5.

Each topology has its own merits and demerits. Flyback and forward converter is made up of single switch. The cost and losses are low. But it supports unidirectional power flow and suitable for low power applications. PWM scheme employed to control the battery charging.

Whereas the half-bridge and full-bridge converter topologies shown in figure 20 and 21 respectively support high power transfer at both directions. Bidirectional power flow eliminates the demagnetizing circuit unlike Flyback/forward converter topologies. Soft switching is also supported by the half-bridge/full-bridge topologies. The main drawback of the half-bridge/full-bridge topology is the more number of power semiconductor devices. It increases the charger cost cum losses associate with it. PWM scheme and phase angle control used to control the battery charging.

Figure 18. DC-DC Flyback converter topology

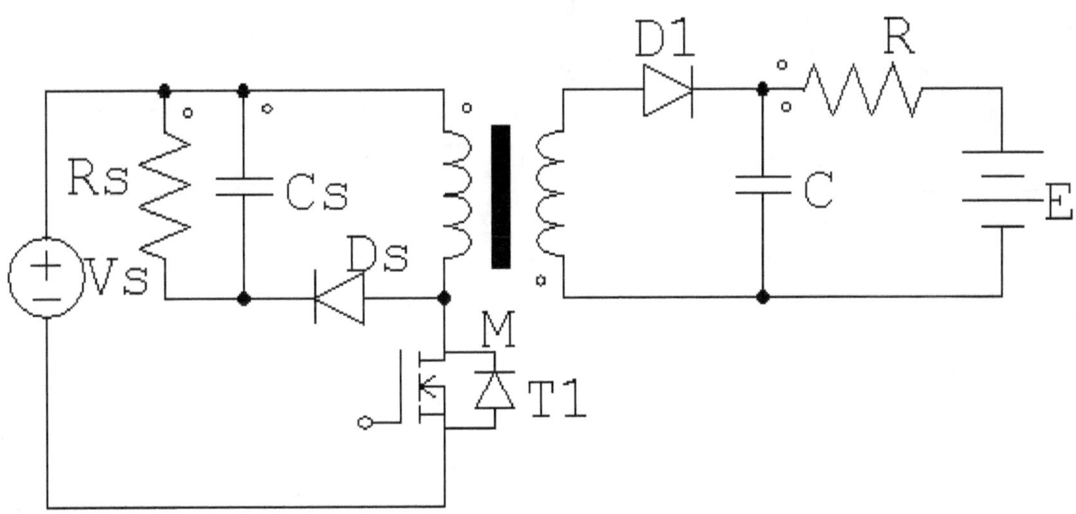

Figure 19. Transistor and diode operating current waveform

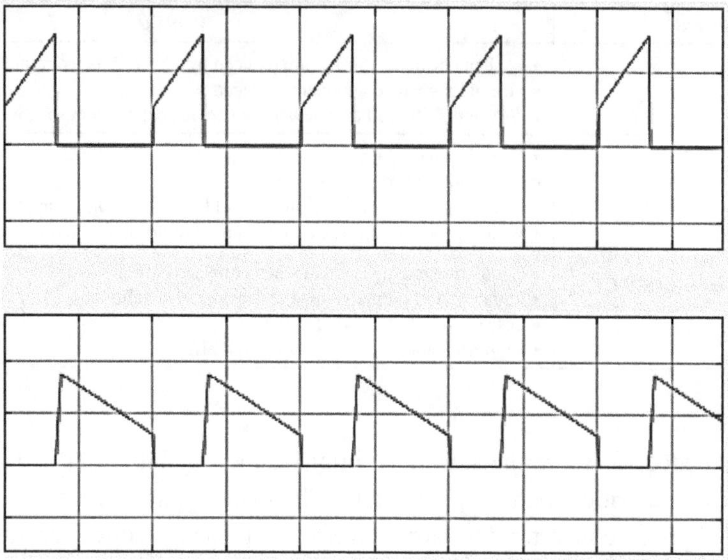

Half–Bridge DC-DC Converter

Figure 20. Half-bridge DC-DC converter

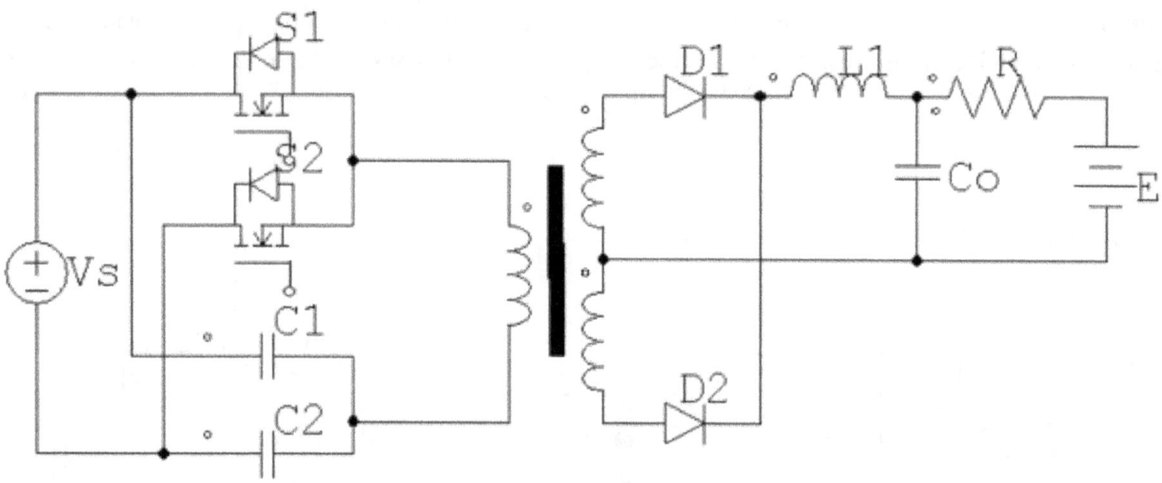

Full-Bridge DC-DC Converter

Figure 21. Full bridge DC-DC Converter

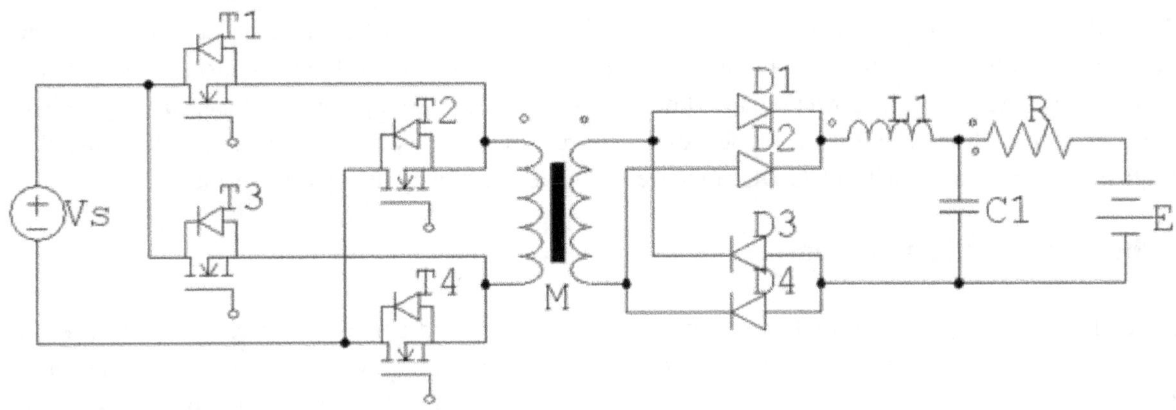

Power Factor Correction Stage

The source harmonics introduced by diode rectifier is reduced by power factor correction (PFC) stage. PFC circuit improves the power factor and stabilizes the DC link voltage. PFC is made up of a boost converter as shown in figure 22. The bridgeless PFC circuit also used to reduce the source harmonics

Figure 22. PFC stage in HEV charger

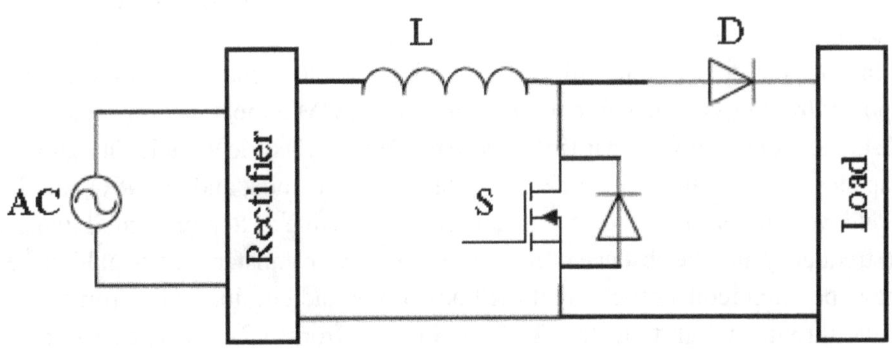

Table 6. Power and efficiency of front end converters for battery charging

Converters Configuration	Conventional Converter with PFC	Bridgeless Converter with PFC	Interleaved Converter with PFC	Bridgeless Interleaved Converter with PFC
Power Ratings (W)	Less than 1000	Less than 2000	Less than 3000	Greater than 3000
Efficiency	Low	High	High	Very high

in battery chargers. The power and efficiency comparison of front end converters with PFC is shown in table 6.

Bidirectional Battery Chargers

The bidirectional battery chargers allow the power flow between the battery and any source or load. In-vehicle to grid mode, the power available from the battery transferred to the grid during peak demand hours. Similarly, during the power outage at home or office, the battery power can be used with the help of bidirectional battery chargers. The safety of the bidirectional battery charger is achieved through either high-frequency transformer in DC-DC converter or power frequency utility transformer.

Other Charger Topologies

Apart from single phase conventional battery chargers discussed above, quick chargers are used in commercial transportation and recharge stations. It is supplied from three-phase AC voltage and saves battery charging time great. It is suitable for only the batteries that allow quick charging. Because the high current charging affects the long term capacity or life battery. The traditional AC charger works with DC input also. In the specially designed DC charger, there is no need for rectifier and PFC stage.

INDUCTIVE AND WIRELESS CHARGING

From the safety aspect, the normal hard-wired charger is not recommended for high power battery charging. There is a need for a trained person and the precautions must be strictly followed when the hard-wired charger used in practice. Otherwise, it causes a spark and electrical hazardous. Especially during charging, mishandling of electrical contact on the outlet leads to potential injury. If there is any physical damage in the power plug and cable also causes electrical hazardous. So, frequent monitoring of power cable and plug also required against wear and tear.

To overcome the problem faced in hard-wired charging, inductive charging proposed. In the inductive charger, the power transfer is done without physical contact. GM adopts inductive charging in the year 1990. In 1995, inductive charging standard introduced by US SAE Electric Vehicle Charging System Group along with Japanese manufacturers including Toyota, Nissan, Honda, and DENSO, and IEC/ISO. The construction and working of inductive power transfer are the same as that of normal transformer except the switching frequency and the absence of iron core. The power transfer is done at high frequency. The high-frequency operation reduces the size of the transformer and eliminates the iron core. The primary coil is energized through a high-frequency DC link inverter from AC grid. The secondary coil placed inside the vehicle and it is used to charge the battery through a full bridge rectifier.

However the inductive charging neglects the physical contact between the transmitter coil and receiver coil, there is a need for the power cable and plug interface to connect the transmitter coil to the source. It is eliminated in wireless charging. In wireless power transfer, the power transfer is done at the resonant frequency. It supports more distance when compared with inductive charging. In wireless power transfer, the transmitter coil and receiver coils are tuned at the resonant frequency. So, the maximum amount of power is transferred from the transmitter coil to the receiver coil. The amount of power transfer and efficiency is affected by the distance between the coils.

Table 7. Rating of SiC devices

SiC Devices	Commercially available device rating	
	Voltage rating	Current rating
DMVTJFET	1200V	48A
EMVTJFET	1700V	30A
MOSFET	1200V	48A
BJT	1200V	20A

Resonant inductive power transfer and capacitive power transfer are the two different types of the wireless power transfer method. This wireless power transfer can be performed when the vehicle in motion by embedding the primary coils on road. This mode of wireless charging scheme is known as dynamic wireless power transfer otherwise it is known as static wireless power transfer.

MODELING AND SIMULATION OF HEV POWER ELECTRONICS

In the design and implementation of any engineering model or system, the simulation plays a vital role. It saves more time and cost before real-time implementation and optimizes the model during the design. The major parts of hybrid electric vehicle involve electrical, mechanical and power electronics. The design of electrical and power electronics circuits requires spice based simulation tools. Whereas, the design of the electromagnetic circuit needs FEA/FEM based analysis. The simulation can be performed in two ways. They are chip/device level simulation and model/system-level simulation. In chip/device level simulation, almost all the real-time attributes are taken on the account. It gives the accurate steady-state and dynamic response and almost equal to real-time implementation. In the case of model/system-level simulation, fewer attributes are taken on the account. It saves simulation time when compared with chip/device level simulation and gives approximate response when compared with real-time implementation.

EMERGING POWER ELECTRONIC DEVICES AND THERMAL MANAGEMENT OF HEV POWER ELECTRONICS

The operating boundary of the power semiconductor needs to increase beyond the present silicon (Si) technology. The device operating voltage, current, power density, efficiency and temperature should be high. More reliable operation, compact size and light weight required in a hybrid electric vehicle application. All are achieved through the new power semiconductor device like silicon carbide (SiC). Hence SiC-based power converter topologies are preferred for HEV applications. The voltage and current rating of SiC devices(Han et al., 2014) are shown in table 7.

The combination of high device switching frequency and high current produces the electrical noise and causes EMI. This EMI affects the nearby circuits and results in malfunction of control circuitry which is undesirable. To reduce the EMI, there is a need for careful design of PCB and components placing. The control circuitry is required to detect the abnormal behaviour of power electronic converter and to protect against voltage and current fluctuations.

Table 8. Power electronic converters cost, power density and efficiency forecasting

	2017		2025		2035	
	Inverter	OBC/ DC-DC	Inverter	OBC/ DC-DC	Inverter	OBC/DC-DC
Cost (Rs/kW)	370	2200	300	1100	220	590
Power density (kW/L)	12	2	15	4	60	15
Efficiency (%)	96	94	97	96	98	97

In any power converter topology, the losses are converted into heat on power semiconductor. The device operating voltage, current and switching frequency determines the junction temperature. In HEV applications, a significant amount of heat produced on the powertrain motors, AC/DC drives and bidirectional battery chargers. Even 98% of efficiency also causes 2kW losses in a 100kW converter. The amount of power loss increases with increase in converter power rating. Thermal runaway causes an increase in loss and decreases efficiency. It results in destroying the power semiconductor device. So the device temperature should be maintained as constant. Based on the device power rating and its type, the following cooling methods such as forced convection and two-phase boiling techniques recommended. The plan of the inverter, onboard charger (OBC) DC-DC converters for HEV applications (Blaabjerg et al., 2020) is shown in table 8.

CONCLUSION

This chapter examines the power electronics circuits for HEV applications. The new power semiconductor device and power converter topology extends the performance of hybrid vehicles. The optimization algorithm of control techniques increases the efficiency increase the life span. The fault-tolerant algorithm and topology improve reliability. The current research involves the continuous improvement of the power semiconductor device, fault-tolerant converter topologies, artificial control techniques and innovative thermal management techniques.

REFERENCES

Alaoui, C., & Salameh, Z. M. (2005). A novel thermal management for electric and hybrid vehicles. *IEEE Transactions on Vehicular Technology*, *54*, 468–476.

Amrhein, M., & Krein, P. T. (2005). Dynamic simulation for analysis of hybrid electric vehicle system and subsystem interactions, including power electronics. *IEEE Transactions on Vehicular Technology*, *54*, 825–836.

Blaabjerg, F., Wang, H., Vernica, I., Liu, B., & Davari, P. (2020). Reliability of Power Electronic Systems for EV/HEV Applications. *Proceedings of the IEEE.* doi: 10.1109/JPROC.2020.3031041

Bose, B. K. (2000). Energy, environment, and advances in power electronics. *Proceedings of the International Symposium on Industrial Electronics, 1*, TU1–T14. 10.1109/ISIE.2000.930467

Ehsani, M., Rahman, K. M., & Toliyat, H. A. (1997). Propulsion system design of electric and hybrid vehicles. *IEEE Transactions on Industrial Electronics, 44*(1), 19–27. doi:10.1109/41.557495

Emadi, A., Ehsani, M., & Miller, J. M. (1999) Advanced silicon-rich automotive electrical power systems. *Proceedings of the Digital Avionics Systems Conference, 2*, 8.B.1-1–8.B.1-8.

Emadi, A., Williamson, S. S., & Khaligh, A. (2006). Power electronics intensive solutions for advanced electric, hybrid electric, and fuel cell vehicular power systems. *IEEE Transactions on Power Electronics, 21*(3), 567–577. doi:10.1109/TPEL.2006.872378

Gargies, S., Wu, H., & Mi, C. (2006) Isolated bidirectional DC/DC converter for hybrid electric vehicle applications. *6th Intelligent Vehicle Symposium.*

Habib, S. (2020). Contemporary trends in power electronics converters for charging solutions of electric vehicles. *CSEE Journal of Power and Energy Systems, 6*(4), 911–929.

Hamilton, D. B. (1996). *Electric propulsion power system - overview*. Power Electronics in Transportation. doi:10.1109/PET.1996.565906

Han, D., Noppakunkajorn, J., & Sarlioglu, B. (2014). Comprehensive Efficiency, Weight, and Volume Comparison of SiC- and Si-based Bidirectional DC-DC Converters for Hybrid Electric Vehicles. *IEEE Transactions on Vehicular Technology, 63*(7), 3001–3010.

Jahns, T. M., & Blasko, V. (2001). Recent advances in power electronics technology for industrial and traction machine drives. *Proceedings of the IEEE, 89*(6), 963–975. doi:10.1109/5.931496

Johnson, R. W., Evans, J. L., & Jacobsen, P. (2004). The changing automotive environment: High-temperature electronics. *IEEE Transactions on Electronics Packaging Manufacturing, 27*, 164–176.

Miller, J. M. (2003). Power electronics in hybrid electric vehicle applications. *Proceedings of the 18th Applied Power Electronics Conference, 1*, 23–29. 10.1109/APEC.2003.1179171

Onoda, S., & Emadi, A. (2004). PSIM-based modelling of automotive power systems: Conventional, electric, and hybrid electric vehicles. *IEEE Transactions on Vehicular Technology, 53*, 390–400.

Rahman, M. F. (2004). Power electronics and drive applications for the automotive industry, *Proceedings of the Power Electronics Systems and Applications Conference*, 156–164.

Rajashekara, K. (2003). Power electronics applications in electric/hybrid vehicles. *Proceedings of IECON'03, 3*, 3029–3030.

Traci, R. M., Acebal, R., & Mohler, T. (1999). Integrated thermal management of a hybrid electric vehicle. *IEEE Transactions on Magnetics, 35*, 479–483.

Vezzini, A., & Reichert, K. (1996). *Power electronics layout in a hybrid-electric or electric vehicle drive system*. Power Electronics in Transportation. doi:10.1109/PET.1996.565910

White, S. B., Gallego, N. C., & Johnson, D. D. (2004). *Graphite foam for cooling of automotive power electronics*. Power Electronics in Transportation.

Williamson, S., Lukic, M., & Emadi, A. (2006). Comprehensive drive train efficiency analysis of hybrid electric and fuel cell vehicles based on motor-controller efficiency modelling. *IEEE Transactions on Power Electronics*, *21*, 730–740.

Chapter 8
Real-Time Optimization of Regenerative Braking System in Electric Vehicles

B. Prasanth
Department of EEE, Amrita School of Engineering, Bengaluru, Amrita Vishwa Vidyapeetham, India

Deepa Kaliyaperumal
Department of EEE, Amrita School of Engineering, Bengaluru, Amrita Vishwa Vidyapeetham, India

R. Jeyanthi
Department of ECE, Amrita School of Engineering, Bengaluru, Amrita Vishwa Vidyapeetham, India

Saravanan Brahmanandam
Alstom Transport India Limited, India

ABSTRACT

In the present era, electric vehicles (EV) have revolutionized the world with their dominant features like cleanliness and high efficiency compared to that of the internal combustion (IC) engine-based vehicles. To crave for the higher efficiency of the EV during the braking, the kinetic energy of the EV is converted into electrical energy, which is harvested into storage system, called regenerative braking. Various techniques such as artificial neural network (ANN) and fuzzy-based controllers consider factors like state of charge of the battery and supercapacitor and brake demand for calculating the regenerative braking energy. A force distribution curve is designed to ensure that the braking force is distributed and applied on the four wheels simultaneously. In real-time optimization, an operating area is formed for maximizing the regenerative force which is evaluated by linear programming. It is proved that the drive range of the vehicle is increased by 25.7% compared to the one with non-RBS. In this work, RTO-based control loop for regenerative braking system is simulated in MATLAB/Simulink.

DOI: 10.4018/978-1-7998-7626-7.ch008

INTRODUCTION

With the increasing demand of fossil fuels, there has been an increase in deterioration of environmental pollution. So, electric vehicles (EV) have been an evident solution for the problem of strident environment and depletion of the energy resources (Wu & Luo, 2017). Apart from these, there are lot more advantages like torque, efficiency. But electric vehicles are limited by its energy storage. It takes a longer duration for the energy storage to be recharged which is not feasible. So, there is a significant research in developing batteries that are compatible with electric vehicles (Grbovic et al., 2010). In order to increase the drive range of the EV, there is requirement of secondary energy storage to support the primary storage. The secondary storage is decided based on the energy density and power density. Energy density is defined as the amount of energy per unit volume whereas power density means the amount of power stored per unit volume. Batteries come with high energy density which signifies that it can supply the vehicle without any support and low power density which implies that it takes longer duration for recharging. So, ideally an energy storage with high power density and high energy density is required which is not available. Therefore, a dual layer module with high power density called supercapacitor has been designed. But due to its lower energy density, it cannot alone supply the vehicle. So, a compound energy storage system (CESS) has been designed for craving the advantages of the supercapacitor and battery (Shabarish et al., 2020). Another advantage of the vehicle is the motor has the capability to harvest the energy back to storage system which is called regenerative braking. Thereby, increasing the regenerative braking energy can effectively increase the overall efficiency and the drive range of the vehicle. Significantly, 60% of the energy during braking is utilized in the urban driving cycles. With addition of the motor braking, a composite braking system is formed which enhances the energy recovery capacity (Gao & Ehsani, 2014).

To enhance the energy harvested during regeneration, a detailed understanding of regenerative braking system is required. However, it is more complex compared to that of the conventional braking system design. A model for regenerative braking has been designed mathematically and with the use of sliding mode controller, torque is generated (Fazeli et al., 2012). A detailed investigation is carried out on the factors like state of charge (SOC) of the battery, SOC of the supercapacitor, characteristics of the battery, supercapacitor, motor of the EV, road's adhesive coefficient which limits regenerative braking energy capability (Crolla & Cao, 2012).

Nevertheless, the aim is to design Regenerative Braking System (RBS) with performance of the braking to be stable and the energy recovery to be efficient. However, these two have a strong contradiction in most of the braking situations (Cao et al., 2005; Hua & Kao, 2011; Jiaqun & Haotian, 2015; Ko et al., 2015; Lu et al., 2014; Seki et al., 2009; Şengör et al., 2018; Xiao et al., 2016; Xu et al., 2016; Yoong, 2010). The factors like speed of the vehicle, braking forces and adhesive coefficient of the road which mostly affect the stability and performance is estimated by a fuzzy-based logic with consideration of the braking safety and wheel slide (Juan et al., 2017; Sasikumar & Deepa, 2019; Xiao et al., 2017). Three Different braking strategies for the RBS has been developed, the first strategy was based on fuzzy logic control to estimate the regenerative braking force by considering the factors like SoC of battery, supercapacitor, speed of the vehicle and brake demand and second, third strategies were based on improving the conventional braking technique to estimate the front wheel brake force. Also, a comparative analysis has been done and fuzzy logic control-based strategy is accurate, reliable, robust and can easily adapt to all condition of the road. It also advances in savings of energy and proved drive range of the vehicle is increased by 25.7% compared to the one with non-RBS (Akhila & Ratnan, 2016; Ko et al., 2015; Simoes et al., 1997; Takahashi & Seki, 2009; Xu et al., 2011; Yao et al., 2006; Zhang, Xu, Li et al, 2012). But as fuzzy calculates with the help of the rule base, it takes longer time than usual to predict which is not feasible and safer with respect to automobile braking. Further to improve the speed and response to the transients, an ANN based controller has been designed. It uses multi-layer perceptron network for the estimation of regenerative force due to its unique feature to take multiple inputs and define the logic. Neural networks predict the output based on the previously analyzed data. Sigmoid activation function is used to compute the output. The neural network consists of 4 neurons at the input layer (i.e. SoC of Battery, Supercapacitor, Speed and brake demand), 8 neurons at the hidden layer and one neuron at the output layer. Backpropagation algorithm is used to compute the gradient of the cost function and fminunc (i.e. advanced optimization algorithms) is used to minimize the cost function. The fuzzy and ANN controllers are designed based on the ideal power, torque vs speed curve of the motor (Gao et al., 2001; Liu et al., 2018; Rahman & Hoque, 1998; Mladenov et al., 2014; Wu et al., 2002). There have been interesting developments done by adapting the traffic conditions. However, regenerative braking alone cannot support the required (applied through brake pedal) braking force. So, there is a blending requirement of frictional/mechanical braking force to support the total braking force. Therefore, a hybrid control strategy has been developed to effectively utilize the frictional and regenerative braking force simultaneously. It is proved that the algorithm can increase the energy and simultaneously have control over the frictional brake force distribution on the rear wheels to ensure vehicle stability and safety (Thiruvonasundari & Deepa, 2020a). However, the regenerative braking force is limited by the SoC of the battery and supercapacitor, speed of the vehicle and brake demand. So, an algorithm has been designed to effectively harvest the regenerative braking energy based on the aforementioned parameters. The algorithm is integrated with fuzzy/ANN based estimators, so that the dynamic performance is achieved. Also, an ideal braking force distribution curve is designed to ensure uniform distribution among the front and rear wheels (Ehsani et al., 2018; Foiadelli et al., 2006; Frilli et al., 2016; Prasanth et al., 2021; Prasanth et al., 2020; Thiruvonasundari & Deepa, 2020b; Zhang, Kong, Chen et al, 2012).

In this chapter, a detailed analysis of the regenerative braking system is done and effectively optimize the process of estimating the regenerative braking force. The whole analysis of the system is discussed in the section II. To estimate the regenerative braking force, there is requirement of complete understanding of the brake system used in electric vehicles. To ensure the vehicle safety and stability, the applied braking force is distributed among the front and rear wheels according to the ideal braking force distribution curve which is discussed in the section III. A complex analysis has to be carried out for the applied braking force to follow the ideal braking force distribution curve. However, the process has been made simple with the advancements of electronics in the field of automotive industry. Nevertheless, ideal braking performance cannot be achieved practically. So, a threshold has been set on the rear braking force called ECE regulation which is explained in the section IV. There are other factors like coefficient of adhesion which influence the braking performance and various scenarios of braking and the problems faced by the vehicle at different scenarios is discussed in the section V. So, after mathematically designing the brake system, a control strategy has been developed to estimate the regenerative braking force for the maximum energy recovery which is discussed in the section VI. The main objective of this chapter is to optimize the process of estimating the regenerative braking force which is done by the optimization techniques discussed in the section VII. The implementation of the system is done in MATLAB/Simulink and the results are evaluated for an urban drive cycle named NEDC ECE R15 to evaluate the braking force scenarios at different speeds which is explained in the section VIII. Finally, the real time optimization has been compared with the existing fuzzy/ANN logic controller to put forth the robust and better responsive control system, which has been detailed in the section IX.

ANALYSIS OF THE SYSTEM

The Block diagram of the electric vehicle is shown in the Fig. 1 which posits the control structure employed for harvesting regenerative energy back into storage device. The vehicle is a front wheel driven (i.e. motor is equipped on the front axle). Generally, a primary storage device like battery is used to support the EV during traction mode, when vehicle goes into braking mode of operation, Force applied by the driver is distributed among the front and rear wheels with help of real time optimization techniques.

Figure 1. Overall block Diagram of the Electric Vehicle

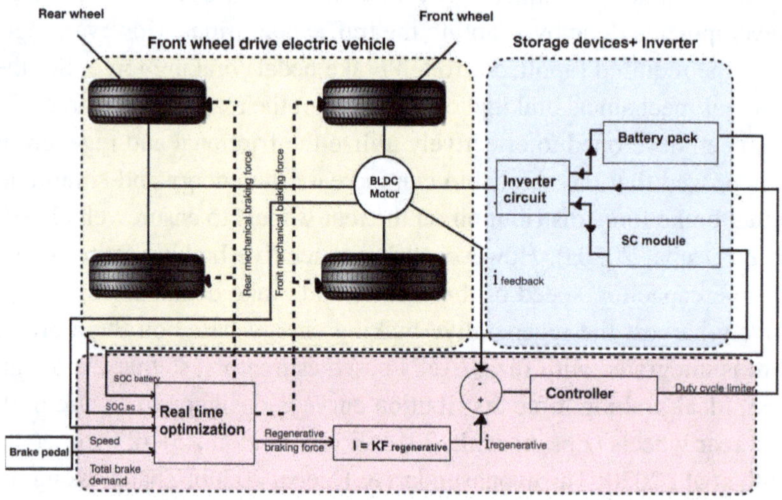

The in detailed analysis of real time optimization is shown in the Fig. 2. With respect to the present state of charge of the battery and state of charge of the supercapacitor, the regenerative braking force for the electric vehicle is determined.

Figure 2. Block Diagram of the System

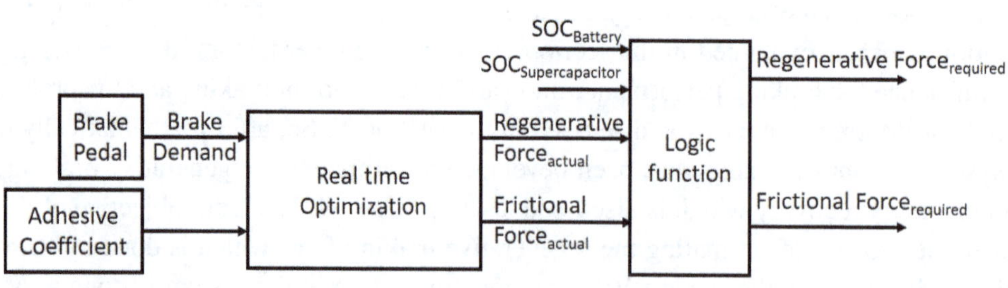

Real time optimization techniques are utilized to evaluate the required regenerative braking force of the system. In order to implement the optimization technique, there is a need to understand the distribution of the applied braking force between the front and the rear wheels so that vehicle stops accurately and safely.

DISTRIBUTION OF BRAKING FORCE AMONG THE FRONT AND REAR AXLES

When the vehicle is moving on a flat road, the forces acting on it during the braking process is shown in the Fig. 3. The opposing forces like rolling resistance and aerodynamic drag are comparatively small with respect to the braking forces and hence they are not considered for analysis in Fig. 3. The vehicle deceleration during braking, j, is given by equation (1).

$$j = \frac{F_{bf} + F_{br}}{M} \tag{1}$$

Where F_{bf} and F_{br} are forces acting during braking on the front and rear axle respectively.

Figure 3. Forces acting when the vehicle is moving on a flat road (Cao et al., 2005)

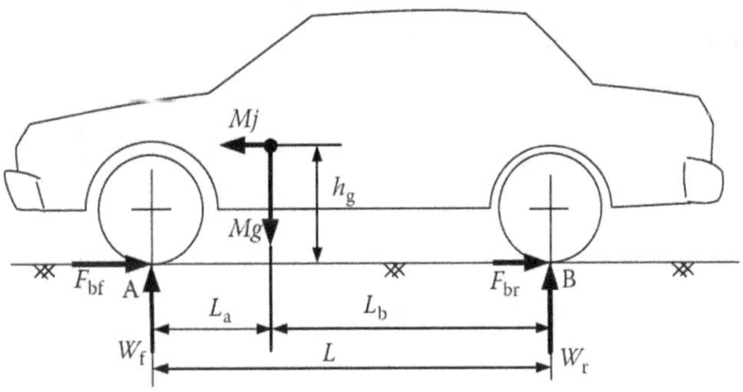

The friction between the tyre and road opposes the maximum braking force and it varies directly to the weight of the vehicle. To obtain maximum braking capacity from the front and rear axles, the actual brake torque from the pedal also should vary proportional to that of the weight of the vehicle. During the process of braking, there will be transfer of load from rear to front axle. With the centre of gravity of the vehicle as a reference and the two equilibrium positions of the point of the friction A and B, as shown in the Fig. 3, the normal load on the front and rear wheels, W_f and W_r, with a vehicle deceleration (j) are given by the equations (2).

$$\left.\begin{aligned} W_f &= \frac{Mg}{L}\left(L_b + h_g \frac{j}{g}\right) \\ W_r &= \frac{Mg}{L}\left(L_a - h_g \frac{j}{g}\right) \end{aligned}\right\} \tag{2}$$

Where M= Total mass of the vehicle (kg),

g= Acceleration due to gravity (9.8m/s^2),
L_a= Distance between centre of mass and front axle (m),
L_b= Distance between centre of mass and rear axle (m),
L=L_a+L_b= Axle Spread (m)
h_g= height between the centre of mass of the vehicle and ground

The braking force on the front and rear wheels is proportional to the normal load on the front and rear wheels, given by the equation (3),

$$\frac{F_{bf}}{F_{br}} = \frac{W_f}{W_r} = \frac{L_b + h_g j / g}{L_a - h_g j / g} \tag{3}$$

Combining equations (1) and (3), the ideal braking force distribution curve (I-curve) is obtained as depicted in the Fig. 4. The front and rear wheel force on the vehicle should exactly follow the curve to halt the vehicle safely and accurately.

Figure 4. Ideal Braking Force Distribution (I-curve)

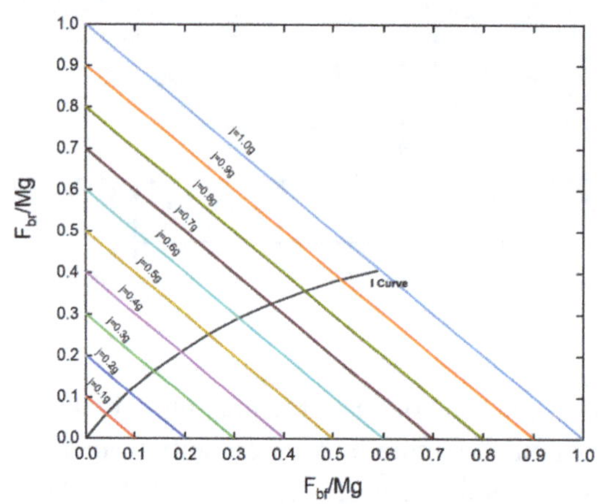

When the applied braking force is strong, there will be a maximum braking force applied on the front and the rear wheel, but the amount of braking force applied to the front and rear wheels cannot be beyond a threshold which depends on the friction between the tyre and ground. It is represented as adhesion coefficient (μ). Therefore, for the particular situation, the rate of deceleration is given by the equation (4).

$$\left| j_{max} \right|_{\mu} = \frac{F_{bf-max} + F_{br-max}}{M} = \frac{\left(W_f + W_r\right)}{M} = g\mu \tag{4}$$

For the braking force to completely follow the I-curve, the system turns out to be more complex in control and structure. The advancements in the development of Electronic braking systems (EBS) and various technologies has enhanced the performance of braking. The concept of regenerative braking with EBS technology will be discussed in the further sections.

The present technology is designed to have a fixed linear ratio for the original braking force applied to the front and the rear wheels. This ratio is given by the front braking force to that of the total force in equation (5).

$$\beta = \frac{F_{bf}}{F_b} \quad (5)$$

Where F_b is the total braking force of the car (F_b=F_{bf}+F_{br}), β is a design perspective parameter like radius of the vehicles but independent on the vehicle parameters. The original braking force on the front and rear wheel is obtained using β, and is expressed as in the equation (6,7).

$$Fb_f = \beta Fb \quad (6)$$

$$Fb_r = (1 - \beta)Fb \quad (7)$$

Solving equations (6) and (7):

$$\frac{F_{bf}}{F_{br}} = \frac{\beta}{1-\beta} \quad (8)$$

Fig. 5 depicts the ideal braking force with β-line. It is observed that there is only one intersection (i.e. j=0.8g), where the force applied is equally distributed among the front and the rear wheels. The point is specific to the adhesive coefficient of the road (μo)

With refence to equation (3), the brake demand (j/g) is replaced by μo which results in

Figure 5. Ideal braking Force distribution curve with β – line

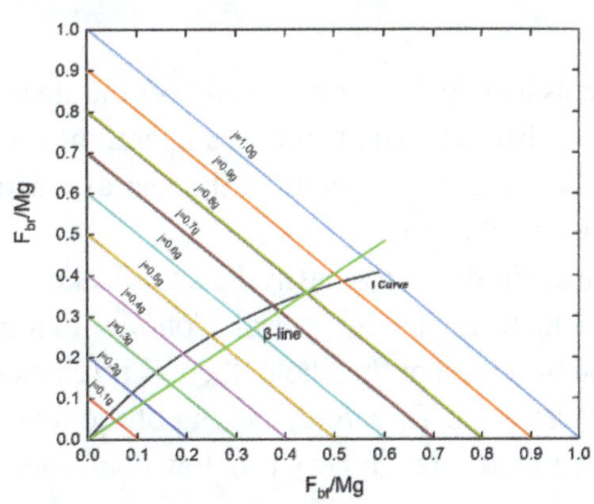

$$\frac{\beta}{1-\beta}=\frac{L_b+\mu_o h_g}{L_a-\mu_o h_g} \tag{9}$$

The values of μo and β are derived using the equation (9)

$$\mu_o=\frac{L\beta-L_b}{h_g} \tag{10}$$

$$\beta=\frac{\mu_o h_g+L_b}{L} \tag{11}$$

When the brakes are applied on the roads with coefficient of adhesion less than μo (From Fig. 5, the region where I curve is above the β-line) the locking of the front wheel is done first followed by the rear, whereas if the coefficient of adhesion is greater than μo ($_F$rom Fig. 5, the region where β-line is above the I curve), the locking of the rear wheel is done first followed by the front.

The direction of the vehicle is unstable if the rear wheels are locked first as shown in the Fig. 6, which depicts the vehicle with two axles when inertia and braking force are acted. As observed from the Fig. 6, the vehicle movement is seen due to the road camber, side wind, or centrifugal force and movement due to the yaw centre of the front axle. As the motion due to yaw increases, the inertia force is increased and results in accelerating the yaw. So, the vehicle is rotated to 180° (i.e. A-E), where rear faces the front end.

Figure 6. Loss of directional stability when the rear wheels lockup first (Cao et al., 2005)

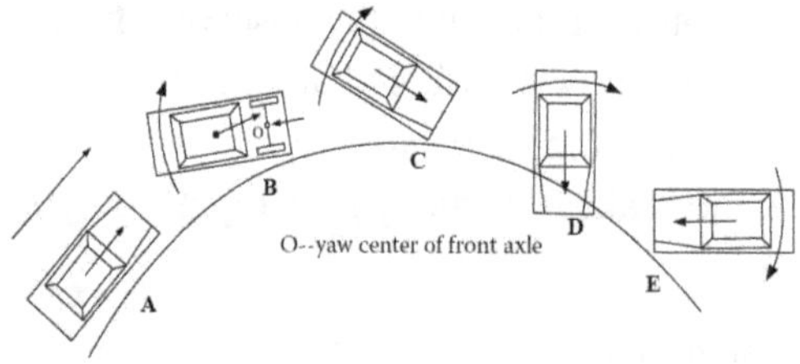

Whereas when the front wheels lockup first, the control over the steering is lost, but the vehicle does not go to unstable position. This is because when there is a lateral movement of front wheel occurs, the movement is self-corrected by the force at yaw centre of the rear axle acted due to inertia which consequently brings the vehicle into a straight path.

When the steering control is lost, the driver can easily detect and take the appropriate action of reducing the pressure on the brake pedal which regains the control. Contrary to this, when rear wheels get locked first, when deviation is exceeded by a limit or threshold, there is no chance of regaining the control even when the brakes are completely released. Therefore, locking of the rear wheels first is considered as a critical situation, especially when roads are slippery (i.e. low coefficient of adhesion), there will high dissipation of kinetic energy which creates a directional instability. For the system proposed, it required to ensure that the rear wheels do not lock first.

BRAKING REGULATION

As described earlier, if the β-line is below the I curve, the front wheels get locked first followed by the rear wheels, which is the stable operation of braking. However, when β-line is much smaller than I curve, large amount of force has to be applied to front wheels and a small amount force is applied to the rear wheel which causes a problem of reduction in the utilization of adhesive capability of the road (i.e. when front wheels only get locked and rear wheels are not locked). So, to prevent this situation, regulations have been developed regarding the brake design by United Nations Economic commission for Europe (ECE). The ECE regulation for passenger vehicle is given by the equation (12).

$$\frac{F_{bf}}{W_f} \geq \frac{F_{br}}{W_r} \tag{12}$$

So, equation (12) dictates that the locking of rear wheels never happens before that of the front wheels (i.e. β-line is always below the I curve). ECE regulation also provides a minimum braking force that has to be applied on the rear wheels, which is given by the equation (13).

$$\frac{j}{g} \geq 0.1 + 0.85(\mu - 0.2) \tag{13}$$

Figure 7. Braking Force Distribution stipulated by ECE regulation

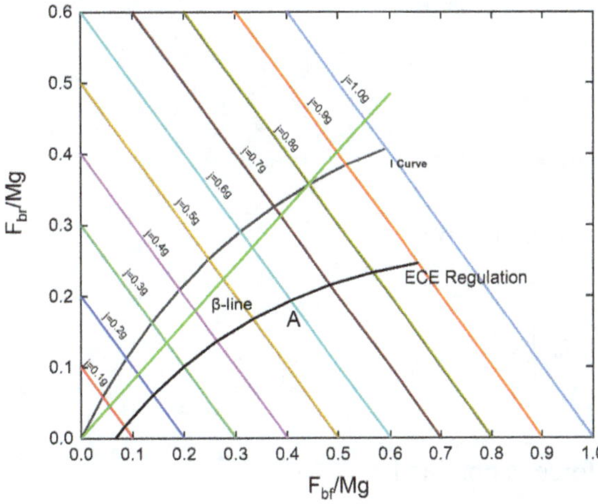

Where j/g is the brake demand of the vehicle. Equation (13) illustrates that after the front wheels are locked, the braking force on the rear wheel should be large so that deceleration value is not less than that mentioned in the equation (13). Using all the above equations, the front and rear wheels braking force is calculated and is shown in the Fig. 7.

From the Fig. 7, it can be observed that at point A of the ECE regulation, the rate of deceleration is j=0.6g, which clearly does not mean that coefficient of adhesion is 0.6 but larger than that because the rear wheels are not locked. Therefore, for the optimization of the braking force distribution, the area of operation lies in between the I curve and ECE regulation.

ANALYSIS OF BRAKING PERFORMANCE

The vehicle with the braking system discussed earlier locks the front and rear wheel simultaneously only at one instant for a road with coefficient of adhesion (μo_j. Whereas for the roads with different coefficient of adhesion, rear or front wheels lock up first followed by the other. So, a further analysis has been to done to illustrate different braking scenarios which is useful for designing the regenerative braking system for EV. There are four different braking scenarios considered, which is as follows:

1. **Unlocked Rear and Locked Front Wheels:** The Braking force when the front wheels are locked with coefficient of adhesion (μ) is given by:

$$F_{bf} = W_f \mu \tag{14}$$

where W_f is the weight of the vehicle. Further, substituting weight of the vehicle from the Equation (2) in Equation (14) results in:

$$F_{bf} = \frac{Mg\mu}{L}\left(L_b + \frac{j}{g} h_g \right) \tag{15}$$

The total braking force is,

$$F_{bf} + F_{br} = Mj \tag{16}$$

So, the front braking force is given by,

$$F_{bf} = \frac{Mg\mu}{L}\left(L_b + \frac{F_{bf} + F_{br}}{Mg} h_g \right) \tag{17}$$

Finally, the rear braking force is obtained,

$$F_{br} = \frac{L - \mu h_g}{\mu h_g} F_{bf} - \frac{MgL_b}{h_g} \tag{18}$$

So, the Equation (18) gives the rear braking force with respect to the front braking force for different coefficient of adhesion (μ) thereby generating group of lines known as f-lines. f-lines represent the locking of front wheels but the rear wheels are not locked, represented by solid lines below the I-curve shown in the Fig. 8.

2. 2. **Unlocked Front and Locked Rear Wheels:** In similar way as case 1, when the wheels on the front are not locked and rear wheels are locked, the rear braking force with respect to front braking force for different coefficient of adhesion is given by,

$$F_{br} = \frac{-\mu h_g}{L + \mu h_g} F_{bf} + \frac{\mu MgL_a}{L + \mu h_g} \tag{19}$$

Thus, Equation (19), generates another group of lines known as r-lines which represents the lockup of rear wheels but not the front wheels represented by dash lines above the I-curve shown in the Fig. 8. The f-lines and r-lines coincide on I-curve.

3. **Analysis of Braking Process:** With the reference to the Fig. 8, which depicts ECE regulation, I-curve, f-lines and r-lines the process of braking in a detailed way is analysed as follows:

    a. For a road whose coefficient of adhesion (μ) is less than μo: For example, take μ=0.6 and μo=$0._{8}$. When the driver applies the brake, the braking force is distributed between the front and rear wheels following the β-line till the point A which is the point of intersection of f-line

Figure 8. Analysis of Braking Force with different coefficient of adhesion

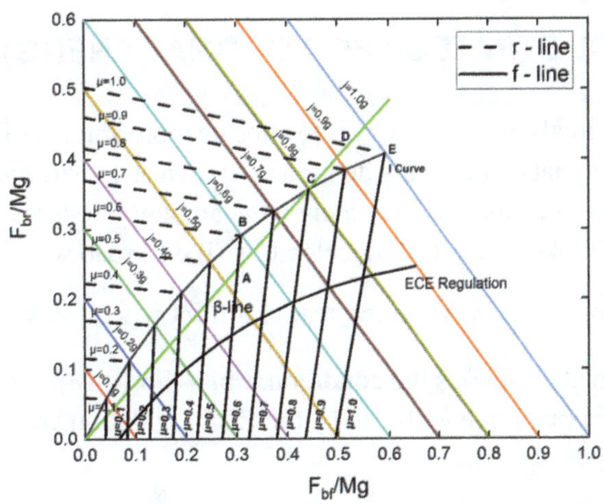

(μ=0.6) and β-line where lockup of front wheels is done but not the rear wheels. Further, with increase in the pressure on the brake pedal, increases the force on the rear wheels to a greater extent and front by a smaller extent. From the Fig. 8, it follows the f-line with μ=0.6 till point B is reached and the front and rear wheels are locked up and the maximum rate of deceleration j=μg=0.6g. However, with this case there is no point of instability in the vehicle as the front wheels are locked up first.

b. For a road whose coefficient of adhesion (μ) is more than μo: Assume μ=1.0 and μo=0.8 as the case of analysis. In the similar way, when is pedal is engaged, the braking force is distributed among the front and rear by following the β-line till the point D which is the point of intersection of r-line with μ=1.0 and β-line where lockup of rear wheels are done but not the front wheels. Further increment of pressure on the pedal makes the braking force follow the r-line with μ=1.0 and reaches point E, where the lockup of front and rear wheels is done and also vehicle achieves its maximum deceleration rate (i.e. j=μg=1.0g). As already discussed earlier, this case causes vehicle to instability due to the lockup of rear wheels first.
c. For a road whose coefficient of adhesion (μ) is equal to μo: Assume μ=0.8 and μo=0.8. However, in this case, the front and rear wheels are locked up simultaneously at point C and the vehicle decelerate at j=μg=0.8g. This case is considered as the ideal case.
    4. Front Wheels with **Maximum Available Braking Force: In electric vehicl**es, the motor is equipped onto front wheels which means during the braking for the regeneration to happen there should be maximum amount of force available on the front wheels compared to that of the rear to harvest maximum amount of braking energy into the storage devices like battery, supercapacitor. So, in order to have maximum energy on the front wheels, the braking force should abide to its norms like ECE regulation, I curve which if not followed may cause instability in the vehicle.

The Existing strategy is explained considering three cases, in which two cases are stable whereas when $\mu>\mu$o, there is instability in the vehicle which is not desirable. To overcome this problem, a control strategy is proposed in such a way that maximum energy is harvested and abides to safety concerns.

PROPOSED CONTROL STRATEGY FOR OPTIMAL ENERGY RECOVERY

The principle behind the optimal control strategy is to allocate more amount of force during braking to the wheels on the front side than that on the rear side for a road with any coefficient of adhesion. Allocating large front wheel braking force results in adequately large amount of energy available for regeneration.

When driver applies the brake with rate of deceleration (j) on a road with coefficient of adhesion (μ), and $\frac{j}{g} < \mu$, the force during the braking on the rear and front wheels can be allocated pertaining that applied force by the brake pedal satisfies the condition, Fbf + Fbr = Mj. As already discussed, the strategy has to ensure the rear braking force is abided by the ECE regulation and gives its maximum braking performance. Therefore, the front and the rear braking force have a certain range to operate which is dependent on the coefficient of adhesion of the road and rate of deceleration of the vehicle.

The whole strategy is illustrated by considering three different cases, assume j=0.7g with the coefficient of adhesion μ=0.9, From the Fig. 9, it is observed that the maximum braking force is at the point of intersection of the f-line with coefficient of adhesion μ=0.9 and j=0.7g which is denoted by F. F is also on the ECE regulation curve, which illustrates that the maximum force is at the point F. However, with respect to the design perspective, the braking force varies on the line j=0.7g denoted by F-H, but the line G-H is range above the I-curve which denotes the lockup of the rear wheels first and then front wheels which causes vehicle instability. So, the region of operation for braking force with rate of deceleration j=0.7g on a road with coefficient of adhesion is F-G.

Considering the case where rate of deceleration is j=0.5g and coefficient of adhesion is μ=0.7, in this case the point of intersection of the line j=0.5g and f-line with coefficient of adhesion μ=0.7 is at point P which is below the ECE regulation which violates the regulation. So, the maximum braking force is dictated by the ECE regulation which is denoted by the point D in the Fig. 9. The braking force on the front and rear wheels vary between the E-D ignoring the region above the I-curve.

Consider the case where the point of intersection of f-line and deceleration line is above the ECE regulation which is j=0.3g and μ=0.4 at point B in the Fig. 9. The point B is above the ECE regulation thus denoting the maximum braking force to be at B and the operating range is between B-C.

The above analysis is the control strategy to determine the operating range for maximum energy recovery in electric vehicles. So, after deciding the operating range for the allocation of braking force on the front and the rear wheels, the regenerative force has to be determined. The regenerative force is dependent on the motor capability. The motor capability is limited by its specifications. As shown in the Fig. 9, the maximum motor capability is denoted by F-motor. Consider the case where j=0.2g and coefficient of adhesion μ=0.3, the point of intersection of f-line and deceleration rate is at R which is above the ECE regulation and is within range of F-motor, which denotes the total braking force on the front wheel is given by the motor thereby ensuring regenerative braking to its full capacity and mechanical braking is not required on the front wheels.

Figure 9. Control Strategy for optimal energy recovery

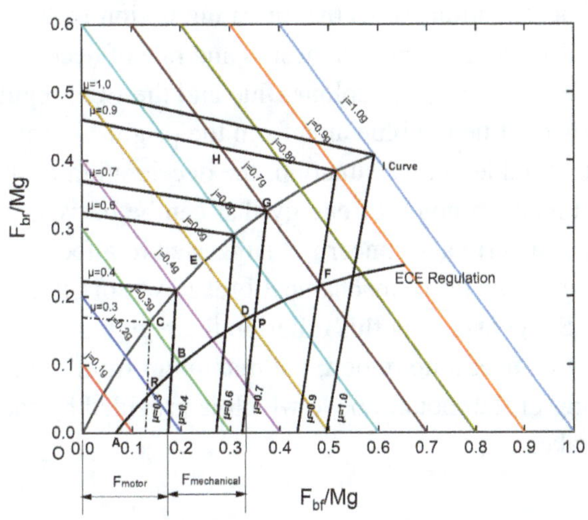

Figure 10. Region of operation for the braking force distribution

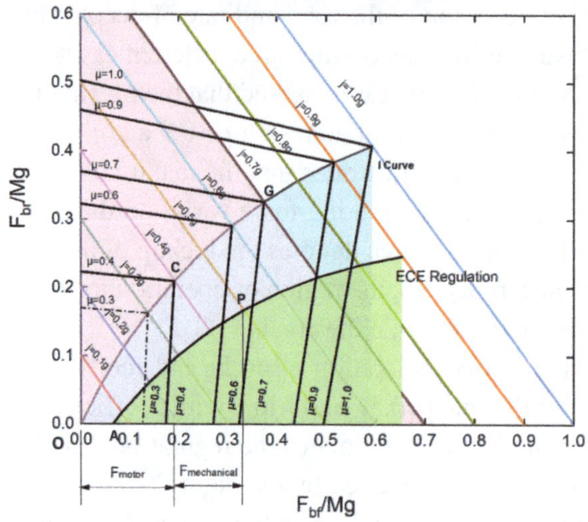

Whereas when there is a situation required Front wheel braking force is greater than F-motor, the rest of the braking force is given by the mechanical brakes. Consider the case where j=0.5g and μ=0.7, as already discussed the maximum braking force lies on the point D, where the braking force required is greater than the F-motor. Therefore, the motor provides its maximum torque and the remaining is provided by the mechanical brakes denoted by F-mechanical. However, for the vehicle with motor equipped on the front side only, the braking force on the rear wheels totally depends on the mechanical brakes or the frictional brake system.

REAL TIME OPTIMIZATION FOR THE PROPOSED CONTROL STRATEGY

In this section, real time optimization techniques are used to calculate the maximum braking force distributed among the front and the rear wheels. As discussed earlier, the operating region is constrained with respect to ECE regulation and the I curve. Also, the braking force is dependent on the coefficient of adhesion (μ) and rate of deceleration (j), so the operating region is decided by the four parameters namely I-curve, ECE regulation, coefficient of adhesion and rate of deceleration. From the Fig. 10, the area that is under the I-curve is given by the colour Blue and the ECE regulation by the colour green. So, the region of operation is confined to blue area from the origin. As safety is primary concern during the braking, when there is sudden brake situation, the deceleration of the vehicle tends to be large (j=1.0g) which in turn inhibits the amount of energy that can be harvested during regenerative braking. However, safety remains the primary concern, it suggested to allocate the braking force above the deceleration curve of j=0.7g gradually to mechanical brake system until j=0.9g and no regenerative brake system above j=0.9g. So, considering this criterion the region of operation converges to the area bounded by OAPGO. With this operating region, the maximum front braking force can be calculated which is limited by the rating of the motor, above which is provided by the mechanical brake system (i.e. basically a frictional brakes).

From the Fig. 10, OPAGO is the operating region which is in purple. The operating region is constrained by a non-linear curve. So, the I-curve and ECE regulation curve can be linearized using linear regression techniques. The I curve is represented by the equation (20).

$$W_r = \frac{1}{2}\left[\frac{M.g}{h_g}\sqrt{b^2 + \frac{4h_g L}{M.g}W_f} - \left(\frac{MgL_b}{h_g} + 2W_f\right)\right] \quad (20)$$

The I curve can be linearized by the breaking the curve into two regions at the point x=0.2.

Figure 11. Region of operation after linear regression

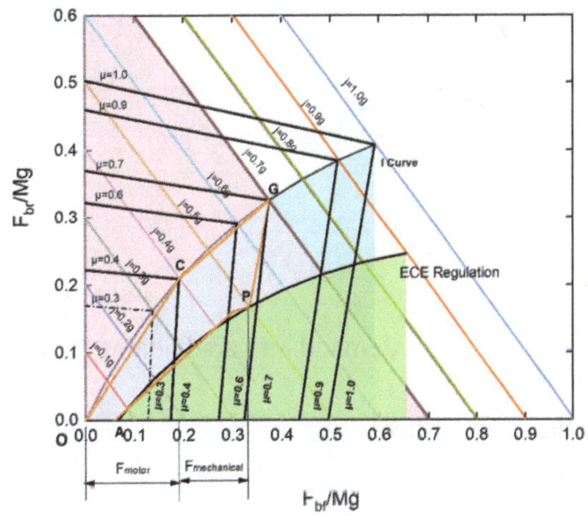

In similar way, the ECE regulation is linearized. As the maximum rate of deceleration under which electrical braking is used is 0.7g, the coefficient of adhesion of the road also is limited by μ=0.7. So, the region operation converges by the area bounded by GPAOG. The area obtained after linearizing I-curve and ECE regulation curves is shown in the Fig. 11. The lines are shown in Orange in the Fig. 11 and the area disclosed by the orange lines (i.e. OCGPAO). The regenerative braking intensity for different scenarios can be evaluated by developing appropriate control logic with respect to the SoC of the battery and supercapacitor. If the required brake demand (j/g) is less (j<0.1g), then only regenerative braking system works. Whereas when the brake demand is medium, regenerative braking works to its full capability and the remaining force is provided by the mechanical braking system. As already discussed, due to safety concerns the regenerative braking is used to its full capability till j/g=0.7, beyond which the capability of regenerative braking decreases and the mechanical braking starts to increase. For sudden heavy brake, (i.e. j/g=0.9 or 1) only mechanical braking system is preferred. The logic for regenerative braking intensity with respect to the brake demand is shown in the Fig. 12. With this strategy, a hybrid brake system which simultaneously makes use of the electrical and mechanical braking system in an effective way.

The main objective of the given optimization problem is to maximize the front braking force in order to harvest maximum braking energy from the motor. So, the objective function for the maximization of the front braking is Fbf/Mg. $Wh_{er}eas$ the required objective function is regenerative braking force which is given by the equation in (21).

Figure 12. Control logic for regenerative braking intensity

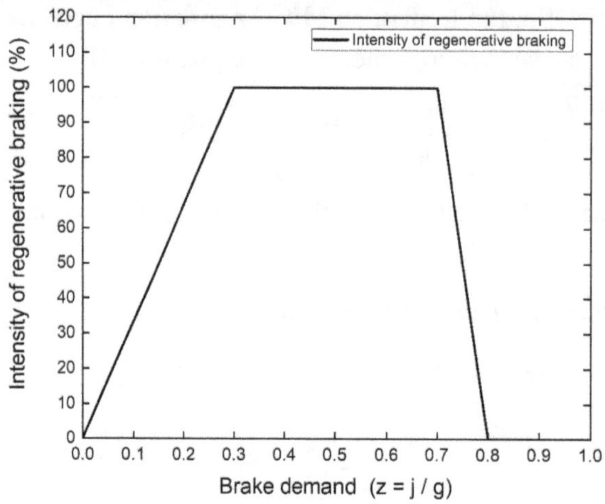

$$F_{regenerative} = \frac{kF_{bf}}{Mg} \quad (21)$$

Where k is a factor dependent on the brake demand and coefficient of adhesion,

Fbf is $_T$otal Braking Force on the Front Wheel
M is the Mass of the Vehicle (kg)
g is the Acceleration due to Gravity (m/s2)
k factor is generalized by analysing different scenarios of road conditions with different deceleration rates and is given by equation (3).

$$k = \begin{cases} 1 & 0 < z < 0.3 \\ \dfrac{F_{motor\_max} L}{\mu\left(L_b + zh_g\right)} & 0.3 \le z \le 0.7 \\ 0 & z > 0.7 \end{cases} \qquad (22)$$

The control strategy is limited by its constraints. The region of operation should lie below the I-curve to avoid vehicle instability as discussed in the previous sections which is given by equation (23).

$$y \le \frac{1}{2}\left[\frac{M.g}{h_g}\sqrt{b^2 + \frac{4h_g L}{M.g}x} - \left(\frac{MgL_b}{h_g} + 2x\right)\right] \qquad (23)$$

Where x and y are the parameters on the X-axis (i.e. F_{bf}/Mg) and Y-axis (i.e. F_{br}/Mg) in the Fig. 11.

When the brake demand is less, there might be a situation where there is no force allocated to the rear wheels and total applied force is applied only on the front wheels which may lose the control of vehicle. So, the minimum rear braking force is given by the ECE regulation which as in Equation (24). As already discussed in the Fig. 7, the ideal case occurs at $\mu=\mu 0=_0.8$.

$$y \geq \frac{L-0.8h_g}{0.8h_g}x-\frac{gL_b}{h_g} \tag{24}$$

However, the brake demand and coefficient of adhesion should abide to braking regulation mentioned in the section IV. The region of operation converges real time based on the required deceleration and the coefficient of adhesion of the road on which vehicle is travelling which is given by the equations (25,26).

$$x+y \leq \frac{j}{g} \tag{25}$$

$$y \leq \frac{L-\mu h_g}{\mu h_g}x-\frac{L_b}{h_g} \tag{26}$$

With respect to the constraints as in Equations (23,24,25,26), the region of operation is constructed. In the regenerative braking system, the force is provided by the motor. The maximum force for regeneration is function of the speed of the motor and the limitation for the maximum force is given by the equation (27).

$$F_{regenerative} \leq \frac{T_{max}i_g}{\eta} \tag{27}$$

Where T_{max} is the maximum regenerative torque during braking, i_g is the drive ratio and η is the efficiency of the braking system. $ig_{,}$ η are totally dependent on the parameters of the machine. The objective function given in Equation (21) is evaluated using linear programming in the specified region of operation which varies with the real time. Therefore, using the optimization technique the maximum braking force on the front wheel is found and the maximum braking force for regeneration so that maximum energy can be recovered back into the storage device.

IMPLEMENTATION OF REGENERATIVE BRAKING SYSTEM

The proposed system is implemented in MATLAB/Simulink based simulation environment. The system is evaluated for the following parameters shown in the Table 1. The system is evaluated for the NEDC ECE R15 drive cycle which is developed by the United Nations Economic Commission for Europe. It is modified version of the ECE R15, named as the new European driving cycle for united nations economic commission for Europe (NEDC ECE R15). The regenerative braking is evaluated based on the state of charge of the battery, state of charge of the supercapacitor. The logic control goes as follows, when the

SoC of both the battery and supercapacitor is very less (i.e. less than 20%) which is an extreme condition, where vehicle halts and as the internal resistance of the battery/supercapacitor is large, there is requirement of very large current to charge, which cannot be harvested with regenerative braking. When the SoC of the either battery or supercapacitor is between 20% and 70%, then the regenerative braking capacity is used to its maximum to charge the battery. Charging of the supercapacitor will be the priority, due to its high-power density which makes it faster to recharge compared to that of the battery. When the SoC of each battery/supercapacitor is high, the total braking force on the front wheel is provided by the frictional brakes and there is no regenerative braking utilized in this case.

Table 1. Simulation Parameters of the system

Parameters	Specification	Value	Units
Body Parameters	Mass (M)	1390	kg
	Axle Spread (L)	2.6	m
	Distance between centre of mass and front axle (La)	1.56	m
	Distance between centre of mass and rear axle (Lb)	1.04	m
	height between the ground and centre of mass (hg)	0.5	m
	Drag coefficient (Cd)	0.335	_
	Radius of the wheel (r)	0.282	m
Motor Parameters	Motor Type	BLDC motor	_
	Maximum Power	50	kW
	Maximum Torque	275	Nm

As the region of operation varies real time with variation of the brake demand. The region of operation for brake demand of j=0.4g and is shown in the Fig. 13. From the Fig. 13, the region of operation for the case has been constructed. The maximum braking force on the front wheel lies on the point which is maximum X-value.

Figure 13. Region of operation for the case when μ=0.5 and j=0.4g

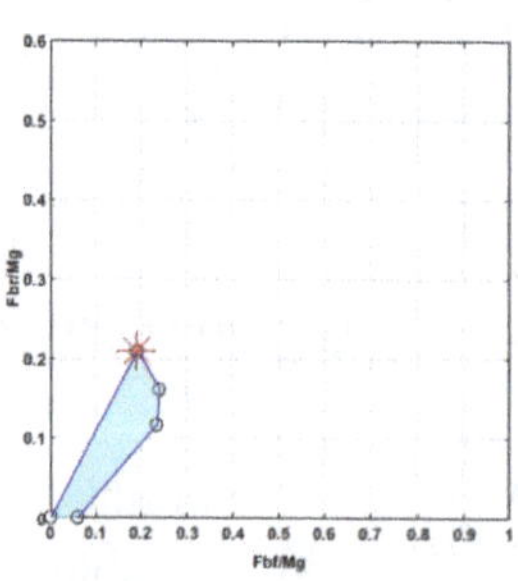

A Case study has been done by considering the vehicle to be moving at constant speed of 100kmph on a road with coefficient of adhesion (μ=0.5). The distribution of braking forces with the variation of brake demand from minimal brake situation (j=0.1g) to sudden brake situation (j=1.0g) are tabulated in the Table-2. Also, a graphical analysis has been shown in the Fig. 14.

Table 2. Case study when a vehicle is moving with a constant speed of 100kmph on a road with coefficient of adhesion (μ=0.5)

Brake demand(j)	Total Braking force	Front Braking force	Rear Braking force	Frictional Braking force	Regenerative Braking force
0.1g	-1362	-1142	-220.4	-220.4	-1142
0.2g	-2724	-1958	-765.9	-765.9	-1958
0.3g	-4087	-2775	-1311	-1710	-2377
0.4g	-5449	-3248	-2200	-2779	-2670
0.5g	-6811	-3370	-3411	-4149	-2662
0.6g	-8173	-3370	-4804	-5610	-2563
0.7g	-9535	-3370	-6166	-7065	-2471
0.8g	-10897	-3370	-7528	-10897	0
0.9g	-12259.8	-3370	-8890	-12259.8	0
1g	-13622	-3370	-10252	-13622	0

From the Fig. 14, It is observed that Regenerative braking increases in the negative direction from j=0.1g to j=0.3g, above which it remains constant (i.e. maximum capacity of the motor). Above j=0.7g, the regenerative braking decreases gradually and becomes zero at j=0.8g because above a threshold of j=0.8g, it is considered to a sudden brake situation. During the sudden heavy brake situations, only frictional braking force acts on the vehicle and no regenerative braking force will be acting.

Figure 14. Distribution of forces with respect to the brake demand

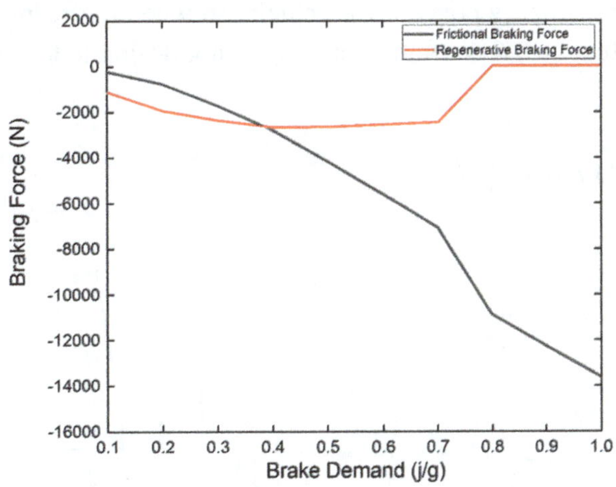

Table 3. Case Study when vehicle is moving at a constant speed of 100kmph on a road with coefficient of adhesion (μ=0.5)

Brake Demand (z)	Effective Braking Time(sec)	Braking Distance (km)	Motor Deceleration (m/s²)	Braking time by mot* (sec)	Braking distance by mot*(km)
0.1g	28.34467	0.787352	0.821583	33.81008	0.821583
0.2g	14.17234	0.393676	1.408633	19.71967	1.408633
0.3g	9.448224	0.262451	1.710072	16.24363	1.710072
0.4g	7.086168	0.196838	1.920863	14.46109	1.920863
0.5g	5.668934	0.15747	1.915108	14.50455	1.915108
0.6g	4.724112	0.131225	1.843885	15.06481	1.843885
0.7g	4.049239	0.112479	1.777698	15.6257	1.777698
0.8g	3.543084	0.098419	0	--	--
0.9g	3.149408	0.087484	0	--	--
1g	2.834467	0.078735	0	--	--

*When the vehicle is halted only with the regenerative effort on the front wheels

Braking distance and braking time also play significant role in designing an efficient controller. For aforementioned case study, the respective braking distances and braking times are calculated for different brake demands and tabulated in Table-3. Consider a case when the vehicle is moving with a constant speed of 100kmph and driver applies a brake which requires brake demand is z=0.3g (i.e. required deceleration = 2.94 m/s^2) which means the vehicle should stop within 9.4 sec. The effective braking time by the controller is computed to be 9.448224 sec. If only regenerative braking is utilized to halt the vehicle, vehicle stops within 16.24363 sec which is not desirable and corresponding braking distance when only regenerative effort is applied computes to be 1km 710m which is not at all desirable for the vehicle whereas when both mechanical and electrical braking are utilized simultaneously, the braking distance is 262.451m.

Braking distance shown in the Table-2 can be called as the drivetrain response time.

The safety is always the primary concern for vehicle, so when the required brake demand is very high (i.e. sudden brake situation), regenerative braking is not desirable to use. From the Table 3, it is

Figure 15. NEDC ECE R15 Drive Cycle

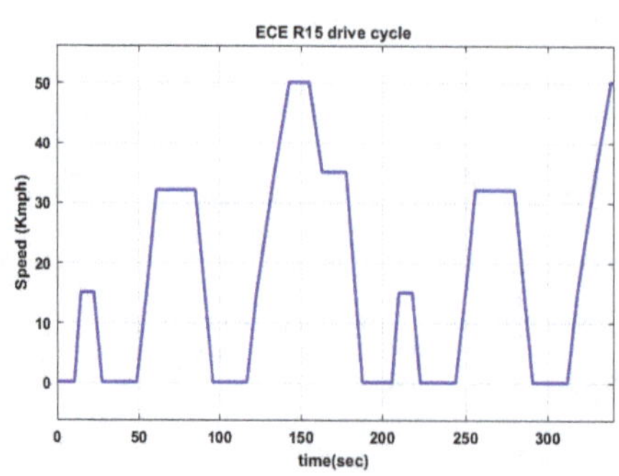

observed that during the situations when brake demand is 0.8g, 0.9g, 1g; the vehicle utilizes only the mechanical brakes to stop the vehicle with braking distances at 98.4m, 87.4m, 78.735m. The decisive logic is achieved as explained in the section VII.

The whole system is evaluated for NEDC ECE R15 drive cycle which is shown in the Fig. 15.

The distribution of braking forces for the NEDC: ECE R15 drive cycle obtained from the real time optimization is shown in the Fig. 16.

Figure 16. Braking Force Distribution in Newtons

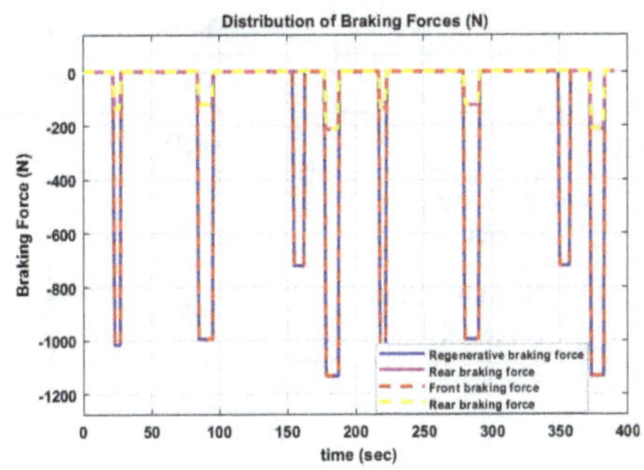

From the ECE R15 drive cycle in Fig. 15, it is observed that the rate of deceleration is varying from 0 to 0.1g as it is a city-based drive cycle. The whole drive cycle is evaluated on a road with coefficient of adhesion being (μ=0.5). When the brake demand is 0.1 (i.e. j=0.1g) which is comparatively less, the total braking force is provided by the motor and there is no need of frictional braking system. However, the rear braking force is always provided by the mechanical brake system, but at some cases when brake demand is 0.05 (j=0.05g), from the region of operation it observed that the total braking is allocated only to front wheels but not to the rear wheels. This is because the ECE regulation curve starts from j=0.1g, and fewer than that even if the rear braking force is zero doesn't cause the vehicle to instability. As the rate of deceleration is always negative, the Distribution of forces obtained in the Fig. 16 is observed to negative.

From the Regenerative force obtained, the regenerative power which can be harvested can be calculated and is shown in the Fig. 17.

Figure 17. Regenerative Power

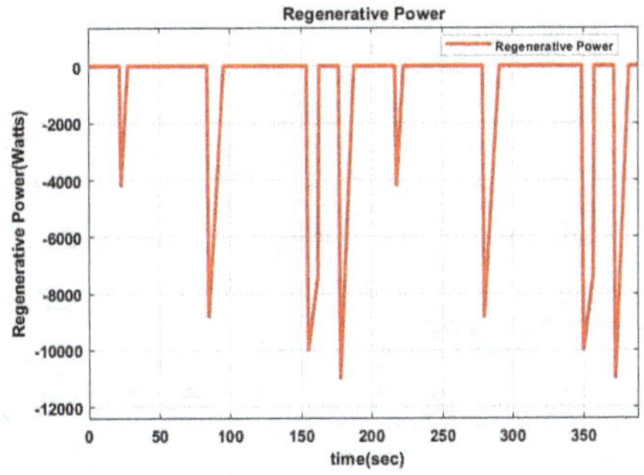

As observed from the Fig. 17, the regenerative power is proportional to the regenerative force. This power is being harvested into battery/supercapacitor based on the control logic designed. However, when both are placed in parallel, the supercapacitor will be of higher priority compared to that of the battery due to its higher power density.

The distribution of forces with fuzzy and ANN are shown in the Fig. 18. Further, analysis on fuzzy and ANN is discussed by the authors in (Prasanth et al., 2020). To evaluate efficiency, robustness and maximum power, a comparative study between Fuzzy, ANN and RTO has been performed and results are shown in the Table 3. The maximum regenerative power for the whole drive cycle is given in the Table 4 which happens between 170-190 s. It is observed that the regenerative power obtained at this interval is higher in RTO compared to that of the Fuzzy or ANN and similar way for the regenerative force. With ANN/Fuzzy controller, when the brake demand is 0.05g, it is not possible to operate with the effort applied only by the motor due to lack of controllability with Fuzzy/ANN, whereas with RTO, it is possible to operate and the rear braking force is totally zero and the total braking force is supported by the front wheels (i.e. Regenerative Braking Force). Another advantage is that when brake demand is j=0.1g, the total braking force on the front wheels is supported by the motor and only rear wheels are provided by the frictional brake system. Therefore, more amount of energy can be harvested with real time optimization compared to that of the Fuzzy/ANN.

Table 4 Comparative Analysis between Fuzzy, ANN and real time optimization

Parameters		Fuzzy	ANN	RTO
Max. Regenerative Power (W)		-8000	-8100	-10000
Max. Regenerative Force (N)		-810	-820	-1180
When the brake demand is 0.05 (between 155-165sec)	**Rear braking force (N)**	-400	-390	0
	Frictional braking force (N)	-390	-390	0

Figure 18. Distribution of Forces with Fuzzy and ANN

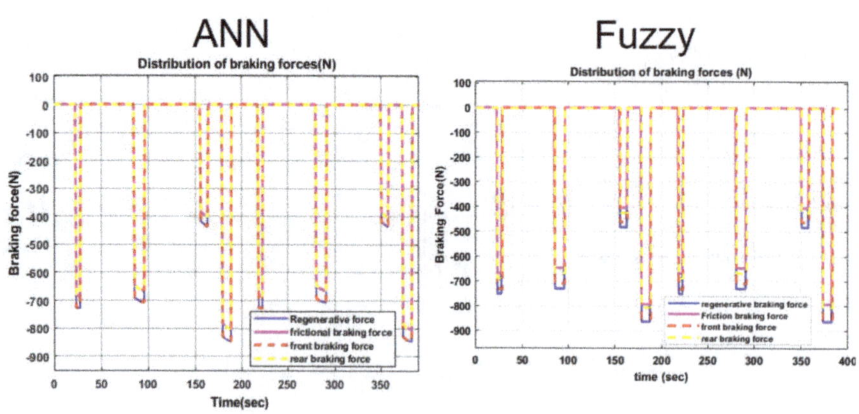

In the view of the practical implementation of the system, there will be high inrush current at the starting of the motor which is controlled by the threshold limiter after the controller in the Fig. 1. The regenerative force regenerated from the real time optimization is converted to the current using scaling factor and parameters of the machine. The threshold current is subtracted from the actual current to produce the error which is given to the controller to produce the duty ratio of the inverter which helps in handling the current into the battery/supercapacitor. The high C-rated charging current is taken care by the real time optimization controller as it generates regenerative force which is further converted to the necessary threshold current required for the battery to charge.

Therefore, based on the above analysis, it can be concluded that RTO based system excels in accuracy with minimal error and capturing huge amount of energy compared to the Fuzzy/ANN. Even the real time optimization is more durable, robust and less computational time compared to Fuzzy/ANN.

CONCLUSION

The estimation of the regenerative braking force is done by using the real time optimization which is implemented by constructing a graph which gives the distribution of the front and rear wheels. When the brake pedal is pressed, the braking force is calculated for different scenarios and formulas regulating those are derived. Further, it has to ensure the brake safety by distributing the braking force among the front and the rear wheels appropriately which is evaluated for different coefficient of adhesion and rate of deceleration and a generalised curve known as I-curve has been obtained. To avoid the rear braking force to be zero, a braking regulation named ECE regulation is defined and has been taken into consideration for evaluating the regenerative braking force. The region of operation is constructed using the I-curve and ECE regulation curve and is limited by the brake demand and coefficient of adhesion. An objective function has been derived for the constructed region of operation such a way that the regenerative braking force is maximized. Also, a comparative analysis has been performed with the fuzzy and ANN which proves that the results obtained with real time optimization are accurate, reliable, robust, time and cost effective. The whole system is evaluated using MATLAB/Simulink. Further, the power management system at the batteries/supercapacitor can be optimized which can be taken as a future scope. Even, a speed control loop can be implemented for efficient energy transfer during cruise control.

REFERENCES

Akhila, M., & Ratnan, P. (2016). Brushless DC motor drive with regenerative braking using adaptive neuro based fuzzy inference system. *2016 International Conference on Electrical, Electronics, and Optimization Techniques (ICEEOT)*, 748-751, 10.1109/ICEEOT.2016.7754784

Cao, B., Bai, Z., & Zhang, W. (2005). Research on control for regenerative braking of electric vehicle. *IEEE International Conference on Vehicular Electronics and Safety*, 92-97. 10.1109/ICVES.2005.1563620

Crolla, D. A., & Cao, D. (2012). The impact of hybrid and electric powertrains on vehicle dynamics, control systems and energy regeneration. *Vehicle System Dynamics*, *50*(sup1), 95–109. doi:10.1080/00423114.2012.676651

Ehsani, M., Gao, Y., Longo, S., & Ebrahimi, K. (2018). *Modern electric, hybrid electric, and fuel cell vehicles*. CRC Press.

Fazeli, Zeinali, & Khajepour. (2012). Application of Adaptive Sliding Mode Control for Regenerative Braking Torque Control. *IEEE/ASME Transactions on Mechatronics, 17*(4), 745-755.

Foiadelli, F., Roscia, M., & Zaninelli, D. (2006). Optimization of storage devices for regenerative braking energy in subway systems. *2006 IEEE Power Engineering Society General Meeting*, 6. 10.1109/PES.2006.1708894

Frilli, A., Meli, E., Nocciolini, D., Pugi, L., & Rindi, A. (2016). Energetic optimization of regenerative braking for high speed railway systems. *Energy Conversion and Management, 129*, 200–215. doi:10.1016/j.enconman.2016.10.011

Gao, H., Gao, Y., & Ehsani, M. (2001). A neural network based SRM drive control strategy for regenerative braking in EV and HEV. *IEMDC 2001. IEEE International Electric Machines and Drives Conference* (Cat. No.01EX485), 571-575. 10.1109/IEMDC.2001.939368

Gao, Y., & Ehsani, M. (2014, December). Electronic Braking System of EV and HEV-Integration of Regenerative Braking. *Automatic Braking Force Control and ABS, 3*. Advance online publication. doi:10.4271/2001-01-2478

Grbovic, P. J., Delarue, P., Le Moigne, P., & Bartholomeus, P. (2010, October). A bidirectional three-level dc-dc converter for the ultra-capacitor applications. *IEEE Transactions on Industrial Electronics, 57*(10), 3415–3430. doi:10.1109/TIE.2009.2038338

Hua, C., & Kao, S. (2011). Design and implementation of a regenerative braking system for electric bicycles based on DSP. *2011 6th IEEE Conference on Industrial Electronics and Applications*, 703-707. 10.1109/ICIEA.2011.5975676

Jiaqun, X., & Haotian, C. (2015). Regenerative brake of brushless DC motor for light electric vehicle. *2015 18th International Conference on Electrical Machines and Systems (ICEMS),* 1423-1428. 10.1109/ICEMS.2015.7385262

Juan, J. C. A., Javier, P. F., Juan, M. V. G., & Juan, A. C. C. (2017). Regenerative Intelligent Brake Control for Electric Motorcycles. *Energies, 10*(10), 1648. doi:10.3390/en10101648

Ko, J., Ko, S., Son, H., Yoo, B., Cheon, J., & Kim, H. (2015, February). Development of Brake System and Regenerative Braking Cooperative Control Algorithm for Automatic-Transmission-Based Hybrid Electric Vehicles. *IEEE Transactions on Vehicular Technology, 64*(2), 431–440. doi:10.1109/TVT.2014.2325056

Liu, X., Su, C., Li, Z., & Yang, F. (2018, April). Adaptive Neural-Network-Based Active Control of Regenerative Chatter in Micromilling. *IEEE Transactions on Automation Science and Engineering, 15*(2), 628–640. doi:10.1109/TASE.2017.2667709

Lu, S., Weston, P., Hillmansen, S., Gooi, H. B., & Roberts, C. (2014, December). Increasing the Regenerative Braking Energy for Railway Vehicles. *IEEE Transactions on Intelligent Transportation Systems, 15*(6), 2506–2515. doi:10.1109/TITS.2014.2319233

Prasanth, B., Sai Surya, G., Sai Vinay, G., Deepa, K., Sailaja, V., & Manitha, P. V. (2021). *Placement of Distribution generators in IEEE 14 bus system with consumer benefit maximization* (Vol. 711). Advances in Electrical and Computer Technologies.

Prasanth, B., Sanju Srivasthav, B., Srinivas Reddy, Y. S., & Deepa, K. (2020). Estimation of regenerative Braking Force in Electric Vehicles for maximum energy recovery. *17th IEEE India Council International conference*, 1-4.

Rahman, M. A., & Hoque, M. A. (1998, December). On-line adaptive artificial neural network based vector control of permanent magnet synchronous motors. *IEEE Transactions on Energy Conversion*, *13*(4), 311–318. doi:10.1109/60.736315

Sasikumar, S., & Deepa, K. (2019). *Comparative Study of LCL-S and LCC-S Topology of Wireless EV charging System. In Innovations in Power and Advanced Computing Technologies (i-PACT).* Vellore.

Seki, H., Ishihara, K., & Tadakuma, S. (2009, May). Novel Regenerative Braking Control of Electric Power-Assisted Wheelchair for Safety Downhill Road Driving. *IEEE Transactions on Industrial Electronics*, *56*(5), 1393–1400. doi:10.1109/TIE.2009.2014747

Şengör, İ., Kılıçkıran, H. C., Akdemir, H., Kekezoglu, B., Erdinc, O., & Catalao, J. P. S. (2018, July). Energy Management of a Smart Railway Station Considering Regenerative Braking and Stochastic Behaviour of ESS and PV Generation. *IEEE Transactions on Sustainable Energy*, *9*(3), 1041–1050. doi:10.1109/TSTE.2017.2759105

Shabarish, P. R., Krishna, R. P., Prasanth, B., Deepa, K., Manitha, P. V., & Sailaja, V. (2020). Fuzzy based Approach to Enhance the Wireless Charging System in Electric Vehicles. *2020 4th International Conference on Electronics, Communication and Aerospace Technology (ICECA)*, 176-181.

Shetty, S. S., & Karabasoglu, O. (2014). Regenerative Braking Control Strategy for Hybrid and Electric Vehicles Using Artificial Neural Networks. In V. Mladenov, C. Jayne, & L. Iliadis (Eds.), *Engineering Applications of Neural Networks. EANN 2014. Communications in Computer and Information Science* (Vol. 459). Springer. doi:10.1007/978-3-319-11071-4_10

Simoes, M. G., Bose, B. K., & Spiegel, R. J. (1997, January). Fuzzy logic based intelligent control of a variable speed cage machine wind generation system. *IEEE Transactions on Power Electronics*, *12*(1), 87–95. doi:10.1109/63.554173

Takahashi, Y., & Seki, H. (2009). Fuzzy logic based regenerative braking control system of electric wheelchair for senior citizen. *2009 IEEE International Conference on Rehabilitation Robotics*, 725-730. 10.1109/ICORR.2009.5209589

Thiruvonasundari, & Deepa. (2020a). Electric vehicle battery modelling methods based on state of charge-review. *Journal of Green Engineering*, *10*, 24–61.

Thiruvonasundari & Deepa. (2020b). Active cell balancing for electric vehicle battery management. *International Journal of Power Electronics and Drives System*, *11*(2), 571-579.

Wu, J., & Luo, Z. (2017). A new uncertain analysis method and its application in vehicle dynamics. *Mechanical Systems and Signal Processing*, *50–51*, 659–675.

Wu, J.-F., Hu, N.-S., Hu, S., & Zhao, Y. (2002). Application of SOM neural network in fault diagnosis of the steam turbine regenerative system. *Proceedings. International Conference on Machine Learning and Cybernetics*, 184-187. 10.1109/ICMLC.2002.1176735

Xiao, B., Lu, H., Wang, H., Ruan, J., & Zhang, N. (2017, November). Enhanced Regenerative Braking Strategies for Electric Vehicles: Dynamic Performance and Potential Analysis. *Energies*, *10*(11), 1875. doi:10.3390/en10111875

Xiao, B., Zeng, X., Zhang, Z., Li, B., & Yang, Y. (2016). Present status and prospect of the regenerative-braking control strategies of electric automobiles. *Auto Electr. Parts*, *12*, 1–3.

Xu, G., Li, W., Xu, K., & Song, Z. (2011). An Intelligent Regenerative Braking Strategy for Electric Vehicles. *Energies*, *4*(4), 1461–1477. doi:10.3390/en4091461

Xu, G., Xu, K., Zheng, C., Zhang, X., & Zahid, T. (2016, March). Fully Electrified Regenerative Braking Control for Deep Energy Recovery and Maintaining Safety of Electric Vehicles. *IEEE Transactions on Vehicular Technology*, *65*(3), 1186–1198. doi:10.1109/TVT.2015.2410694

Yao, J., Zhong, Z., & Sun, Z. (2006). A Fuzzy Logic Based Regenerative Braking Regulation For a Fuel Cell Bus. *2006 IEEE International Conference on Vehicular Electronics and Safety*, 22-25. 10.1109/ICVES.2006.371546

Yoong, M. K. (2010). Studies of regenerative braking in electric vehicle. *2010 IEEE Conference on Sustainable Utilization and Development in Engineering and Technology*, 40-45. 10.1109/STUDENT.2010.5686984

Zhang, H., Xu, G., Li, W., & Zhou, M. (2012). Fuzzy logic control in regenerative braking system for electric vehicle. *2012 IEEE International Conference on Information and Automation*, 588-591. 10.1109/ICInfA.2012.6246881

Zhang, J., Kong, D., Chen, L., & Chen, X. (2012). Optimization of control strategy for regenerative braking of an electrified bus equipped with an anti-lock braking system. *Proceedings of the Institution of Mechanical Engineers. Part D, Journal of Automobile Engineering*, *226*(4), 494–506. doi:10.1177/0954407011422463

Chapter 9
Solar Powered Electric Vehicle Through Wireless Power Transfer

Balamurugan M.
Jain University (Deemed), India

Raghu N.
https://orcid.org/0000-0002-2091-8922
Jain University (Deemed), India

Kamala N.
Jain University (Deemed), India

Trupti V. Nandikolmath
Jain University (Deemed), India

Sarat Kumar Sahoo
Parala Maharaja Engineering College, India

ABSTRACT

Solar powered wireless electric vehicle charging technology functions independently without interface with the utility grid. Wireless power transfer (WPT) technology is incorporated for wireless charging, which brings the benefits of safe operation, less pollution, and little maintenance cost. WPT technology necessitates no physical connection between the charging device and vehicle, thus hazards and inconvenience produced by conventional charging methods have been minimized. WPT in electric vehicle can be used to reduce the charging time, range, and cost. In this chapter, the various configurations of WPT like inductive, capacitive, resonant, and roadway power transfer techniques have been presented. The small-scale prototype of wireless charging has been developed in the laboratory by incorporating inductive power transfer technique. The experimental results have been presented to validate the feasibility of the system in real time.

DOI: 10.4018/978-1-7998-7626-7.ch009

INTRODUCTION

In recent years of the energy transition, electric mobility has become an essential part. These significant changes will imply for manufacturing of electric vehicles. During the last few decades, petroleum-based transportation has a huge environmental impact, along with fear of degradation of fossil fuel at the peak. For these reasons, electric transportation has gained a renewed interest. Electric vehicles (EV) are anticipated to surge from 2% of international share in 2016 to 22% in 2030.

In traditional vehicles, the consumed electricity has been produced from various resources of non-conventional energy. Depending upon the choice of technology and fuel adopted for the production of electricity, there is a variation in carbon footprint and other emissions of EV. Supercapacitors, flywheel, or the battery can store the electricity onboard.

An EV may be driven through a collector arrangement by electrical energy from off-vehicle resources or could be self-sufficient with a fuel cell, solar panels, battery, or an electric generator to convert fuel to electrical energy. EV has been employed in two different configurations which are purely electric and hybrid. Pure electric vehicles have an electric motor, controller and rechargeable battery. The same technology has been used in hybrid vehicles but alongside a small gasoline engine runs a generator (Parise, 2002).

It is essential to charge an EV on regular basis. The major disadvantage of electric vehicles is the longer duration of charging typically it requires 8 hours of charging to be fully charged. To obtain the most benefit, electric vehicle arrangement necessitates four synchronized strategies (i) Vehicle electrification; (ii) facility of adequate charging apparatus; (iii) decarbonization of electricity production, (iv) EV integration to the grid (Kabalo, Berthold, Blunier, Bouquain, Williamson, & Miraoui, 2014).

Like any other new technology, EV has created potential economic development challenges and opportunities. EVs are suitable for shorter trips but not limited to rail and road transport, underwater and surface vehicles, electric spacecraft and aircraft.

The idea of wireless power transmission has been a goal from the time when electric engineering concepts were derived based on Faraday and Ampere law. Even in radio systems, power has been transmitting power for small distances with good efficiency but when the distance has been increased which results in low efficiency (Javier Vázquez, 2018). The same concept has been applied for IPT systems in electric vehicles.

Wireless Power Transfer (WPT) is a technology in which the transmitters generate electromagnetic waves that are processed by the receivers without any conductor the power has been extracted from the wave. Unlike the wireless communication systems, the electromagnetic wave generated in WPT systems is utilized by the receiver to store energy in a power electronic device or battery. It is well known that the first experiment on WPT was performed by Nikola Tesla in the 19th century. He has transmitted power between two objects 48 km apart by using the microwave. In another experiment, he powered 200 bulbs without the inclusion of cables, from a power source situated 25 miles apart (Musavi, Edington, & Eberle, 2012).

The major benefit of WPT is to eliminate high tension power transmission line cables, substations and towers connecting the interconnection with generating stations. Another advantage is that power might be easily transferred to remote locations where transmission through wires is not feasible. It could also avoid the theft of power. The major disadvantage of wireless power transmission is the high capital cost required for practical implementation and also microwave radiations received at high doses are not suitable for human health. There are also chances of interferences with a wireless communication system(Qiu, Chau, Liu, & Chan, 2013).

Wireless power transmission consists of three major components which are resonance, solar cells and microwaves. Microwaves are utilized in electric equipment to transfer electromagnetic radiation from transmitter to receiver. As the name suggests electrical power could be transmitted from a source to a receiver without the use of wires.

An electric vehicle that is powered using direct solar energy is called the solar vehicle. It consists of solar panels that convert solar energy directly into electrical energy. Solar energy is utilized to power completely or a portion of vehicle propulsion. Solar energy might also be utilized to deliver power for communication, control or other auxiliary operations (Shubhangi Das, 2018). Convex solar cells have been added on the rooftop of hybrid electric vehicles which provides power to an aeration system while parked to offer to cool. Further, solar cells are utilized in electric vehicles either for motive power or auxiliary power units. Because of the limited space accessible in each vehicle either range or speed or both are restricted while utilized for motive power (Huang, Iam, Hoi, Lam, Mak, & Martins, 2019).

Solar Panel charging connections for EVs can be generally classified into two categories. First one, mobile solar charging systems which consist of rooftop solar panels are utilized to charge the batteries of an EV. The second one, static solar-powered charging connection, a ground-based installation (off-grid or grid-based) which comprises direct charging installations/devices accessible at EV charging locations. The major constituents of a solar energy charging system for EVs are solar panels, microcontrollers, motor to control solar panel axis rotation, DC-DC converter with MPPT, communication boundary between electric vehicle battery system and charging method, interface and electric vehicle battery charging equipment (Adel El-Shahat, 2019).

The arrangement of the solar panel structure and EV charging system has numerous profits and offers a cost-effective method to yield and make usage of solar energy. It's tough to charge inherent solar panels during the winter period. The result is a solar-powered electric vehicle is sustained with a charging cable that plugs into the vehicle and a 230v wall socket. EVs have inbuilt structures like drive guidance system, android application support, route detection, Wi-Fi, Battery and security system. Therefore Solar powered electric vehicle (SPEV) consequences in less pollution transportation. More than one vehicle at a time can be charged using solar energy. Two configurations of charging representations are considered. One is the real-time charging model, which specifies that EVs could be charged at any time as long as solar power is produced and the other exploits a mobile charge truck can be furnished with a mobile charging location(S, 2020).

The objective of this chapter to review the various WPT technologies used for solar-powered electric vehicles. The Chapter has been divided into six sections. The second section has been discussed the basic principles of solar-powered WPT. The third section discusses the various types of WPT like Inductive WPT, capacitive WPT, resonant inductive power transfer and roadway/online power transfer. The lab-scale prototype of inductive WPT and its testing has been presented in the fourth section. The comparison of various WPT technologies has been presented in the fifth section and the challenges of WPT are discussed in the sixth section.

WPT Configuration

With the advent of power converters, WPT operations employ the frequency and power in kHz and kW range. The key function of WPT is to pass the electrical devices to be charged without using a power cord

(N.UthayaBanu, 2018). WPT can be employed in an extensive range of applications and environments. It is commonly supported in commercial elements such as electric toothbrushes and chargers for EV (Anon Namin, 2018). IHS Markit has predicted that the WPT market segment may raise to more than 2.2 billion units by the year 2023. Navigant Research has estimated that the income from wireless chargers will be approximately 17.9billion dollars by 2024 (Grazian, Shi, Dong, Duijsen, Soeiro, & Bauer, 2019).

Figure 1 illustrates the WPT system block diagram which comprises of power source (Can be battery or renewable energy source), one or more transmitters produce an electromagnetic wave, which has been managed by one or more receivers where the power has been extracted from wave without any conductor. The power transmitter and power receiver are detached by a distance *d*, which is usually called a gap. Electromagnetic waves are categorized by their frequency *f* or wavelength λ.

Figure 1. Block Diagram of a WPT system

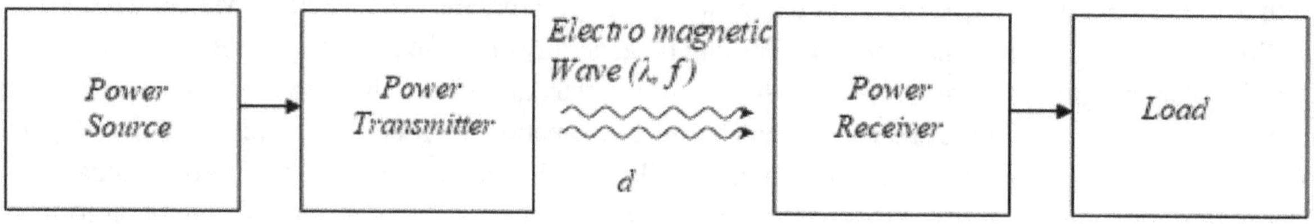

Electrical vehicles (EV) are powered by clean electricity, and it can be produced from non-conventional energy sources like wind, solar and biomass. The growth of photovoltaic technology has been a widespread approach to extract power from the solar panel for charging EV. Two configurations of solar power extraction for charging EV stations are centralized and self-contained type. In a centralized configuration to stabilize the source and demand power, it has been connected to the utility grid which resulted in high capital cost and limited infrastructure facility. To overcome the drawbacks in centralized configuration, the self-contained configuration has been introduced whereas the power extracted through the solar panel is accountable for charging EV. Self-contained configuration for EV charging has been especially used in remote locations where the grid power is unavailable (Huang, Iam, Hoi, Lam, Mak, & Martins, 2019).

The configuration of the self-contained solar-powered for charging EV is shown in Figure 2. It consists of a solar panel, MPPT controller, Inverter and transmitter coil on the ground side. On the other side, the vehicle consists of the receiver coil, rectifier and battery. Inductive power transfer (IPT) technique is integrated with self-contained configuration through wireless charging, to get benefits of low maintenance cost and safe operation.

Solar panel drives the IPT converter directly and it is loaded by the battery on the vehicle side. The resistance at the input of the IPT converter determines the output power of the solar panel and the load resistance determines the efficiency of the IPT converter. Maximum Power Point Tracking (MPPT) is used to provide the maximum power of the panel accessible at a particular time. The small-scale prototype of this self-contained configuration has been set up in the laboratory. A brief discussion of the prototype and its results has been presented in the next section.

Figure 2. Solar Powered Self-Contained Configuration for charging EV

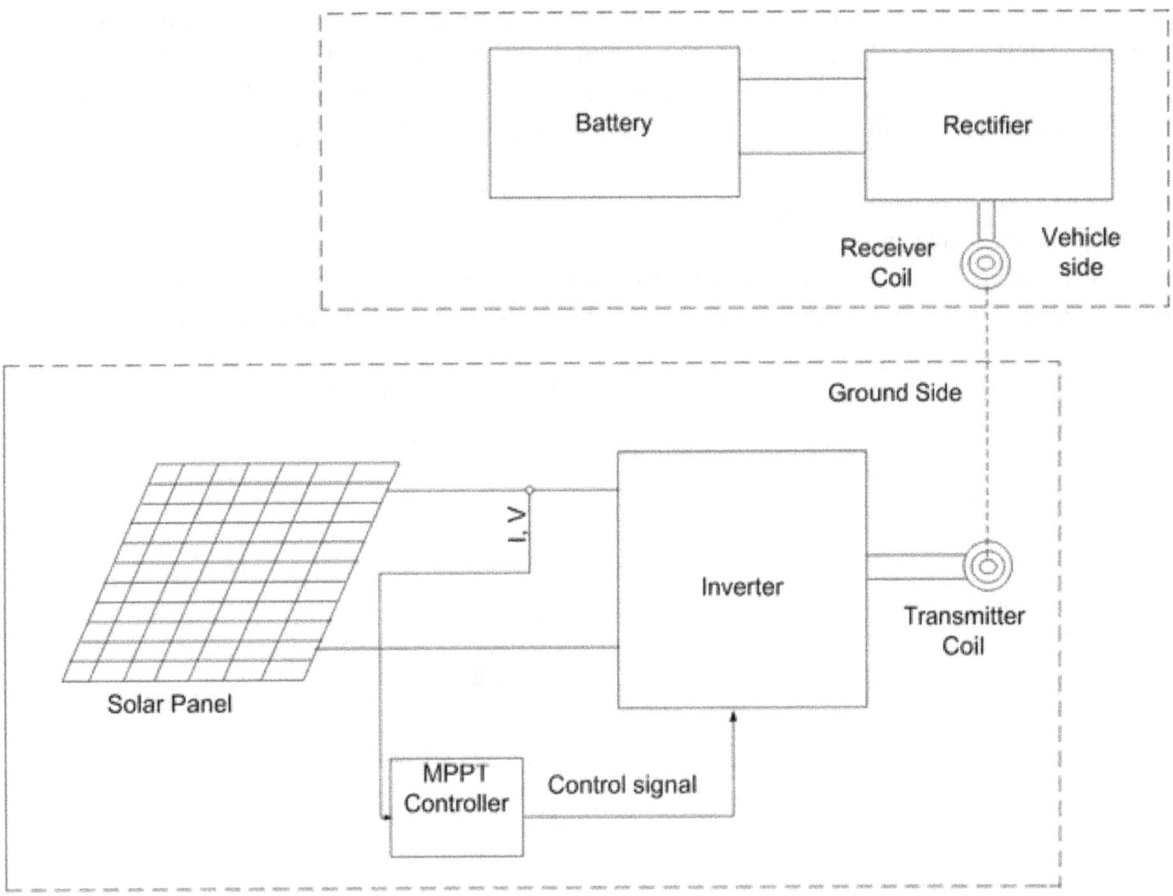

Different WPT Technologies

The different WPT technologies discussed in this section are Inductive Power Transfer (IPT), Capacitive Power Transfer (CPT), Resonant Inductive Power Transfer (RIPT) and Roadway/Online Power Transfer (RPT) systems.

Inductive Power Transfer (IPT) System

Inductive Power Transfer (IPT) technology is the most commonly used WPT systems. IPT technologies are extensively used in numerous applications and it is especially used in electric vehicles and also this technology offers more benefits when compared to other WPT systems for electric vehicles applications. Even though the technology is promising still there are challenges in design (Covic & T.Boys, 2013).

In the IPT system, the power transfer is enhanced from a transmitter (track) to a receiver (pickup coil) is tuned to operate at the desired frequency. Moreover, the IPT system offers hands-free charging methods it has been not affected by weather, chemicals or also it can be extended to dynamic charging systems though the vehicle is in motion alongside the road (Nama, Verma, Srivastava, & Tomar, 2020). The major problem in all IPT systems is the uncontrolled frequency in the resonant circuit. Because of this problem, the system will become unstable when too many circuits are switched on at the same time for light load conditions (Mou, Gladwin, Zhao, Sun, & Yang, 2020). By introducing the decoupling control in the IPT system the instability in the system can be eliminated and also multiple loads operated at the single track without interference (Vishvendra Singh, 2014). By including the decoupling control the IPT system has been designed with the following essential components.

- Power supply system to energize the primary circuit (track).
- Frequency Compensation circuit.
- A secondary circuit (pickup coil) to receiver power magnetically from the primary circuit (track).
- A controller circuit to control the power transfer from primary to the secondary circuit.

The general circuit arrangement of an IPT system is shown in Figure 3.

Figure 3. General Configuration of an IPT system

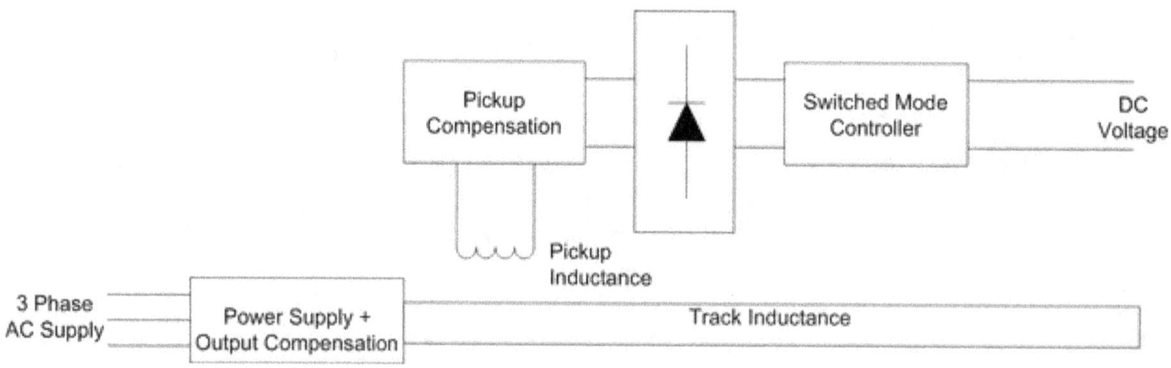

From utility, power is taken from the power supply which energizes a primary circuit (track) and the secondary circuit (pick up coil) is magnetically connected to the primary of the system.

The resonant circuit for the IPT system is presented in Figure 4. It comprises two thyristors, two transistors, DC voltage source, transformer, inductor and capacitor. These types of circuits will use the transistors to switch only the real power and the reactive power is large on the track. This resonant circuit does not meet the requirements of an IPTS system and at the same time, it produces very low distortions and also maintains efficiency. The track current is usually 5 to 10 times greater than the switch current. A drawback of this circuit it is very tough to start as well as stop the motor and it is also a challenge to protect the circuit from overload conditions. Further, this circuit is not isolated from the utility grid and this will create a major hazard during the circuit operation (Sergeant & Bossche, 2008).

An isolated fixed frequency LCL tuned circuit has been introduced to overcome the drawbacks of the current source resonant converter is shown in Figure 5. In the WPT system, it is very difficult to operate with variable frequency because the significant power will be lost but the fixed frequency will provide a better result without much difficulty. Further, the filter at the input side used to prevent electromagnetic interference and also having a three-phase diode bridge rectifier reduces the generation of electromagnetic interference. In LCL tuned resonant circuit, the frequency is high when compared to

Figure 4. Resonant Circuit of IPT System

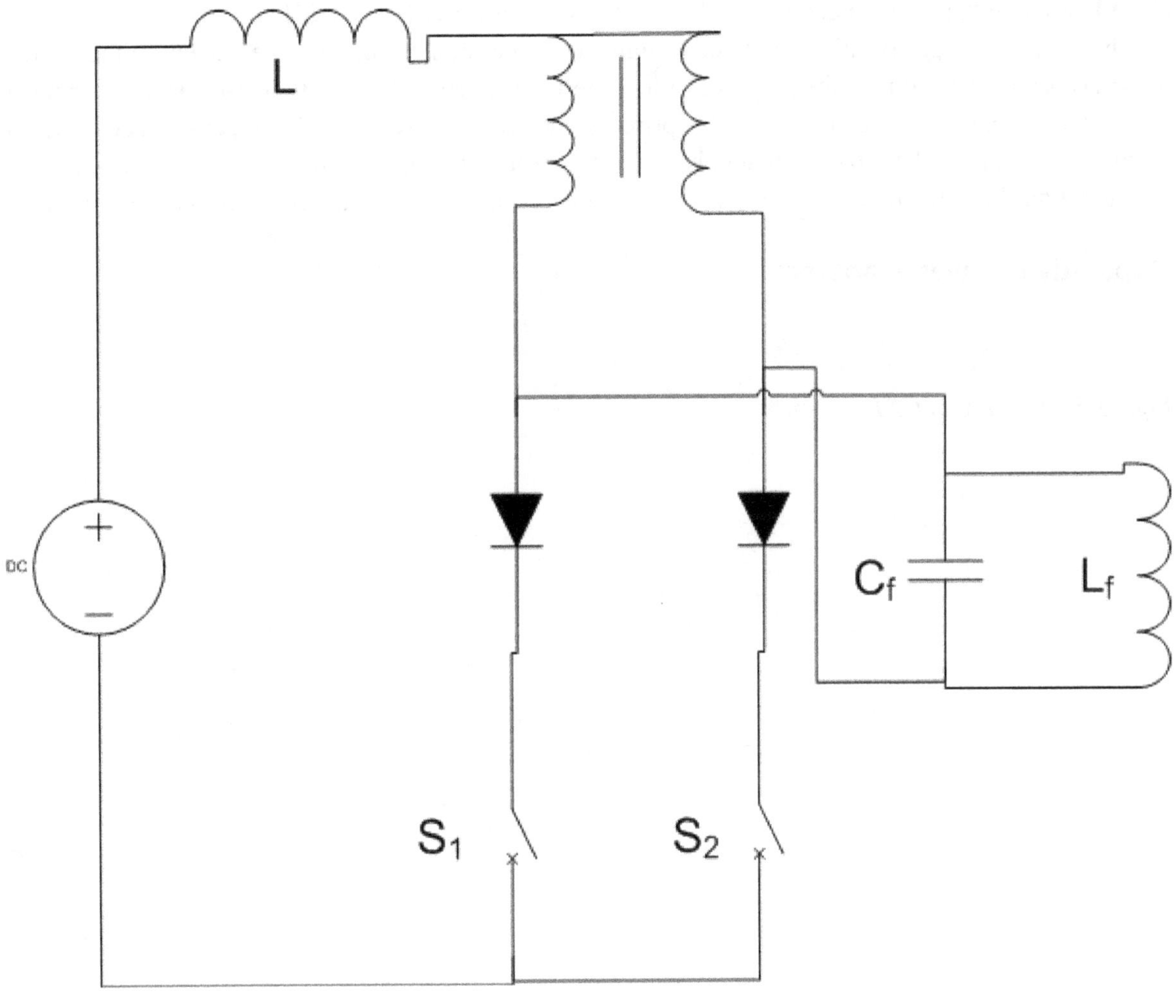

normal circuits and the filter capacitor value is small which will lead to a small inrush current in addition to that the efficiency and power factor of the circuit has been improved (Ludios, 2015).

From the rectifier circuit, the pulsating DC is given to the H-bridge inverter through the DC link capacitor. Then the H-bridge is connected to the isolation transformer through the leakage inductor which forms the primary side of the transformer (track). An LCL filter is coupled to the second portion of the

Figure 5. Fixed Frequency LCL Tuned Circuit with isolation

transformer which filters ripple-free voltage with fewer harmonics. The current in the track has been altered based on the firing angle provided to the H-bridge inverter (J.Huh, 2011).

The main advantage of this circuit is to eliminate the harmonics and also the current in the source track has been created due to the presence of leakage inductance in the isolation transformer. A typical IPT system is composed of inductance in the primary track and it must be series compensated determined by the power supply. The track and supply provide power to the loads using a pickup inductor. Then the output has been rectified and delimited to provide a controlled output voltage for the chosen load.

Capacitive Power Transfer:

Figure 6. Structure of CPT System

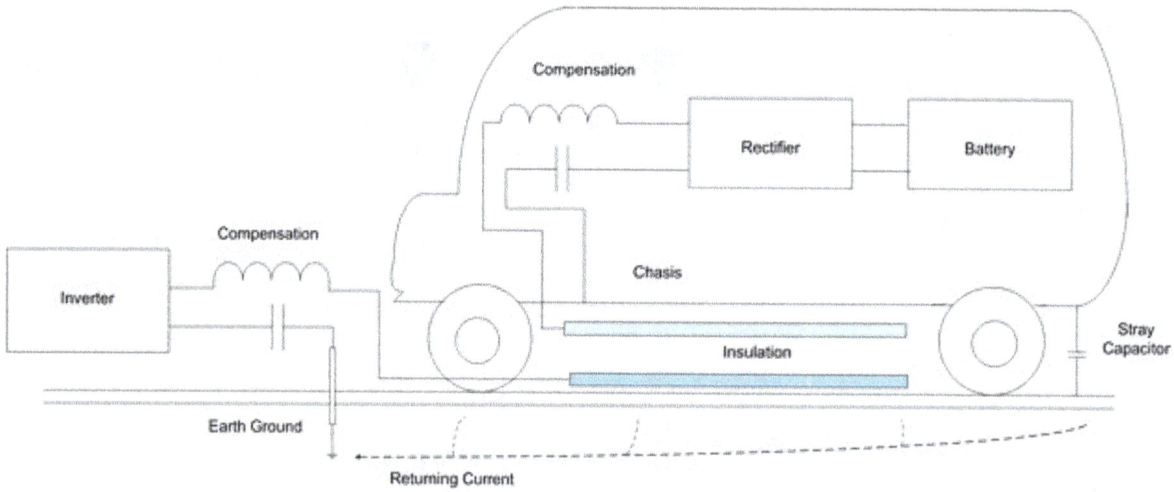

One of the most common techniques in WPTS is Capacitive Power Transfer System (CPTS). The implementation of CPTS will replace the materials used in the inductive power transfer system (IPTS) like copper which provides a shield to fossil surfaces to afford a cost-effective system and also to maintain the capability of power transfer is efficient. Due to the increase of wireless charging stations WPT technology has been developed for both IPTS and CPTS. However, IPTS experiencing a power ranging from few watts to kW and the distance in the air gap is relatively large when compared to CPTS. Recently, in CPTS huge air gap systems are established which was modest with IPTS on a throughput power basis. Due to this development, CPTS has been widely utilized in EV applications and the arrangement of the CPT system is displayed in Figure 6.

Inverter followed by compensation circuit forms the primary side where the metal plate and earth ground acts as transmitter and also an insulation layer is placed between them. The return current path has been provided by the parasitic capacitance between earth ground and vehicle chasis. In practical applications, to reduce the conductive loss the earth ground can be connected with metal plate. As long as the connection is good enough the size of the vehicle chasis must be higher than the ground plate. Since the compensation circuit has been directly connected with the ground plate, all the displacement current from chasis to earth ground will be accumulated to this ground plate and then fed back to the compensation circuit on the secondary side. A rectifier circuit preferably full bridge and a filter circuit on the secondary side to provide a DC current to the battery.

The ability to deliver the power of the vehicle should be high and also the breakdown field strength should be around 3 MV/m. Inside the vehicle caveat is provided for commercial or industrial applications with electric field exposure, IEEE also provides the regulation to provide the electric field within a safe operating limit. IEEE recommends the value of the operating area is 614 V/m for the frequency range of 0.03 – 1.34 MHz and it reduces as the frequency is increasing 823.8/f V/m for 1.34 – 30 MHz. In addition to that, CPTS used capacitance of a large air gap which results in voltage to be high across the coupling capacitors (Lu, Zhang, & Mi, 2018).

Figure 7. Schematic Diagram of CPTS

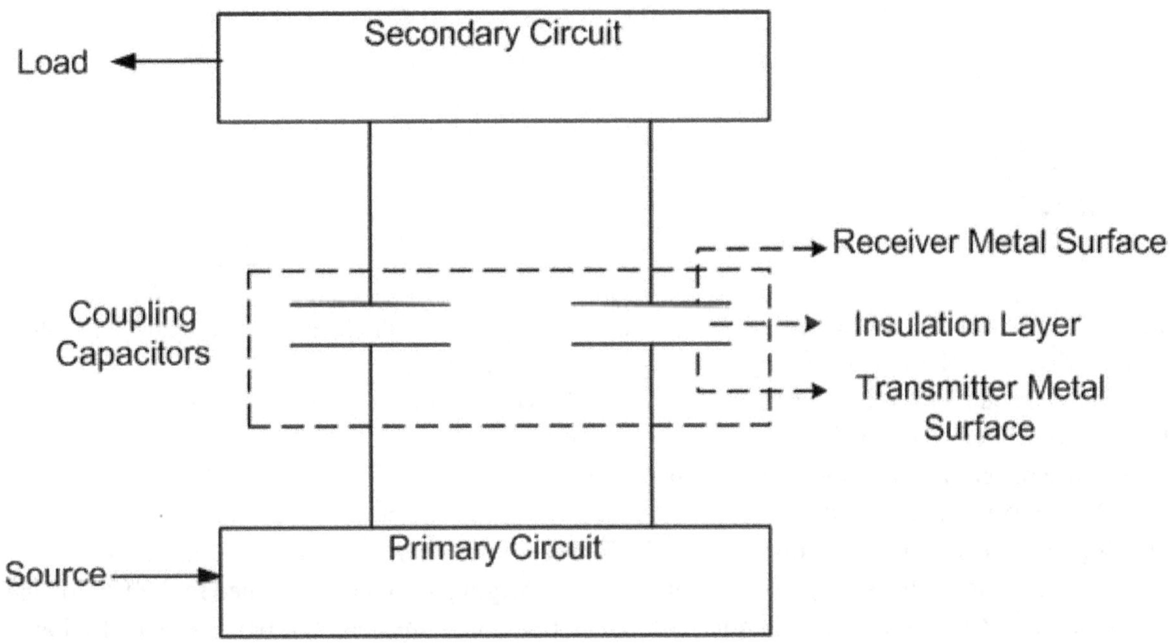

The schematic diagram of the CPTS is shown in Figure 7. It consists of primary and secondary circuits along with the source, coupling capacitors and load. The throughput power of CPT has been maximized to impose the safety constraints of humans beyond the limit and the air gap should be less so that the electrical field is mostly confined in caveat to mitigate the risk of electromagnetic interference. Since the air gap of CPT is small, it allows maximizing the throughput power which makes the system is applicable for stationary charging. If the copper metal surfaces are replaced with permeable material then it will be the potential application of dynamic charging and the purpose of the insulation layer is to protect the metal surface of the transmitter to contact with the receiver. Corrosion can be eliminated with the help of galvanic isolation (Dai & Ludois, 2016).

To prevent the flow of high voltage between the receiver and transmitter and also the efficient power transfer capability nanofarad coupling capacitance is required for CPT. A dielectric medium is inserted between the metal plates of the capacitor to increase the volume of effective permittivity of capacitance. A dielectric is inserted to raise the effective permittivity between the metal plates of the capacitor to provide a high value of coupling capacitance. To ensure the galvanic isolation between the transmitter and receiver it is coated with the dielectric material. For a high value of permittivity, titania coatings can be used for better dielectric strength.

In an electric vehicle application, a receiver is always a mobile object and it will charge only when the transmitter surface comes in contact and it will always create an air gap. This air gap may neutralize the merits of a dielectric coating having higher relative permittivity on coupling surfaces. Two layers of coupling present in a capacitor which are the air gap surface and the coating surface. The capacitance and equivalent permittivity will be determined with the help of equation (1) and (2).

$$C_v = \frac{A_v}{\frac{d_{ag}}{\varepsilon_o} + \frac{d_{cg}}{\varepsilon_o \varepsilon_r}} = \frac{\varepsilon_{eq}\varepsilon_o A_v}{d_{ag} + d_{cg}} \tag{1}$$

$$\varepsilon_{eq} = \frac{d_{ag} + d_{cg}}{d_{ag} + \frac{d_{cg}}{\varepsilon_r}} \tag{2}$$

where C_v = capacitance, A_v = Area, d_{ag} = thickness of air gap, d_{cg} = thickness of coating, ε_o = absolute permittivity, ε_r = relative permittivity, ε_{eq} = equivalent permittivity.

From the equations, it is clear that based on the thickness of the air gap is higher than the coating, the air gap will lead the capacitance. Therefore to reduce the size of the air gap the only way is to insert a dielectric material to raise the effective capacitance.

CPT system is implemented for electrical vehicles with a high degree of field containment. The copper windings have been replaced by foil plates for CPTS because of its cost-effective solution. Alignment and coupling structure are the vital parameters of electrical charging. CPTS is best suited for higher voltage lower current loads, it is used for charging series batteries and on the contrary, IPTS is suited for lower voltage higher current loads and it is used for fast charging in parallel batteries. Therefore, depending upon the voltage and current ratings the suitable WPT technologies have been selected.

Resonant Inductive Power Transfer (RIPT)

The main principle behind the integration of resonance with IPT is the oscillation transformer. To raise the coupling and performance of the system the resonant circuits are implemented along with IPT. Resonant circuits are employed to increase the transmitting distance of IPTS (Darab, 2019). The energy transfer rate has been improved with the help of resonant circuits. A combination of capacitor and inductor creates a resonant circuit which has been called magnetic resonance coupling. The interaction between the objects has been very strong by enabling the magnetic resonance coupling. Using this technology, energy might be transferred efficiently from transmitter to receiver and it offers better operation when it is compared with conventional IPTS. Magnetic resonance coupling is safe to the environment because of its resonance property. An IPT system without resonance can cause reduction in the power transfer between the primary and secondary if there is any misalignment between the transmitter and receiver coils. The inclusion of the resonant circuit with IPT provides more flexibility in the orientation between the load and source during the operation (Daida, 2019).

The interesting feature of resonant inductive power transfer is that it can transmit the energy from one transmitter to many receivers and also it is used to deliver the energy for different power requirements.

Resonant frequency has been defined as the resonance occurs at a particular frequency where the imaginary parts of admittance and impedances cancel out each other in an electric circuit. Resonance can take place in any circuit that has an arrangement of a capacitor and an inductor can store the energy.

In a resonant circuit, the energy will be fluctuating between both the capacitor and inductor and the rate of transfer of the energy between these two components. Because of these energy transfer oscillations in the circuit has been produced. If this circuit is ideal, and if there is absence of any kind of resistive elements, these fluctuations will continue endlessly. In reality, though, all the circuits will have some resistance involved in them. Because of the presence of this resistance, there will be depreciation in these oscillations. To maintain these oscillations, supply must be provided from an external source with the same frequency to this LC circuit. This way the oscillations in the circuit have been maintained. Parallel and series RLC circuits are the two main basic resonance circuits. In resonance circuits, the sharpness of the resonance can be measured by a term called Quality factor (Alberto Reatti, 2018). The formula to calculate the quality factor has been mentioned below.

Q = 2π * {Net energy stored in circuit /Dissipation of energy at resonance during one period}

where Q is the quality factor.

For parallel connected resonant circuits, the quality factor can be calculated by the following equation.

$$Q = \frac{R_p}{\omega L_p} = \omega R_p C_p = R_p \sqrt{\frac{C_p}{L_p}} \tag{3}$$

For the series-connected resonant circuit, the quality factor can be calculated by the following equation.

$$Q = \frac{\omega L_p}{R_p} = \frac{1}{\omega C_p R_p} = \frac{1}{R_p} \sqrt{\frac{L_p}{C_p}} \tag{4}$$

Where R_p, C_p and L_p are resistance, capacitance and inductance of the resonant circuit respectively and ω is the resonant frequency. The model of the system has been derived with the help of basic resonance concepts. The merits of resonant circuits make them suitable for WPT application and depending upon the resonant topology voltage and current of the system has been increased (Mou, Gladwin, Zhao, & Sun, Survey on magnetic resonant coupling wireless power transfer technology for electric vehicle charging, 2019). The voltage across the inductor of the series circuit can be calculated with the formula given below.

$$V_{LL} = \frac{v_m}{R_p} = \omega L_p \tag{5}$$

where V_{LL} voltage across the inductor and v_m is the maximum voltage and by replacing the quality factor $Q = \omega L_p/R_p$ then equation 5 can be rewritten as

$$V_{LL} = v_m Q \tag{6}$$

From the above equation, it is clear that the voltage across the inductor will be greater than the supply voltage and hence this circuit can also call as a voltage amplifier. By using the same concept in a parallel circuit the voltage through the capacitor can be determined by properly designing the value of capacitor and inductor in the resonant circuit (Zhu & Gao, 2020).

The block diagram of RIPT is presented in Figure 8.

Figure 8. Block Diagram of RIPT

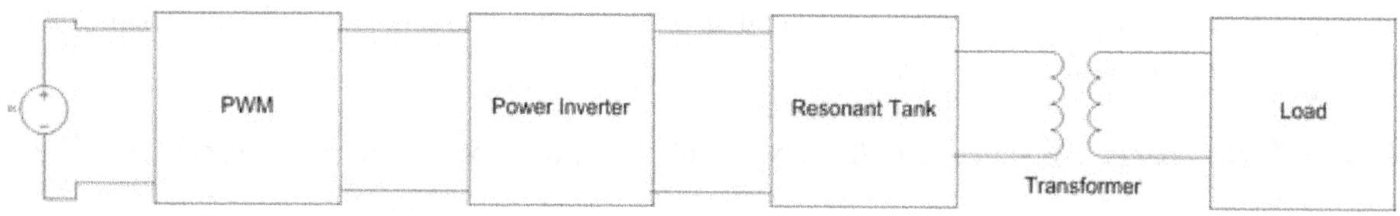

Two-stage conversion AC-DC-AC is followed in most of the WPT models. But a single-stage DC-AC conversion is more efficient when compared to a two-stage conversion in terms of cost and efficiency. Whenever a two-stage conversion is used it requires more switches and gate driver circuits which will lead to more losses. But in single-stage conversion, the only inverter circuit is used and also switching frequency ranges from 20 kHz to 100 kHz and this type of topology will provide more benefits to WPT systems.

The major components of the circuit are the PWM module, Power Inverter, Resonant tank and transformer. PWM module is used to provide the gate pulses to control the inverter switches. A conventional single phase inverter circuit which consists of four switches is used for this configuration. Inductor and Capacitor form the resonant circuit preferably class E or class D resonant inverter and transformer has been used for isolation (Moustapha Elwalaty, 2020).

In the resonant inductive power transfer system transformer turns ratio plays a significant part in determining the performance of the system. So, utmost care to be taken to select the transformer turn ratio which ensures the performance of the system is good for resonant circuits. Full bridge inverter design is preferred when compared to other topologies because of its simple design and easy implementation. Because of the presence of resonant circuit in the system the transferring ability of the system is increased and also attained 90% efficiency.

Roadway/Online Power Transfer

Roadway/online power transfer comprises two subsystems, one portion is connected to the bottom of the vehicle and another portion is connected under the road or on the surface of the road. The subsystem on the roadway has a power supply coupled to the high-frequency inverter through the rectifier (Saransch Chhawchharia, 2018). Another subsystem attached to the bottom of the vehicle consists of the capacitor bank and battery regulator and also having a pick-up coil connected via a rectifier. The subsystem on the roadway should be cheap and robust to withstand the disturbance in the road for a long duration and this system is economic when it is installed for long-distance (Choi, Gu, Jeong, & Rim, 2015). On the other side, the subsystem which is connected to the vehicle should be lightweight and compact size to adopt into the vehicles has been shown in Figure 9.

Figure 9. Configuration of Wireless Power Transfer System for RPT

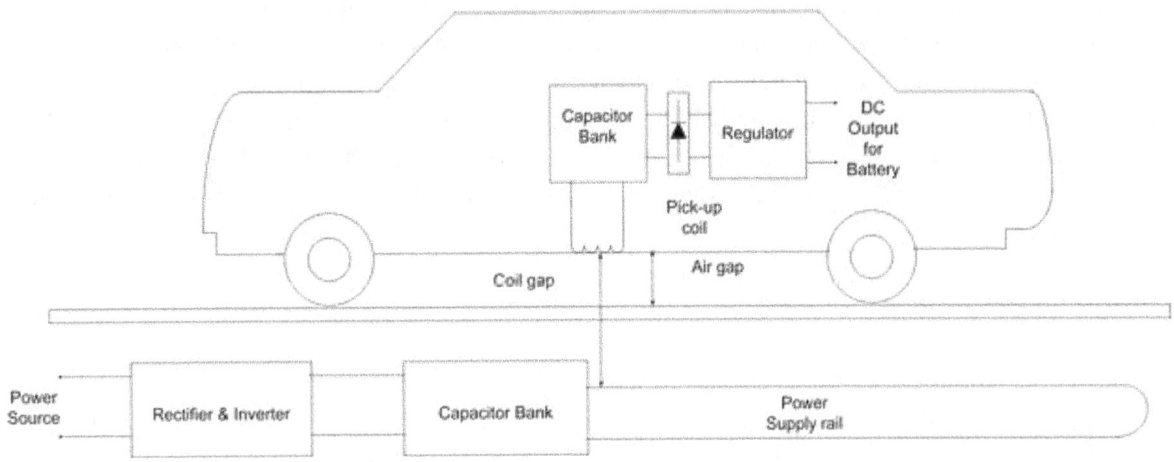

In general, there are three different configurations of WPT used in RPT Systems which are IPTS, CPTS and Coupled Magnetic Resonance System (CMRS) (Michela Longo, 2018). However, the operations of these three different configurations are quite different from one another. CMRS is found suitable for short term distance and power delivery is highly efficient with a high value of quality factor Q. Later, IPTS configuration provided the same purpose with a low value of Q. CPTS is not the right choice for RPT system because to transfer the small power from road subsystem to vehicle subsystem it requires huge space area of conductors which is more than the space of the vehicle. Therefore IPTS is the most appropriate technology for RPT (M. Takagi. Y. Iwafune, 2013).

Principles of IPTS for RPT

The IPTS for RPT is mainly designed on the principle of Ampere's law and Faraday's law among Maxwell's equations.

- Generation of magnetic flux from power supply rail of AC in accord with Ampere's law.

- The power supply rail is coupled with a pick-up coil where the voltage is induced by Faraday's law.
- Through magnetic coupling, power is delivered wirelessly where the inductive reactance is nullified with the help of capacitor banks.

An inverter with a high switching frequency to deliver the high-frequency current to the power supply rail and also a rectifier is included to pick-up the coil side to attain the direct current voltage to provide power in the battery of RPT.

Essential Requirements of IPTS for RPT

The requirements of IPTS for RPT are different from the requirements for traditional IPTS because RPEV vehicles require dynamic charging and also the air gap must be high and lateral tolerance is large and the cost of construction is low. Further, the IPTS required for RPEV must persist from extremely difficult road conditions such as low and high temperatures and humidity must be high and it should withstand mechanical shocks for around ten years (Koufakis, Rigas, Bassiliades, & Ramchurn, 2020).

The subsystem under the road must be mechanically and electrically well protected, but during the rainy conditions the road becomes wet and it is not compatible. Therefore, it is a challenging task to build a subsystem under the road. On the other side, the subsystem under the vehicle also withstands the road conditions and it should be able to operate under vibrating conditions of RPEV (Mi, Buja, Choi, & Rim, 2016).

Design Issues of IPTS for RPT

- The magnetic coupling has to be increased to obtain the induced voltage.
- The power efficiency has to be increased for given device ratings, power capacity and cost.
- The module should be compact to accommodate the given weight and space.
- Electromotive Force should not be increased.
- The variation in resonance frequency has to be managed and also the change in the coupling parameter due to pick-up position is misaligned and variation in the air gap.

Figure 10. Wireless power transmission block diagram

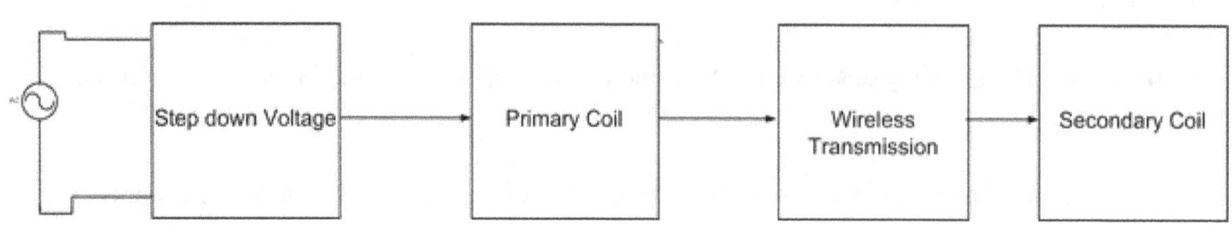

LABORATORY SCALE PROTOTYPE OF INDUCTIVE WPT

The self-contained solar-powered wireless EV charging setup has been developed in the laboratory on a small scale by replacing the solar panel with an AC voltage source in the laboratory. The proposed system developed by using modern and new technologies, which can make the vehicles charge faster than the existing one as shown in Figure 10.

Figure 11. Process of a Flow chart to Charge the battery

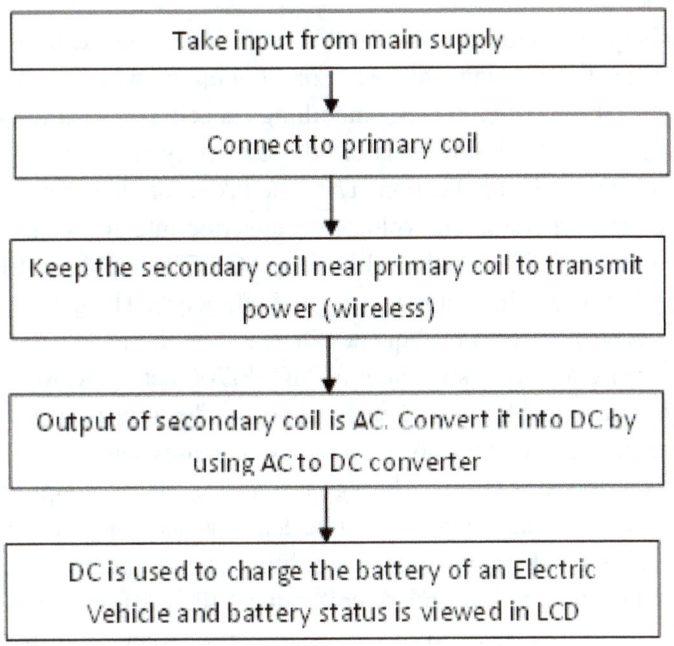

Figure 12. Block diagram of the proposed system.

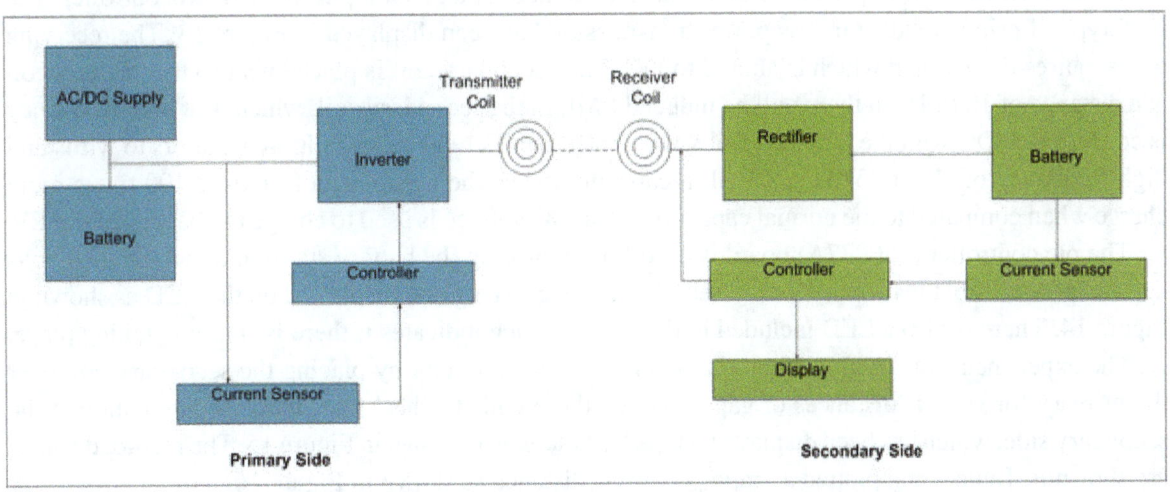

Instead of using wires in this system, the coils can be used to charge the vehicle battery using primary and secondary coils. This uses inductive power transmission, where the primary coil can be used in the charging stations and the secondary coil used to charge the battery of electric vehicle as shown in flow chart Figure 11. The battery status at the secondary side such as voltage and power can be obtained and displayed on the LCD screen.

Power has been transmitted from the primary to the secondary coil by using the mutual induction principle and the charge has been stored in the battery, the status of the battery has been displayed in LCD as shown in Figure 12. The 230V AC source is reduced to 12V using a step-down transformer. The AC supply has been given to the primary transmission coil used in the circuit in which the EMF is induced in the primary coil is transferred to receiving coil by mutual induction to energies the secondary circuit of Electric vehicle.

The primary side AC supply is stepped down to 12V using a center-tapped high-frequency transformer. This 12V voltage is rectified to DC using a full-wave rectifier circuit which uses 1N4007 diodes as they are superior compared to other IN400x series, smoothing capacitor is used to obtain a distortion-free output for storing the DC at the batteries which is a lead-acid battery of 12V and 4.5A rating. The current rating of the battery depends on the application. The time taken for charging can be reduced by using a V-taper charge control and the stored DC voltage is converted into AC with the help of an inverter. The inverter circuit uses MOSFET"s which is driven by SG 3525 PWM IC. MOSFET"s as they are sensitive, fast switching can be obtained along with good efficiency. The gate pulse is controlled using IC SG 3525 which is a control IC whose frequency increases with increases with increase in load. It is basically used in inverter circuits for switching of MOSFETs that is DC to AC conversion. This IC provides improved performance with lowered external ports. The high-frequency AC voltage is given to the primary coil through an inverter, whose frequency increases with an increase in load. The AC power source has been transferred wirelessly through resonant inductive coupling. The inductor can be measured, and then joined with a capacitor to be adjusted to a frequency. The frequencies can be accorded by both the transmitter and receiver coils, permitting them to interconnect together much more efficiently. At the secondary side, the secondary coil induces the EMF, when placed near the primary coil. The received AC voltage is converted to DC using rectifiers and high-frequency capacitors are used to withstand at high frequency. A PIC controller programmed circuit is used to sense the current and voltage in the secondary coil then it has been displayed at LCD. The regulator has been used in the PIC board, which helps to produce a constant five voltage for a given input to the microcontroller. The prototype of primary side wireless power transmission has been displayed in Figure 13. The receiving coil acquires the AC and which is altered to DC. The secondary coil is placed near to the primary coil at a distance of 10 to 15cm there will be induced EMF in the secondary coil, which is at high frequency is converted to DC with the help of a full-wave bridge rectifier circuit and ultra-capacitors to withstand high frequency of about 25KHz. The ultra-capacitors have the capability of holding 100 times more charge when compared to the normal capacitors. The DC voltage is used to charge the battery of the EV.

The pic controller (16F877A) board is used for monitoring the battery status and the obtained voltage and power from the primary for the vehicle battery charging and displays it on the LCD as shown in Figure 14. There is also a LED included in the circuit which indicates if there is charging taking place.

The experiment was conducted on the wireless charging circuit by placing the secondary coil over the primary for various distances or gaps between these coils to check the charge accumulation at the secondary side, which has been displayed on the LCD screen as shown in Figure 15. The recorded values are shown in Table 1 and plotted a graph power v/s distance as shown in Figure 16.

Figure 13. A prototype of wireless power transmission at the primary end

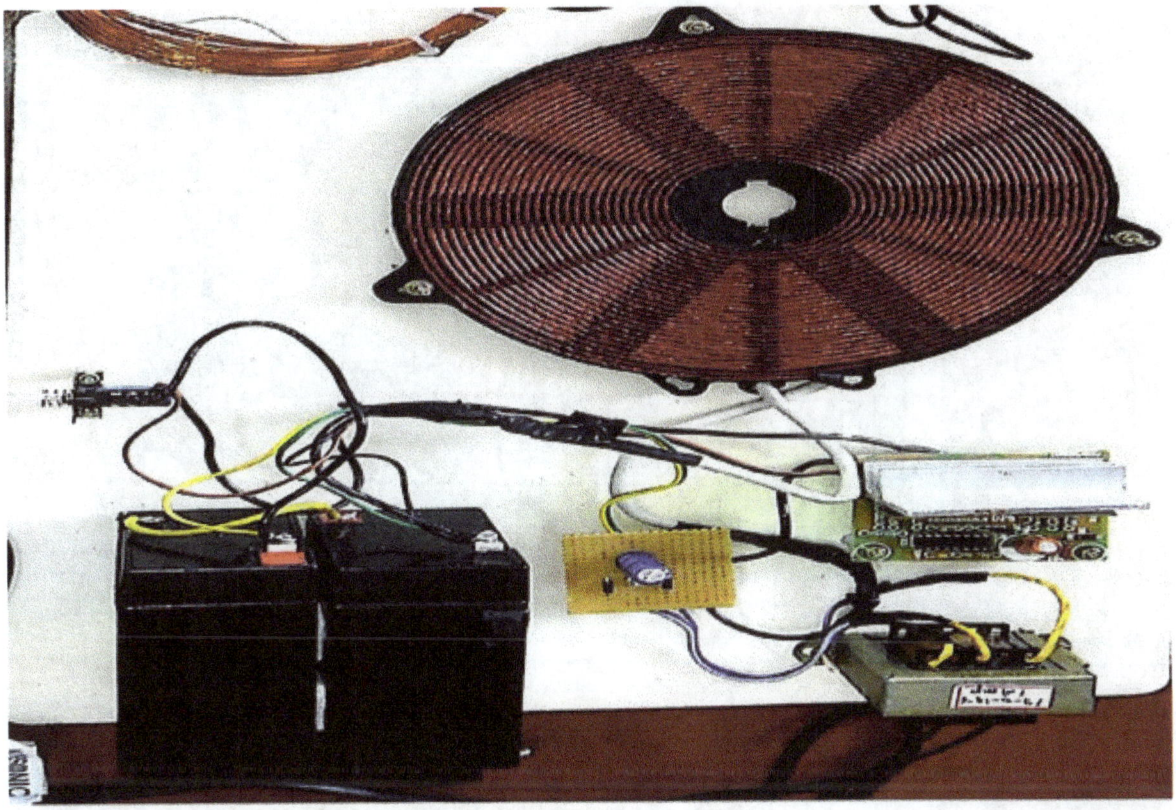

The battery status is shown in figure 16 clearly states that if the distance is less the power transfer is more. From table 1, it has been clear that the efficiency of the wireless power transmission system increases as the distance or gaps between the primary and the secondary coils decreases. Battery status at the secondary circuit has been shown in LCD. If this technology can be included for the EV's to charge the battery. If the secondary circuit can be incorporated in the electric vehicles and the primary circuit can be used in the charging stations. Like the petrol bunks, if there are charging stations included, EV's will be able to replace the present vehicles which use fuels such as petrol and diesel vehicles where the fuel cost is high and it is a non-renewable source of energy.

COMPARISON OF VARIOUS WPT TECHNOLOGIES

The comparison of different WPT technologies based on size, frequency, EMI, efficiency, cost and complexity have been summarized in Table 2.

For low power applications, the CPT system is the most popular technology because of size, cost and power level are low. For medium power applications, IPT and RIPT are the most suitable technologies because of the range of frequency and conversion of power from the primary and secondary sides. The most promising WPT is technology is RPT for charging applications in electric vehicles because in this technology, one part of the system is connected to the vehicle and the other part of the system is pro-

Figure 14. A prototype of wireless power transmission at the secondary end

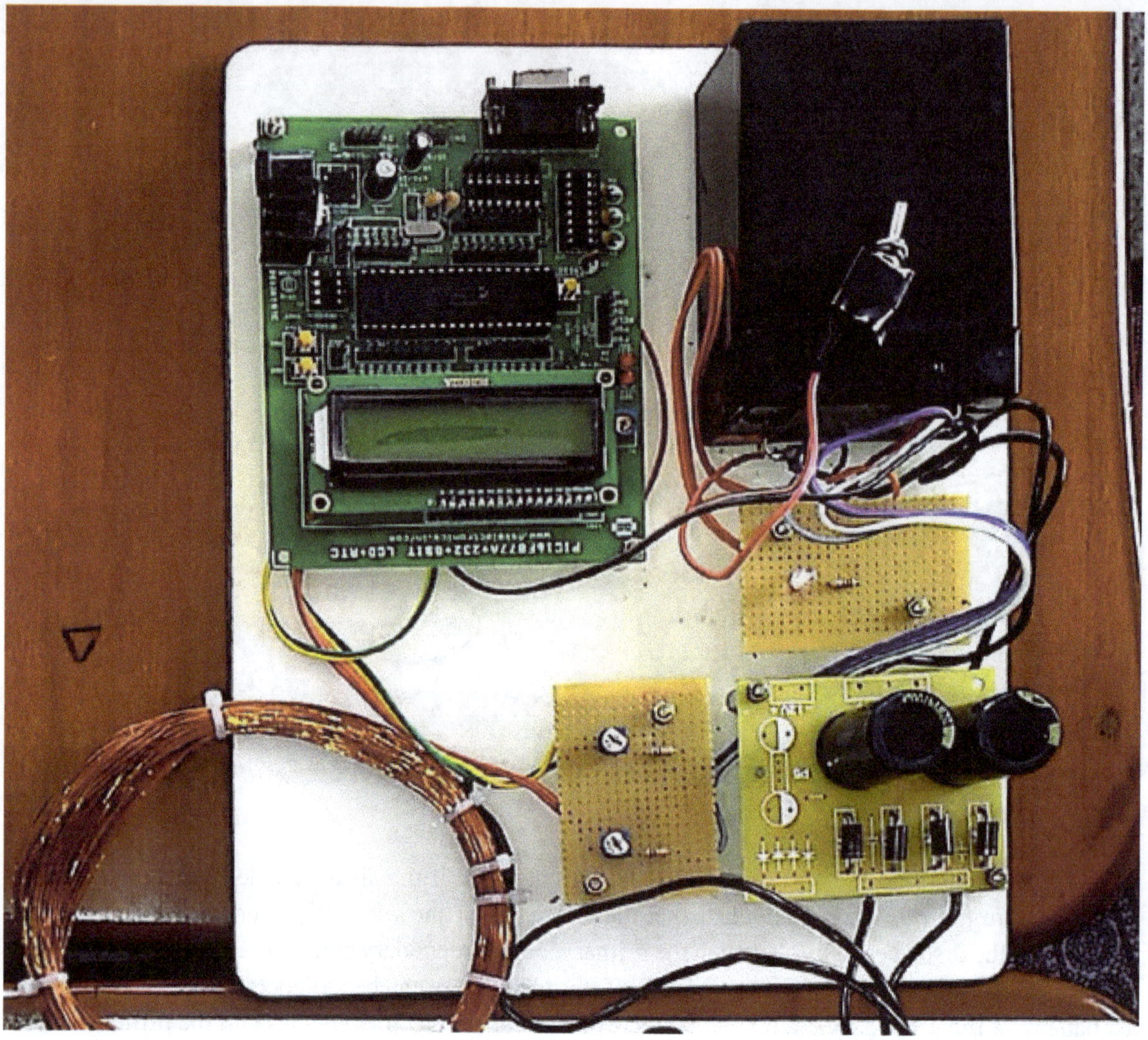

vided on road. In addition to that, the conversion efficiency of this technology is more when compared to other technologies.

CHALLENGES IN WPT TECHNOLOGY

The most significant drawback is the low conversion efficiency during the power transfer in all WPT system and the losses will also be more when the conversion of power from the primary coil to the secondary coil. The cost of installation of the WPT system is also high when compared to the conventional plugin charging methods. In terms of cost perspective, the customer will think the WPT will be a disadvantage for them in terms of initial capital cost (Musavi, Edington, & Eberle, 2012).

Figure 15. Prototype for wireless power transmission circuit for electric vehicle charging

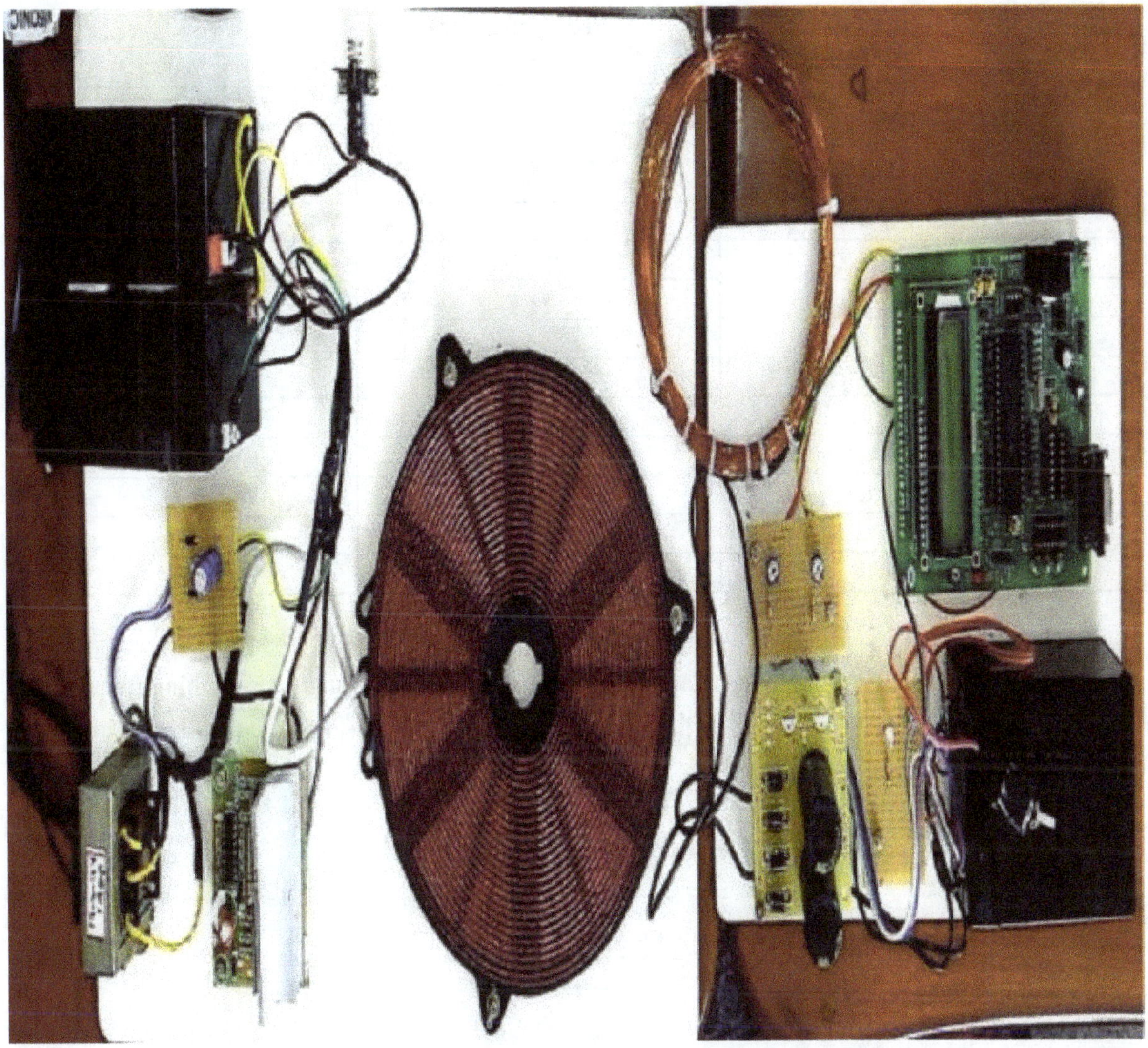

One more significant concern with WPT technologies is the exposure of radiation on humans will create health hazards. The limits on radio frequency radiations will be varied from one country to other and each country has to follow its own national standards/ For example, in Canada EV manufacturers have to follow Canadian Safety Code. IEEE C95.1 standards have been followed by many countries including the United States. International Commission on Non-Ionizing Radiation and Protection (ICNIRP) will set the rules and standards for European countries and the Australian Radiation Protection And Nuclear Safety Range (ARPANSA) will set the bounds on the exposure of radiofrequency in Australia.

In WPT technologies, standards should be set for magnetic field exposure in addition to radiofrequency limits. The magnetic field exposure criterions set through the International Committee on Non-Ionizing Radiation Protection (ICNIRP). The guidelines for the emission of time fluctuating electromagnetic fields on humans have been established by ICNIRP. The standards set by ICNIRP for exposure of radiation

Table 1. Battery status for different gaps between the Primary and the secondary coils.

SL no	Distance cm	Power Watts	Voltage Volts
1	1	12	6.17
2	2	12	6.17
3	3	11	6.17
4	4	10.5	6.17
5	5	10.5	6.17
6	6	10.05	6.17
7	7	9.6	6.16
8	8	9.6	6.17
9	9	9.5	6.17
10	10	9.2	6.17
11	11	9.2	6.17
12	12	9.1	6.16
13	13	8.95	6.16
14	14	8.75	6.16

on human-based on switching frequency and average flux density. The range of switching frequency is 0.8 to 150 kHz and the average flux density should not be more than 6.25μT.

Figure 16. Distances vs Power and Voltage below 15cm.

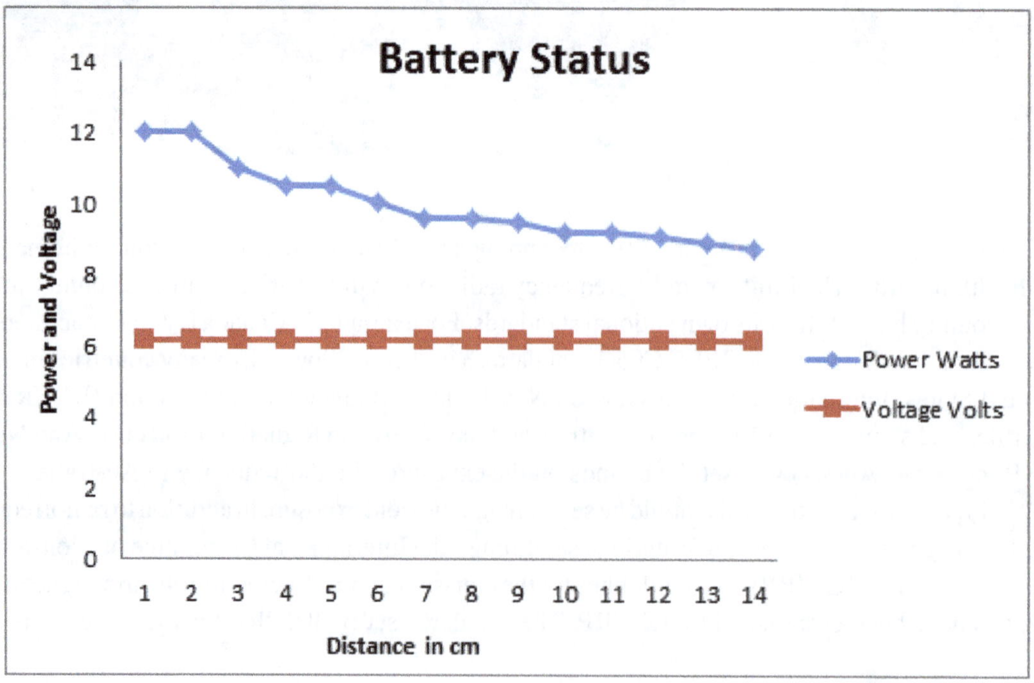

Table 2. WPT Technologies Comparison (Musavi, Edington, & Eberle, 2012)

Technology	Frequency	EMI	Efficiency	size	Cost	Recommended power level	Complexity
IPT	10 – 50 kHz	Intermediate	Intermediate	Intermediate	Intermediate	Intermediate	Intermediate
CPT	100 – 500 kHz	Intermediate	Low	Low	Low	Low	Intermediate
RIPT	1 – 20 MHz	Low	Intermediate	Intermediate	Intermediate	Intermediate	Intermediate
RPT	10 – 50 kHz	Intermediate	Intermediate	High	High	High	Intermediate

CONCLUSION

The WPT systems are becoming progressively popular in research area, in particular the charging application of electric vehicles. In this chapter, the comprehensive survey of various WPT Technologies has been presented.

The focus is provided on the design of solar powered WPT along with various topologies of WPT has been discussed in detail along with advantages as well as challenges. IPT is the most effective topologies for intermediate and high power applications compared to all other topologies. Efficiency is also better in IPT systems. By considering the merits of IPT systems, a laboratory prototype of IPT has been designed and tested with various distances or gaps between these coils to check the charge accumulation on the secondary side. From the observed results, it has been concluded that the efficiency of the wireless power transmission system increases as the distance or gaps between the primary and the secondary coils decreases. Further, the comparisons of various topologies have been presented in terms of cost, size, EMI and efficiency. It should be noted that this chapter does not cover all the details about the WPT. However, the basic principles of the WPT have been discussed precisely.

REFERENCES

Adel El-Shahat, E. A. (2019). Electric Vehicles Wireless Power Transfer State-of-The-Art. *Energy Procedia, 162*, 24–37. doi:10.1016/j.egypro.2019.04.004

Alberto Reatti, M. K. (2018). Resonant Compensations in Inductive Wireless Power System. In A. T.-C. Aguado (Ed.), Emerging Capabilities and Applications of Wireless Power Transfer (p. 383). IGI Global.

Anon Namin, E. C. (2018). Solar Tricycle with Lateral Misalignment Maximum Power Point Tracking Wireless Power Transfer. In *15th International Conference on Electrical Engineering/Electronics, Computer, Telecommunications and Information Technology* (pp. 656-659). IEEE.

Choi, S. Y., Gu, B. W., Jeong, S. Y., & Rim, C. T. (2015, March). Advances in Wireless Power Transfer Systems for Roadway-Powered Electric Vehicles. *IEEE Journal of Emerging and Selected Topics in Power Electronics, 3*(1), 18–36. doi:10.1109/JESTPE.2014.2343674

Covic, G. A., & Boys, T. (2013). Inductive Power Transfer. *Proceedings of the IEEE*, *101*(6), 1276–1289. doi:10.1109/JPROC.2013.2244536

Dai, J., & Ludois, D. C. (2016). Capacitive Power Transfer Through a Conformal Bumper for Electric Vehicle Charging. *IEEE Journal of Emerging and Selected Topics in Power Electronics*, *4*(3), 1015–1025. doi:10.1109/JESTPE.2015.2505622

Daida, S. R. (2019). *A Design of Inductive Coupling Wireless Power Transfer System for Electric Vehicle Applications* (Master of Science thesis). Electrical/Computer Engineering, Old Dominion University.

Darab, C. (2019). Electric Vehicles in Smart Grids. In H. H. Hayek (Ed.), Handbook of Research on Smart Power System Operation and Control (p. 489). IGI Global. doi:10.4018/978-1-5225-8030-0.ch009

Grazian, F., Shi, W., Dong, J., Duijsen, P. v., Soeiro, T. B., & Bauer, P. (2019). Survey on Standards and Regulations for Wireless Charging of Electric Vehicles. In *AEIT International Conference of Electrical and Electronic Technologies for Automotive (AEIT AUTOMOTIVE)* (pp. 1-6). IEEE. 10.23919/EETA.2019.8804573

Huang, Z., Iam, I. W., Hoi, I. U., Lam, C.-S., Mak, P.-I., & Martins, R. P. (2019). Self-Contained Solar-Powered Inductive Power Transfer System for Wireless Electric Vehicle Charging. In *IEEE PES Asia-Pacific Power and Energy Engineering Conference (APPEEC)* (pp. 1-6). Macao: IEEE. 10.1109/APPEEC45492.2019.8994554

Huh, J., Lee, S. W., Lee, W. Y., Cho, G. H., & Rim, C. T. (2011, December). Narrow-Width Inductive Power Transfer System for Online Electrical Vehicles. *IEEE Transactions on Power Electronics*, *26*(12), 3666–3679. doi:10.1109/TPEL.2011.2160972

Javier Vázquez, P. R.-S. (2018). An Inductive Power Transfer System for the Wireless Charging of Electric Vehicles: Determination of the Magnetic Coupling Factor. In A. T.-C. Aguado (Ed.), Emerging Capabilities and Applications of Wireless Power Transfer (p. 383). IGI Global.

Kabalo, M., Berthold, F., Blunier, B., Bouquain, D., Williamson, S., & Miraoui, A. (2014). Efficiency comparison of wire and wireless battery charging: Based on connection probability analysis. In *IEEE Transportation Electrification Conference and Expo (ITEC)* (pp. 1-6). IEEE. 10.1109/ITEC.2014.6861783

Koufakis, A.-M., Rigas, E. S., Bassiliades, N., & Ramchurn, S. D. (2020). Offline and Online Electric Vehicle Charging Scheduling With V2V Energy Transfer. *IEEE Transactions on Intelligent Transportation Systems*, *21*(5), 2128–2138. doi:10.1109/TITS.2019.2914087

Lu, F., Zhang, H., & Mi, C. (2018). A Two-Plate Capacitive Wireless Power Transfer System for Electric Vehicle Charging Applications. *IEEE Transactions on Power Electronics*, *33*(2), 964–969. doi:10.1109/TPEL.2017.2735365

Ludios, J. a. (2015). *Wireless electric vehicle charging via capacitive power transfer through a conformal bumper. In IEEE Applied Power Electronics Conference and Exposition (APEC)*. IEEE.

Mi, C. C., Buja, G., Choi, S. Y., & Rim, C. T. (2016, October). Modern Advances in Wireless Power Transfer Systems for Roadway Powered Electric Vehicles. *IEEE Transactions on Industrial Electronics*, *63*(10), 6533–6545. doi:10.1109/TIE.2016.2574993

Michela Longo, M. B. (2018). Focus on OLEV: On Line Electric Vehicles. In A. T.-C. Aguado (Ed.), Emerging Capabilities and Applications of Wireless Power Transfer (p. 383). IGI Global.

Mou, X., Gladwin, D. T., Zhao, R., & Sun, H. (2019). Survey on magnetic resonant coupling wireless power transfer technology for electric vehicle charging. *IET Power Electronics*, *12*(12), 3005–3020. doi:10.1049/iet-pel.2019.0529

Mou, X., Gladwin, D. T., Zhao, R., Sun, H., & Yang, Z. (2020). Coil Design for Wireless Vehicle-to-Vehicle Charging Systems. *IEEE Access: Practical Innovations, Open Solutions*, *8*, 172723–172733. doi:10.1109/ACCESS.2020.3025787

Moustapha Elwalaty, M. J. (2020). Modeling, Analysis, and Implementation of Series-Series Compensated Inductive Coupled Power Transfer (ICPT) System for an Electric Vehicle. *Journal of Electrical and Computer Engineering*, *2020*, 1–10. doi:10.1155/2020/9561523

Musavi, F., Edington, M., & Eberle, W. (2012). *Wireless power transfer: A survey of EV battery charging technologies. In IEEE Energy Conversion Congress and Exposition (ECCE)*. IEEE.

Nama, J. K., Verma, A., Srivastava, M., & Tomar, P. S. (2020). An Efficient Inductive Power Transfer Topology for Electric Vehicle Battery Charging. *IEEE Transactions on Industry Applications*, *56*(6), 1–8. doi:10.1109/TIA.2020.3018419

Parise, R. (2002). Model to predict performance of all electric transportation with wireless power beams. In *37th Intersociety Energy Conversion Engineering Conference* (pp. 1-6). IEEE.

Qiu, C., Chau, K., Liu, C., & Chan, C. (2013). Overview of wireless power transfer for electric vehicle charging. In *2013 World Electric Vehicle Symposium and Exhibition (EVS27)* (pp. 1-9). Spain: IEEE. 10.1109/EVS.2013.6914731

S, P. S. (2020). Wireless Power Transfer Topologies used for Static and Dynamic Charging of EV Battery: A Review. *International Journal of Emerging Electric Power Systems, 21*(1), 1-34.

Saransch Chhawchharia, S. K. (2018). Investigation of wireless power transfer applications with a focus on renewable energy. *Renewable & Sustainable Energy Reviews*, *91*, 888–902. doi:10.1016/j.rser.2018.04.101

Sergeant, P., & Bossche, A. V. (2008). Inductive coupler for contactless power transmission. *IET Electric Power Applications*, *2*(1), 1–7. doi:10.1049/iet-epa:20070059

Shubhangi Das, K. P. (2018). Wireless Power Transfer in Electric Vehicles. *International Journal of Applied Environmental Sciences*, *13*(7), 643–659.

Takagi, M., Iwafune, Y., Yamaji, K., Yamamoto, H., Okano, K., Hiwatari, R., & Ikeya, T. (2013). Economic value of PV energy storage using batteries of battery-switch. *IEEE Transactions on Sustainable Energy*, *4*(1), 164–173. doi:10.1109/TSTE.2012.2210571

Uthaya, N., & Banu, U. A. (2018). Wireless Power Transfer in Electrical Vehicle by Using Solar Energy. *Asian Journal of Electrical Sciences*, *7*(1), 6–9.

Vishvendra Singh, A. P. (2014). Introduction to Wireless Power Transmission. *HCTL Open International Journal of Technology Innovations and Research*, *8*, 1–10.

Zhu, G., & Gao, D. (2020). High-Frequency Voltage and Current Sense Circuits for Inductive Power Transfer Systems. *IEEE Transactions on Power Electronics*, *35*(11), 11352–11362. doi:10.1109/TPEL.2020.2989922

Chapter 10
Transformation of the Turkish Automotive Sector:
The Electric Vehicle Challenge

Yigit Aydogan
https://orcid.org/0000-0002-1823-0352
Kirklareli University, Turkey

ABSTRACT

As an experienced automotive manufacturer, Turkey is following the global trend of the EV transition and on an enthusiastic start. A government-led action group for EV production called TOGG was initiated in 2018. Automotive sector in Turkey yields (but is not limited to) the following titles at the same time: a strength, a public revenue source, a trade dilemma, an (national) ambition. To tackle power grid problems due to increasing demand, renewable energy usage seems both challenging and necessary. There is a potentially favorable demand for EVs. Promoting the EVs for the customers requires strong infrastructure which seems lacking in Turkey. The EVCSs have been scarce against large land area of the country. On the contrary, consumers in Turkey buy too much BEVs against PHEVs compared to the EU which needs to be directed by the government accordingly to make a healthy transition in the future.

INTRODUCTION

A global pursuit for transforming the economies into a lower environmental-impact model has been going on for a long time. The automotive sector is the usual suspect for such quest. Improving efficiency in production facilities, squeezing engine volumes and significantly decreasing consumption per mileage have always been on the spot in this manner. But the primary element of lowering carbon emissions is decreasing the fossil fuel dependence of the motor vehicles. If all vehicles would have no carbon emissions at all, engineers who have been endlessly working on progressively smaller and more complex engines might finally get some rest.

Despite the contemporary struggle by the environmentalists for bringing down the petroleum powered cars, the invention of the Electric Vehicles (EVs) goes back in time. The early versions of an EV

DOI: 10.4018/978-1-7998-7626-7.ch010

had been built in Hungary, the Netherlands and the US during the 1830s. By end of the first decade of the twentieth century, almost one third of all vehicles in the US had been EVs. Moreover, even a hybrid car was invented by Ferdinand Porsche in 1901. The EV era was over when Ford Model T had been dominated the automotive sector in just a few years as an affordable transportation solution (U.S. Department of Energy, n.d.).

There has been much research and development (R&D) done by almost all of the big automotive manufacturers around the world in order to adapt EVs into their product lines. The most important EV producer in the contemporary car market can be seen as Tesla Motors for several reasons (Thomas and Maine, 2019). In the middle of automotive giants, surprising rise of Tesla Motors raise questions about its future in the global car market as a major player. However, the current global market for EVs remains extremely limited of scope. Global firms with millions of cars sold each year seem like taking their time. In 2019, global EV sales accounted to only 2.6% of total car sales and the EV share is only 1% in total car stock (International Energy Agency, 2020). This can be associated with the same reason as disappearing of EVs in the first place after the first decade of the twentieth century.

Production and maintaining conditions can still be considered as unappealing for both producers and customers. Expectations on the EV transition of active big firms stood unclear in the midst of Covid-19 pandemic. There have been discussions about which part of the world might be the future leader in the EV market, China, Europe or the US (e.g. Jetin, 2020). In this context, however, there can be no healthy inferences about the future players in the EV market due to its proximity to a possible technological and methodological shift. The EV share in global car market can still be considered as a frictional side market by looking at it accounting perspectives. On the other hand, it gathers much more attention from all related parties that taking into account its future impact. Nilsson and Nykvist (2016) expect a total EV transition by 2050 if a breakthrough happens during the development period.

As an experienced automotive manufacturer, Turkey is following the global trend of the EV transition and on an enthusiastic start. The country is one of the top 10 largest car exporters in the world, yet without a single national brand. The ambitious efforts for a Turkish car had started in 1961, but Turkey's first 100% nationally produced passenger car brand *Devrim* did not succeed to mass production (Tülomsaş A.Ş., n.d.). FIAT, Renault, Ford, Toyota, Honda and Hyundai currently produces passenger vehicles in Turkey and there have been producers of several other types including military vehicles and trucks, e.g. BMC, Man and Mercedes. Fortunately, a government-led action group called *TOGG* has been initiated in 2018 and the country is hoping to embrace this new line of transportation with a Turkish product in the near future.

Automotive industry is a multidirectional element in Turkish economy. In a nutshell, it yields (but not limited to) the following titles at the same time which are about to be discussed in detail: a strength, a public revenue source, a trade dilemma, an (national) ambition. It is a strength because of the large scale and varied production capacity of the sector. The positive impact of domestic automotive and fuel sales via extreme taxation has become indispensable for the government. The high demand for automotive imports and related fossil fuel to run the contemporary vehicles pose challenges to the sector's future due to the considerable amount of current account deficit. Most importantly, 97.6% of the people supports the TOGG project and 89% of people would like to buy a "nationally branded" car according to official surveys (Anatolian Agency, 2020). In this chapter, current outlook and possibilities regarding Turkish EV adventure will be discussed in detail.

BACKGROUND

Establishing long-term R&D traditions by automotive producers and universities in Turkey have led to a considerably big pile of publications concerning many aspects of the local automotive sector, also with ties to the electric powered future of it. Ertuğral (2011) asserts that joining the Customs Union with the European Union in 1996 has been highly influential on the Turkish automotive sector, enabling its development towards export orientation with better products. Furthermore, Taymaz and Yılmaz (2017) conclude that Turkish automotive industry had reached to the current development level thanks to the organizational and competitional skills of the producing firms, rather than a well-planned and visionary governmental policy. Görener and Görener (2008) assert that Turkish automotive industry complies with the high standards of the global market having strong research and development focus relying on the availability of the high-quality human capital in Turkey. They also point out the first three problems of the Turkish automotive industry as increase in imports, lack of cooperation between main and the supplier industries, high sales taxes. Çoban (2007) states that as firm diversity in the automotive industry in Turkey increases, competitiveness of Turkish automotive industry has been improving steadily. Moreover, it was also put forward that firms in the Turkish automotive sector effectively benefits from scale economies (Yaylalı and Çalmaşur, 2014).

Yılmaz et al. (2017) claims that Turkish consumer behavior indicates that neither they are indifferent of whether the car had been produced in Turkey, nor they connect the national policymaking with the automotive demand. In line with the contradictory evidence from recent surveys, they suggest a more focused research to find out the underlying dynamics.

Akçomak and Bürken (2019) claims that the Turkish automotive industry could be seen in a middle-technology trap. The main pillars of such trap have been found to reside in weak innovation system and partial role in Global Value Chain of the sector. Firm structure as joint ventures between local and global manufacturers, Turkey's imported input dependencies for production and absence of complementarities have been pointed out to form the trap together. On the other hand, Özatağan (2011) finds that such power asymmetries depending on Global Value Chain has been dynamic and suppliers in Turkey have chances to break free from their struggles against developing.

Yılmaz and Ustaoğlu (2013) investigates the early EV transformation in Turkish automotive sector. At the time of the research, main demands of the manufacturers had been favorable taxation and governmental incentives on new investments. They conclude that competitiveness of the Turkish automotive industry is crucial for constituting a production center for EVs in the global market. According to the interviews conducted by Ustabaş and Simav (2018) primary steps to be taken in order to adapt this new line of transportation have been noted by the sector representatives as maintaining a positive public opinion with trust, supporting consumers physically and financially, taxation and export-oriented production incentives for the producers. They also note the participants expect that the transformation to higher technology using and more environmentally friendly vehicles might contribute to the economic growth in Turkey. On the other hand, Paşalıoğlu and Cengiz (2019) find that Turkish consumers' demand for EVs have been affected by their lifestyle through their attitudes related with innovativeness, uniqueness and exploratory behavior.

Research on renewable energy and EVs have been increased recently in Turkey. Ustabaş (2014) posits that locally produced spare parts and especially batteries of EVs would benefit Turkey much on the international trade accounts. Moreover, the most beneficial phase of the EV transition would be powering them with renewable energy resources of Turkey. Tascikaraoglu, Boynuegri and Uzunoglu (2014) forecasts energy demand with renewable energy systems utilizing EVs advantages, proposing a smart home system in Turkey. Çolak et al. (2015) proposes a solar power-based charging station network for Ankara, Turkey. Erdinc et al. (2017) tackles the managing the optimal size and location of EV charging stations and distributed generation units considering Istanbul case. It was noted that one of the most important problem for the power network system operators is the variation of electric supply and demand when renewable energy and EVs are active.

TURKISH TRANSITION TO ELECTRIC VEHICLES

Analyzing the possibilities regarding the transformation of the somehow traditional automotive sector in Turkey towards EV production could be achieved on four steps. In this part, first, a detailed description of the current automotive industry in Turkey will be provided. Next, the energy importing problem of Turkey will be tied to the efforts related to the EV transition. Third, Turkish car market dynamics and the relation with EVs will be displayed. In the last section of this part, possible effects of the EV transition on the Turkish government budget will be discussed as the automotive maintains several important relations with both revenue and expense of the central government.

Turkish Automotive Industry: General Outlook

Automotive ventures in Turkey goes back in time. The first ever car assembly line was built in Istanbul by Ford Motor Company in 1929 (Kapan, 2018). Unfortunately, there is very little information about this initiative today, and shortly after the Great Depression it ceased to exist, until returning with a local partnership in 1959. Current automotive manufacturers are presented in Table 1 in detail.

As seen in Table 1, Turkey is an important figure in the global automotive sector with a lot of foreign capital investment in several agglomeration areas in the Western Anatolia. Most of the largest global automotive producers have operations in the country for many years. The oldest entry in Table 1 corresponds to the oldest active factory, tractor production of New Holland under the name of Turk Traktor, starting in 1954. For the near future, Honda recently announced plans to shut down their operations in Turkey in 2021 (Kucukgocmen, 2019). Additionally, Volkswagen group had plans to start producing passenger cars in Turkey, but they dump the project as Covid-19 hits the global demand for passenger cars (Rauwald, 2020).

Automotive sector in Turkey has strong links to the international trade, as expected. Sectoral imports and exports correspond to sizeable portions of the national accounts. The summary data of the Turkish Automotive Sector is presented in Table 2.

Table 2 presents the international trade data of Turkish automotive industry. While automotive exports more than double during the subjected period, total exports also increase 68%. On the other hand, automotive imports decreased to 2009 levels in 2019, while total imports of the nation increased 44%. This step-up in automotive provided $16.5 billion trade surplus in 2019, a 28% increase compared to the previous year. In depth analysis of the figures in Table 2 reveals that car imports got hit severely in the last two years. It also highlights the importance of locally produced cars and changes in national demand possibly related to social or economic factors. Sharp increase in export/import ratios fortify the Turkish automotive sector's importance in terms of the persistently high trade deficit of the country.

Table 1. Active Automotive Manufacturers and Factories in Turkey as of 2020

Firm	City	Production Started in	License	Foreign Capital, %	Total Area, 1,000m²
Anadolu Isuzu	Kocaeli	1966	Isuzu	29.73	299
Ford Otosan	Eskişehir	1983	Ford	41.04	1,125
	Kocaeli-1	2001			712
	Kocaeli-2	2014			639
Hattat Traktör	Tekirdağ	2002	Valtra, Hattat	0	182
Honda Türkiye	Kocaeli	1997	Honda Motor Europe. LTD.	100	292
Hyundai Assan	Kocaeli	1997	Hyundai Motor Comp.	70	692
Karsan	Bursa	1966	Hyundai Motor Comp. Menarini Bus	0	237
M.A.N. Türkiye	Ankara	1966	Man Truck&Bus SE	99.9	317
Mercedes Benz Türk	İstanbul	1968	Mercedes Benz	84.99	552
	Aksaray	1985			696
Otokar	Sakarya	1963	Otokar	0	561
Oyak Renault	Bursa	1971	Renault	51	630
Temsa	Adana	1987	Temsa	100	510
Tofaş	Bursa	1971	FIAT	37.8	978
Toyota	Sakarya	1994	Toyota	100	917
Türk Traktör	Ankara	1954	New Holland / Case IH	37.5	257
	Sakarya	2014			402

Source: Automotive Manufacturers Association (2020)

It can be seen in Figure 1 that during the last decade, Turkish automotive industry has been on a favorable trend in terms of international trade. Automotive exports increase within total exports. Moreover, automotive imports have been decreasing significantly as percent of total. On the other hand, car share in automotive exports decreases except a temporary peak in 2017. This interesting fact could mean that Turkish car manufacturers might be losing stance in a transformation regarding global car demand. Perhaps, a contemporary car, e.g. an EV, might boost car exports to maintain a more stable exports composition.

Table 3 provides more evidence regarding the recent change in Turkish automotive trade. Adding to the previous comments, unit value of Turkish automotive imports increases significantly. It means that Turkish automotive market seeks higher value-added products over time. On the other hand, unit value of exports remains stable except significant drops in some years. It can be seen that despite increased demand for Turkish exports, the automotive industry cannot increase supply for high end products neither to the local market nor to the international partners. It is clear that there is room for EVs in the Turkish automotive industry in terms of trade.

Table 4 depicts imports and exports covering all kinds of EVs of Turkey. It can be seen that Turkey runs a consecutive trade deficit on EVs until 2020. It is evident that EV imports decrease in number

Table 2. Turkish Automotive Sector and International Trade, 2009-2019

	2009	2010	2011	2012	2013	2014	2015	2016	2017	2018	2019
Total Imports, $, Bn	140.9	185.5	240.8	236.5	251.7	242.2	207.2	198.6	233.8	223.0	202.7
Total Exports, $, Bn	102.1	113.9	134.9	152.5	151.8	157.6	143.8	142.5	157.0	167.9	171.5
Trade Balance, $, Bn	-38.8	-71.7	-105.9	-84.1	-99.9	-84.6	-63.4	-56.1	-76.8	-55.1	-31.2
Automotive Imports, $, Bn	9.0	13.4	17.2	14.5	16.8	15.7	17.5	17.8	17.4	13.9	9.6
Automotive Exports, $, Bn	12.3	13.8	15.8	15.1	17.0	18.1	17.5	19.8	23.9	26.8	26.2
Automotive Trade Balance, $, Bn	3.3	0.4	-1.4	0.6	0.2	2.3	-0.1	2.0	6.5	12.9	16.5
Automotive Exports/ Imports Ratio	1.4	1.0	0.9	1.0	1.0	1.2	1.0	1.1	1.4	1.9	2.7
Car Imports, $, Bn	4.3	6.8	8.5	7.3	9.1	7.7	9.2	9.8	8.6	5.9	3.4
Car Exports, $, Bn	6.1	6.2	6.5	6.1	6.9	7.3	6.9	8.4	11.8	12.4	11.9
Car Trade Balance, $, Bn	1.8	-0.6	-2.0	-1.2	-2.3	-0.5	-2.3	-1.5	3.2	6.5	8.5
Car Exports/Imports Ratio	1.4	0.9	0.8	0.8	0.8	0.9	0.8	0.9	1.4	2.1	3.5

Source: Automotive Manufacturers Association (2019)

while the unit cost increases. Turkish automotive sector seems importing cheaper EVs relative to exports, which is promising. On the other hand, EV exports rise quickly but decrease in terms of unit price. As of Q1-2020, Turkey runs a trade surplus on EVs but the rest of the year shall be determined by the Covid-19 pandemic's global effects on the international demand.

Stagnant nature of the current account deficit of the Turkish economy have been presented in Table 2. Combined with the international trade data on EVs in Table 4, it can be seen that locally manufacturing EVs is utmost importance for the economy's long-term wellbeing. The European Union is the primary market for Turkish exports, where the market for EVs is highly favored for the future. Turkey is the largest passenger car exporter to the EU, exporting almost a million cars each year (European Automotive Manufacturers Association, 2019). With the market experience at hand, EV transition offers high return for Turkish exporters.

In short, the EV target for Turkish automotive sector seems extremely plausible. While the automotive giant country has been struggling to increase average return of automotive exports in recent years, its imports have been on the rise in terms of unit price. Luckily, EVs' terms of trade are highly favorable for Turkey with high value of exports relative to imports. It could justify the TOGG's primary aim to be an EV.

Car Market Dynamics and Opportunities Regarding EVs

Occasionally, embracing a new product line might be troublesome. When it comes to EVs, the dynamics of the subjected market can be assessed by analyzing the history of automotive production and evolution of the vehicle park in Turkey. Moreover, by comparing several ratios i.e. car ownership per 1000 inhabit-

Figure 1. Shares of Automotive in International Trade and Car Share in Automotive, 2009-2019 (Automotive Manufacturers Association, 2020)

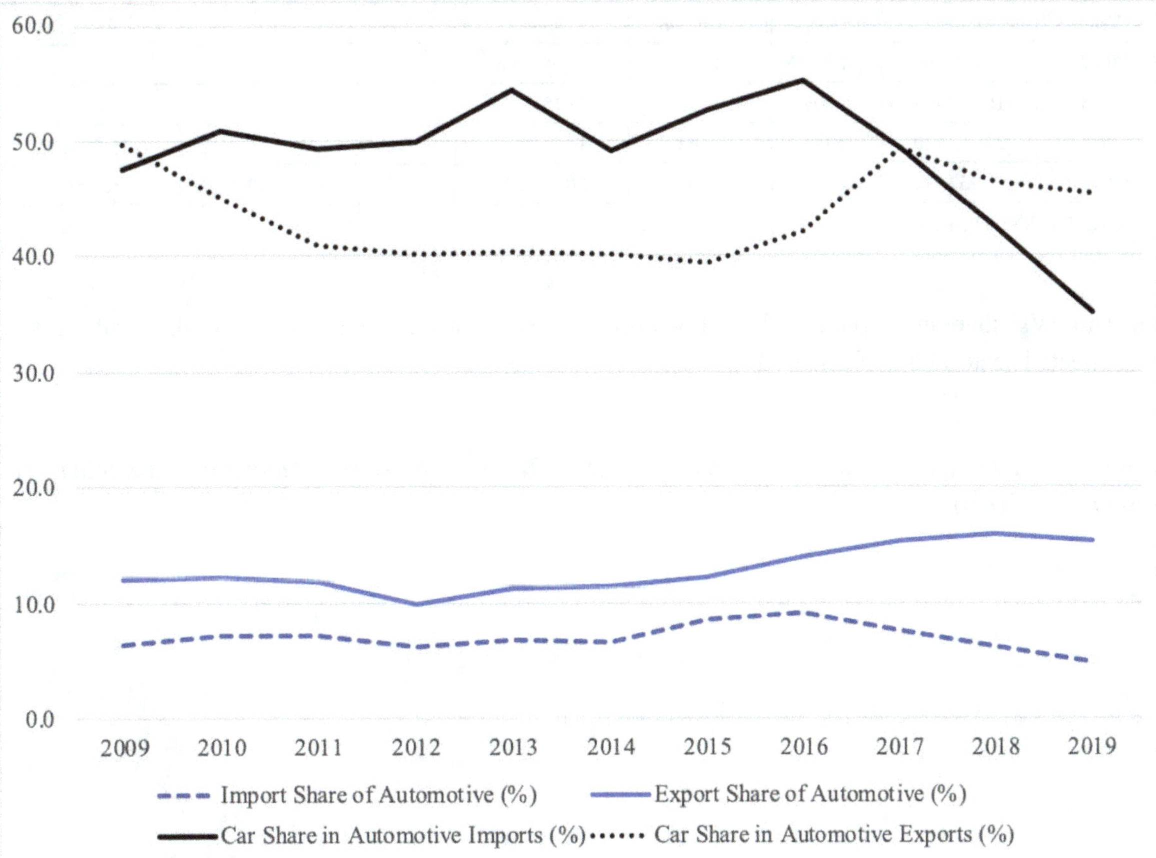

Table 3. Turkish Automotive Market and Trade Value, 2013-2019

Year	Local Market (units)	Production (units)	Imports (units)	Exports (units)	Export Value ($ per unit)	Import Value ($ per unit)
2013	893,124	1,125,534	615,609	828,471	20,519	27,290
2014	807,486	1,170,445	524,831	885,180	20,447	29,914
2015	1,011,194	1,358,796	658,690	992,335	17,635	26,568
2016	1,007,857	1,485,927	681,308	1,141,382	17,347	26,126
2017	980,277	1,673,664	624,409	1,332,794	17,932	27,866
2018	641,550	1,550,260	397,268	1,318,869	20,320	34,989
2019	491,909	1,461,244	277,070	1,252,586	20,916	34,648

Source: Automotive Distributors Association (2020)

Table 4. Turkish International Trade of EVs, 2017-2020

	Exports			Imports			Trade Balance
Year	*Value, $*	*Units*	*$ per Unit*	*Value, $*	*Units*	*$ per Unit*	*Value, $*
2017	5,944,343	1,347	4,413	34,359,006	57,818	594	- 28,414,663
2018	16,337,144	5,003	3,265	29,189,788	44,470	656	- 12,852,644
2019	25,162,140	16,098	1,563	43,739,368	31,393	1,393	- 18,577,228
2020-Q1	12,152,794	4,213	2,884	11,632,276	11,106	1,047	520,518

Source: TURKSTAT (n.d.-a)

ants and EVs' share in car park, with European counterparts one can infer solutions to the challenges of the expected sectoral transformation.

Figure 2. Automotive Production and Vehicle Park in Turkey, 1963-2019 (Automotive Manufacturers Association, 2020)

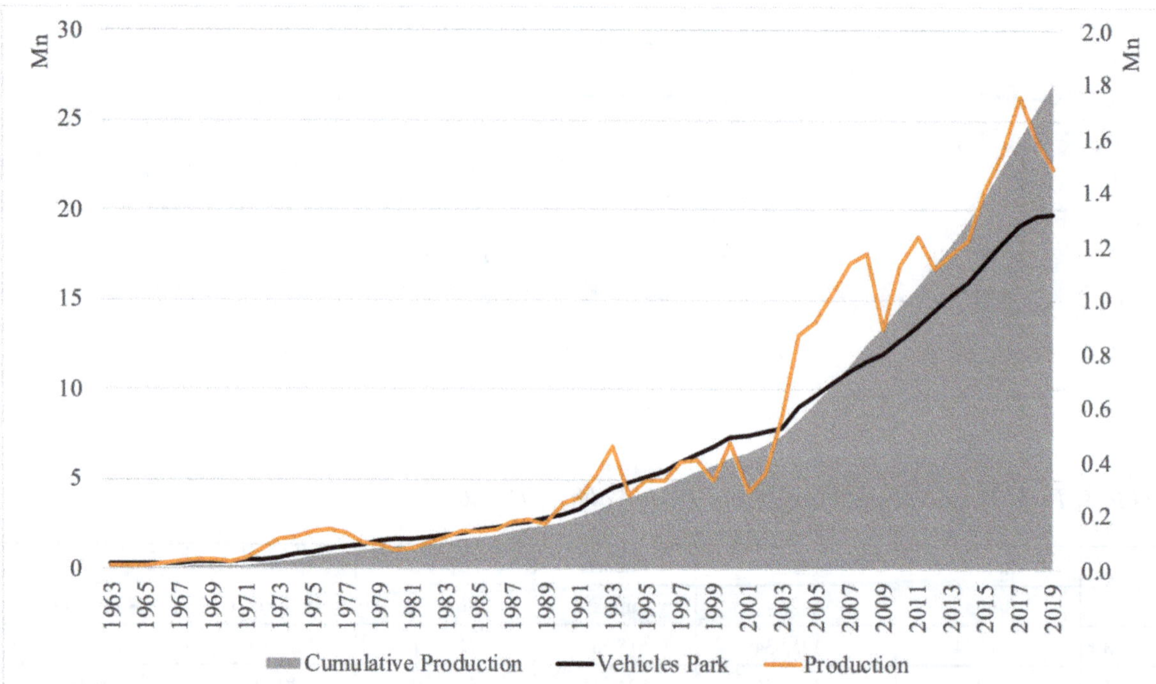

Development of the automotive production and the overall vehicle park in Turkey is presented in Figure 2 for the period 1963-2019. The yearly production data is represented by the right axis. It is evident that there has been always a big demand for automotive products in Turkey. The vehicle park does not stop expanding while production decreases during or after crisis periods e.g. 1994, 2001 and 2008. Therefore, it can be said that demand for automotive products have been more robust against shocks than its production in Turkey. As an interesting turning point, historical Turkish automotive production had surpassed the country's vehicle park in 2007. Turkey could be seen as having a surplus in automotive against the world ever since. It coincides with the Global Financial Crisis, and temporarily lowered production bounced back quickly within a year's duration.

Table 5. Passenger Cars per 1000 Inhabitants (shown as Rate) in European Countries in 2018. Countries above 10 million cars denoted with an asterisk.

Country	Rate	Country	Rate	Country	Rate
Liechtenstein	780	Czechia	540	Denmark	447
Luxembourg	676	Switzerland	539	Ireland	445
Italy*	646	Norway	516	Slovakia	426
Cyprus	629	Portugal	514	Croatia	409
Finland	629	Spain*	513	Bulgaria	396
Poland*	617	Lithuania	512	Hungary	373
Malta	608	Belgium	511	Latvia	369
Germany*	567	Netherlands	494	Romania	332
Estonia	563	France*	478	North Macedonia	200
Austria	562	Sweden	476	Turkey*	151
Slovenia	549	United Kingdom*	473	Average (All)	498

Source: Eurostat (n.d.)

Surpassing the possible export opportunities, future of the domestic demand is appealing for automotive sector in Turkey. Comparing to European counterparts, Turkey has a very low per capita car ownership combined with a population of 83 million. Table 5 presents passenger cars per 1000 inhabitants in European countries and their average. By comparing the figures at Table 5, one can easily recognize the opportunities regarding the Turkish passenger car market. While Turkey is already amongst the largest car markets with over 10 million cars at use, it has only 151 cars per 1000 inhabitants as of 2018. The European average of car ownership is 498 per 1000 inhabitants during the same period. Although this statistic includes several country specific characteristics, e.g. per capita income, life-style, local car production, the gap is too big to justify on any grounds. There is clearly a favorable demand potential if the overall circumstances improve.

In order to increase the per capita car ownership, several measures can be taken. The most important aspect of the car demand in Turkey is extreme taxation on new car purchases. Therefore, it could be effective to modify tax rates in favor of EVs. Accordingly, the government offers lower tax rates for EVs than petroleum run cars. For example, the cheapest fully-EVs has only 3 percent private consumption tax, while the cheapest petroleum powered and hybrid cars has 45 percent instead (Revenue Administration, 2019). Considering the relatively expensive price tag of fully-EVs compared to hybrid cars, current differentiation in tax exemptions can be interpreted as the government is cautious about the hybrids' role in technology introduction to the market.

Table 6 depicts the EV share in passenger cars throughout the European countries. While having a considerable large car stock, EVs in Turkish market are almost non-existent. Only 1 in 10,000 cars in Turkey is powered with electricity. Being a popular activist nation on environmental issues, and also a

Table 6. EV Share of Passenger Cars in European Countries in 2018. Countries above 10 million cars denoted with an asterisk.

Country	EVs	Country	EVs	Country	EVs
Norway	7.10%	Portugal	0.19%	Latvia	0.06%
Liechtenstein	0.79%	Germany*	0.18%	Czechia	0.04%
Austria	0.42%	UK*	0.18%	Italy*	0.03%
Switzerland	0.42%	Estonia	0.17%	Croatia	0.03%
Denmark	0.39%	Belgium	0.16%	Cyprus	0.02%
France*	0.36%	Slovenia	0.11%	Romania	0.02%
Sweden	0.34%	Spain*	0.11%	Poland*	0.01%
Luxembourg	0.33%	Hungary	0.11%	North Macedonia	0.01%
Ireland	0.21%	Finland	0.07%	Turkey*	0.01%
Malta	0.20%	Lithuania	0.07%	Kosovo	0.01%

Source: Eurostat (n.d.)

fossil fuel exporter, Norway by far has the highest share of EVs in Europe with 7.1 percent as expected although its relatively narrow car park. Considering developed car manufacturer nations, France is the leader in terms of EV penetration to car market. Its EV share is twice of what Germany and the UK have. Italy is an outlier to the historic car manufacturers group. Italian market has only 3 in 10,000 cars as EVs. It is surprising when one considers small car preference of Italian drivers stood away from EVs that far. Fortunately, there are fiscal measures taken by the Italian government starting from 2019 to promote EVs and low emission cars (Tomo, 2020). Finally, Poland stands as the same market share of EVs as Turkey. If the global car market is headed towards EVs, then the opportunity for Turkish market can be clearly seen from the Table 6 combined with the previous analysis regarding per capita car ownership.

Impact on the Government Budget

One of the most controversial issues regarding the car market in Turkey is the fact that it has been a solid source of tax revenue. There is an extreme taxation on acquiring, possessing and running motor vehicles in Turkey. An average consumer might pay 45 to 80 percent of the initial price to taxes just for buying a family car, depending on the pre-tax price and engine size (Resmi Gazete, 2020).

Interestingly enough, there is a double taxation on motor vehicles. Turkish consumers pay value added tax over private consumption tax. For instance, if a very modest small car were to be sold by the firm for 100 liras in Turkey, with a 45 percent private consumption tax and 18 percent value added tax, the buyer has to pay 171.1 liras instead of 163 liras because people need to pay the value added tax for the amount including the private consumption tax. Similarly, gasoline has almost 70% tax in its final price. Combined with yearly taxes, automotive sales and usage represents 11% of the total revenue of the central government (Revenue Administration, 2019). With the introduction of EVs, most of this revenue could be at stake. But can it really create a conflict of interest between the public and the private?

One can attribute the severe taxation on automotive products to cut the national demand for imported cars while the country runs a large current account deficit, rather than revenue generation. As shown earlier, Turkey can benefit from the trade surplus of automotive sector and has room for more imports.

However, if the country needs a strict surplus from the automotive industry in order to make up for other sectors' deficits (e.g. energy) there is a deadlock.

Adding to the situation, there are more than enough locally produced vehicles which could be subsidized against imported ones. Local firms have very high idle capacity so they have been prepared for a demand shock. Specifically, Turkish automotive sector utilized 71% of its total capacity in 2019 (Automotive Manufacturers Association, 2020). If the main aim is the tax revenue, then if it could be compensated by other revenue sources then the country may increase per capita car ownership which has been sluggishly low. Both problems can be the solved by the forthcoming national EV project. In the end, it requires a clearer statement of the purpose from the government whether the taxation is important for curbing importing enthusiasm to keep the current account deficit low, or it aims to generate revenue for the central administration. There are solutions applicable for both.

The Energy Nexus

Turkey is a big fossil fuel importer, and certain geographical advantages make the shift towards sustainable energy resources valuable (Kaygusuz & Sarı, 2003). Sharing the similar climate protection vision, EV and renewable energy are closely related and might prosper hand-in-hand. The EU 2050 Climate Commitment aims sharp reduction of greenhouse gas (GHG) emissions up to 95 percent and utilizes the EV transition (European Commission, 2011). Adding to the discussion, sharp changes in oil prices can be considered one of the most important issue regarding Turkish quest for EVs. Despite the sudden drop in oil prices during the Covid-19 pandemic, post-pandemic period might yield prospect for Turkey in terms of EVs nevertheless (Keçeci, 2020). It was shown that oil prices directly affect Turkish current account deficit (Beşel, 2017), consumer confidence (Bildirici & Badur, 2019) and many more macroeconomic aggregates (Al Rasasi & Yilmaz, 2016). Therefore, the EV motivation of Turkey should also be evaluated with a focus on energy needs. Increasing share of renewable energy and EVs against using fossil fuels for electricity generation and transportation methods would mean less dependence on price shocks in the oil market.

With or without the forthcoming EVs, Turkey has a strong power demand as a developing country. To tackle power grid problems due to increasing number of EVs, renewable energy usage seems both challenging and necessary. Higher utilization of renewable energy resources posits different problems for energy planning. Supply and demand for electricity may vary significantly with more renewable energy resources and EVs. Being dependent on natural events such as sunshine or blowing winds, the reliability problem of the renewable energy utilization could reside in EVs. The literature mostly evolves around how to overcome grid planning problems, general systematic suggestions are being coined as "smart" (see Botterud, 2017 and references therein). There are different grid designs proposed including Vehicle to Grid (V2G) and Grid to Vehicle (G2V), former meaning that EVs might serve as energy storage units as well and vice versa (Harighi et al., 2018). An easier solution to fully benefit from having an EV battery at hand could be a Vehicle to Building (V2B) system which serve as a protection against power outages and demand management for the building (Pang et al., 2011). These bilateral systems could provide storage capacity for renewable energy by charging when there is excess supply, cheaper or renewably generated electricity, and playing a regulatory role for the building or grid when necessary (Richardson, 2013).

Figure 3 depicts the electricity generation of Turkey by energy resources between 1970 and 2018. Total generation has been increased substantially during the subjected period, presented on the right axis.

Figure 3. Electricity Generation and Shares by Energy Resources in Turkey, 1970-2018 (TURKSTAT, n.d.-b)

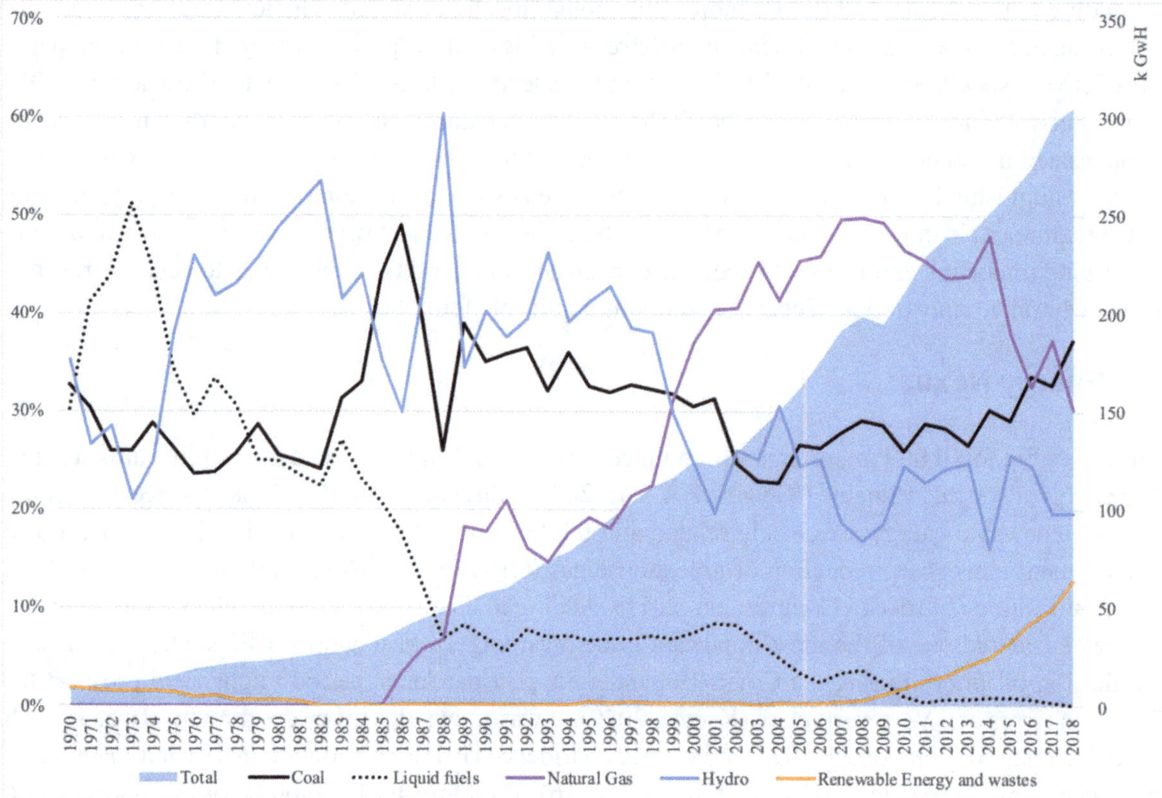

As seen from the shares of energy sources denoted on the left axis, coal consumption has been persistent and on the rise during the last few years. Although Turkey has its own coal reserves, it imports high calorie coal. Most of the national reserves can be classified as low calorie, therefore, cannot be relied on for long-term energy supply (Yılmaz, 2012). Substantial rise and fall of the share of natural gas which is almost fully imported is striking. Despite the geographical proximity to important natural gas exporters, e.g. Russia and Iran, the decrease in the share of natural gas in electricity production observed during the last decade is an important issue for researchers from other disciplines as well. The escape from natural gas has been accompanied by a sizeable increase in the share of renewable energy resources, e.g. wind and solar energy. This trade-off indicates the importance of renewable energy resources in the eyes of the policymakers.

Increasing share of renewable energy in total electricity generation posits great challenges for Turkey. Although there is an appealing geographical potential for solar and wind power, higher grid dependence for such types of energy could be a stability problem. EVs might offer a solution for such problems. For example, if V2B or V2G systems were to work efficiently, the grid management authority could have access to a massive electricity storage capacity. EV batteries could be charged during the day, perhaps with solar power, and returned to the grid at night if available and needed. Thus, incentives on EVs might be seen as an opportunity to evade considerable public spending on grid improvement. People enthusiastically obtaining EVs would inadvertently help the government to reach an environmentally friendly energy composition. It would help reducing GHG emissions further, helping accomplish the EU 2050 climate commitment.

IS INFRASTRUCTURE A SOLVABLE PROBLEM IN TURKEY?

The main issue with the EVs is their relatively limited travel range, especially in a large country with many rural areas. Technological drawbacks regarding EV batteries are being overcome with favorable charging networks in developed nations. The wide and mountainous territory of Turkey and the lack of available capital bring challenges to building the charging infrastructure, but there is an undisputed effort. As of 2019, there have been 11 firms invested in charging stations in Turkey (TEHAD, 2019).

Most of the country is on the wide and mountainous Anatolian peninsula. Moreover, there is a considerable regional inequality across the country. The challenge is real and alive, but the highly populated areas depict prospect. Strong central state and the power of incentives might overcome such problems with careful planning. Supporting the concern, Önder and Kaya (2019) find that urbanization has been the most important socio-economic factor affecting EV sales.

In this regard, there have been a handful of papers promoting well-planned solutions for several important cities of Turkey. Erbaş et al. (2018) maps the capital city of Turkey for efficiently placing electric vehicle charging stations (EVCS). Kaya et al. (2020), proposes innovative EVCS planning for the biggest city of Turkey, Istanbul. They find that southwest of the European side and southwest of the Asian side of the city to be the most appropriate locations for EVCSs. Moreover, Kaya, Alemdar and Çodur (2020) consider Electric Taxis (ETs) as primary component of EV transition and proposes efficient EVCS networks for Istanbul.

According to Table 7, the EV presence in Turkey is extremely low compared to the EU. There are only 15 EVs per 1000 passenger cars in Turkey, while the EU has 405 for 1000 as of 2019. Turkey has only a handful of EVCS compared to the EU. For each EVCS, there are comparable number of EVs in both the EU and Turkey, i.e. 7 and 10 respectively. However, Turkey has only 1 EVCS per 100 km highway while the EU has 20. Moreover, each EVCS in Turkey covers for 4,353 km^2 land area whereas

Table 7. Comparison of the EV and EVCS Statistics of the European Union and Turkey, 2019

Name of the Statistic	European Union	Turkey
Total No. of BEVs (% of Total Passenger Cars)	616,417 (0.229%)	1,319 (0.01%))
Total No. of PHEVs (% of Total Passenger Cars)	474,724 (0.176%)	539 (0.004%)
Total EVs (% of Total Passenger Cars)	1,091,141 (0.405%)	1858 (0.015%)
BEV/PHEV Ratio	1.3	2.4
Total No. of EVCS	164,016	180
EVCS-per-EV Ratio	7	10
Fast Charging Points per 100 km Highway	20	1
Land Area per EVCS, km^2	27	4,353

Source: European Alternative Fuels Observatory (n.d.)

the EU stands for 27 km^2. In contrast, Turkey has 2.4 BEVs for each PHEV while the consumers of the EU possess 1.3 for each.

One real-life project to promote infrastructure for EVs can be planned on the European part of the Turkey. A small part of Turkey resides on the Thrace, including more than half of Istanbul and three cities namely Edirne, Kırklareli and Tekirdağ. Thracian peninsula is rarely mountainous and a relatively well-maintained area in Turkey including high urban and rural population with relatively short distances from each other. Building on the papers presented above, the government might provide subsidies for building infrastructure first in the Thrace peninsula with an open-air laboratory framework. Aside the developed cities of Turkey such as Istanbul and Ankara which could have been equipped with adequate EV network by private agents, the Thrace project can serve as a point of departure for rural planning of the government. It can provide a good insight about the possible challenges covering Anatolia with such technological units where finding investors can be troublesome.

FUTURE RESEARCH DIRECTIONS

As a future research, automotive market modeling could be improved utilizing complex behavior of buyers, sellers and manufacturers which are being forced to adopt a new line of technology. Socioeconomic indicators of EV owners in Turkey could be mapped taking into account the regional differences. The impact of living conditions on EV acquisition in different parts of Turkey can be estimated. Moreover, to propose an optimal charging network throughout the large and mountainous Anatolian peninsula, the characteristics of the small cities and rural areas can be studied.

CONCLUSION

As stated by Meher et al. (2020), sharp oil price movements being witnessed in 2020 might continue in the future. This can be seen as a threat to the long-term planning for renewable energy and EVs. One can say that trade deficits and lack of fossil fuel reserves drive the governments towards such solutions. If the commitment to the green energy and EV investments originate from financial problems, as oil prices decrease, one might see departures from enthusiastic goals. However, as seen in Norway, even if a country has large fossil fuel reserves the EV and climate commitment can be sustained.

EVs might play a key role for a country to achieve better integration of renewable energy resources to the power grid. Serving power storage and demand control mechanisms, EV batteries would be an asset for any country. In this regard, one can see the increasing number of EVs in a country as a privately funded (and owned) public energy security, or regulatory, system. For future research, modeling ownership and control of EV charging mechanisms, i.e. charging car parks, batteries and battery control software, with an economic perspective could be interesting. It would also be interesting to model a private infrastructure investment for a developing country with limited resources for public spending.

Comparing per capita car ownership of Turkey with European counterparts to determine the saturation of demand for cars is one of the core findings of analyzing the automotive data. Corresponding to less than one third of the European average of car ownership, Turkish residents might carry a strong demand for passenger cars in the future if the economic conditions stay favorable. It also means that when the global transition to EVs completed, there will be a "produce or import" deadlock for a long time if the sector is undervalued.

Stimulating demand for EVs is a critical step. Major tax cuts only apply for fully-EVs in Turkey at present. Allowing hybrid cars to the larger tax exemptions would rapidly promote any kind of EV penetration to the Turkish market. It could have further effects on consumers when it comes to domestic fully-EV production. Despite the high retail cost of fossil fuels in Turkey, according to the surveys, Turkish people has a higher tendency towards EVs due to environmental concerns. 47% of the people expressed that they would consider EVs due for environmental protection while 39% aimed cheaper transportation (World Energy Council, 2018). Therefore, demand for EVs in Turkey might be not-so-bounded with retail prices as petroleum run cars, which in turn have favorable marketing consequences which is beyond the scope of this work (for a literature review, see Liao et al., 2017).

Although it does not mean the developed economies are going to have less demand for cars each year than Turkey, which is untrue, it only means that Turkey might become one of those superb markets to distribute EVs. Adoption to EVs is a multi-layered process for car owners as purchasing and running costs, habits and appetites differ (Rezvani et al., 2015). However, Turkish car market should struggle less to adapt due to the obvious reason that many people does not have a car at all. With proper economic development in terms of personal income, Turkey might sell millions of EVs in the near future just to catch up with the European countries' car ownership statistics.

As a nation with no natural resources to export, Turkey runs a current account deficit for a long time. Although there are many export oriented automotive manufacturers, the trade balance is severely affected by the demand for the imported vehicles. On the bright side, domestic cars are rapidly increasing in market share within Turkey.

On the other hand, Turkey is a massive oil importer. When the future demand for cars is considered, it could bring a further burden to the deficit with an increase of petroleum-run vehicles' usage, no matter where it would have been manufactured. With these alarming issues, a domestic EV brand might be a jewel for the country in the long run.

Building on the "national ambition" proposition from the introductory part, all of the problems discussed above might unveil as the government keep up to their announced plans. A strong and decisive public policy equipped with adequate fiscal measures shall make a Turkish EV possible with all the sectoral experience and human capital at hand. It can be seen as a "when" issue instead of "if" considering the automotive expertise of the Republic of Turkey.

Promoting the EVs for the customers requires strong infrastructure which seems lacking in Turkey. The EVCSs have been scarce against large land area of the country. It would bear no value having an EV for a driver if there were no fast charging points during a long journey. With only 1 fast charge station per 100 km highway and considering only a few 100 km range of a typical car, Turkish BEV drivers should be considered very lucky if they make it to their further destinations comfortably. One would expect that without adequate infrastructure, the demand for BEVs should be lower. On the contrary, consumers in Turkey buys too much BEVs against PHEVs compared to the EU. It makes the market transition even more troublesome for the planners. This tendency also needs to be directed by the government according to the infrastructure in order to make a healthy transition to BEVs. If the early adopting consumers complain and leave a bad feedback, it might harm the return of future investments.

The good-looking sports cars of Tesla Motors, some of which travel on their own relying on Artificial Intelligence (AI) might make people believe the future of EVs shall be similar. However, a total market transformation in today's world should rely on strong supply and demand conditions. Imagining a future like it was portraited by The Jetsons might not always be economical in even the long-term. Therefore, analyses on the EV adaptation of national economies should be based on real life circumstances. The near future of automotive industry might still lay in oil and grease, even if electrical power units find place in the market.

ACKNOWLEDGMENT

This research received no specific grant from any funding agency in the public, commercial, or not-for-profit sectors.

REFERENCES

Akçomak, I. S., & Bürken, S. (2019). *The middle-technology trap: The case of the automotive industry in Turkey* (UNU-MERIT Working Paper Series No. 2019-006). https://www.merit.unu.edu/publications/wppdf/2019/wp2019-006.pdf

Al Rasasi, M., & Yilmaz, M. (2016). The effects of oil shocks on Turkish macroeconomic aggregates. *International Journal of Energy Economics and Policy*, *6*(3), 471–476. https://www.econjournals.com/index.php/ijeep/article/view/2416

Anatolian Agency. (2020, January 6). *Bakan Varank: Türkiye'nin Otomobili'nin tasarımları için tescil başvuruları yapıldı.* Anatolian Agency. https://www.aa.com.tr/tr/politika/bakan-varank-turkiyenin-otomobilinin-tasarimlari-icin-tescil-basvurulari-yapildi/1693428# Automotive

Automotive Manufacturers Association. (2019). *Otomotiv Sanayii Dış Ticaret Raporu.* http://www.osd.org.tr/sites/1/upload/files/4.Ceyrek_2019_OSD_Dis_Ticaret_Raporu-5492.pdf

Automotive Manufacturers Association. (2020). *General and Statistical Information Bulletin of Automotive Manufacturers.* http://www.osd.org.tr/sites/1/upload/files/YILLIK_2020-5479.pdf

Beşel, F. (2017). Oil prices affect current account deficit: Empirical evidence from Turkey. *Journal of Applied Research in Finance and Economics*, *3*(2), 13–21. https://www.academia.edu/download/53123395/40._Besel.pdf

Bildirici, M. E., & Badur, M. M. (2019). The effects of oil and gasoline prices on confidence and stock return of the energy companies for Turkey and the US. *Energy*, *173*, 1234–1241. doi:10.1016/j.energy.2019.02.137

Botterud, A. (2017). Forecasting renewable energy for grid operations. In *Renewable Energy Integration* (pp. 133–143). Academic Press., doi:10.1016/B978-0-12-809592-8.00010-X

Çoban, O. (2007). Türk otomotiv sanayiinde endüstriyel verimlilik ve etkinlik. *Erciyes Üniversitesi İktisadi ve İdari Bilimler Fakültesi Dergisi*, *29*, 17-36. https://dergipark.org.tr/en/pub/erciyesiibd/issue/5886/77854

Çolak, I., Bayindir, R., Aksoz, A., Hossain, E., & Sayilgan, S. (2015, October). Designing a competitive electric vehicle charging station with solar PV and storage. In *2015 IEEE International Telecommunications Energy Conference (INTELEC)* (pp. 1-6). IEEE. 10.1109/INTLEC.2015.7572480

Distributors Association. (2020, January). *ODD Genel Değerlendirme*. http://www.odd.org.tr/folders/2837/categorial1docs/2657/Sektörel%20Değerlendirme%20Ocak%202020.pdf

Erbaş, M., Kabak, M., Özceylan, E., & Çetinkaya, C. (2018). Optimal siting of electric vehicle charging stations: A GIS-based fuzzy Multi-Criteria Decision Analysis. *Energy*, *163*, 1017–1031. doi:10.1016/j.energy.2018.08.140

Erdinc, O., Tascikaraoglu, A., Paterakis, N. G., Dursun, I., Sinim, M. C., & Catalão, J. P. (2017, September). Optimal sizing and siting of distributed generation and EV charging stations in distribution systems. In 2017 IEEE PES Innovative Smart Grid Technologies Conference Europe (ISGT-Europe) (pp. 1-6). IEEE. doi:10.1109/ISGTEurope.2017.8260298

Ertuğral, S. M. (2011). Otomotiv Sektörü Ve Gümrük Birliği Sonrasi Gelişmeleri. *Sosyal Bilimler Dergisi*, (2), 75-83. https://dergipark.org.tr/tr/pub/iusosbil/issue/9499/118702

European Alternative Fuels Observatory. (n.d.). *NPF Highlights* [Data Set]. https://www.eafo.eu/countries/european-union/23640/summary

European Automotive Manufacturers Association. (2019). *Economic and Market Report: EU Automotive Industry First Half of 2019*. https://www.acea.be/statistics/article/economic-and-market-report-state-of-the-eu-auto-industry-first-half-of-2019

European Commission. (2011). *Energy Roadmap 2050: impact assessment and scenario analysis*. Brussels. https://ec.europa.eu/energy/sites/ener/files/documents/roadmap2050_ia_20120430_en_0.pdf

Eurostat. (n.d.). *Passenger Cars per 1,000 Inhabitants* [Data Set]. https://appsso.eurostat.ec.europa.eu/nui/show.do?dataset=road_eqs_carhab&lang=en

Görener, A., & Görener, Ö. (2008). Türk otomotiv sektörünün ülke ekonomisine katkilari ve geleceğe yönelik sektörel beklentiler. *Journal of Yaşar University*, *3*(10), 1213–1232. https://dergipark.org.tr/en/download/article-file/179182

Harighi, T., Bayindir, R., Padmanaban, S., Mihet-Popa, L., & Hossain, E. (2018). An overview of energy scenarios, storage systems and the infrastructure for vehicle-to-grid technology. *Energies*, *11*(8), 2174. doi:10.3390/en11082174

International Energy Agency. (2020, June). *Global EV Outlook 2020*. https://www.iea.org/reports/global-ev-outlook-2020

Jetin, B. (2020). Who will control the electric vehicle market? *International Journal of Automotive Technology and Management*, *20*(2), 156–177. doi:10.1504/IJATM.2020.108584

Kapan, K. (2018). İz Birakmadan Kaybolmuş Bir Endüstri Mirasi: Ford'un Otomotiv Montaj Fabrikasi (1929-1944). *Journal of International Social Research*, *11*(59), 384–391. doi:10.17719/jisr.2018.2646

Kaya, Ö., Alemdar, K. D., & Çodur, M. Y. (2020). A novel two stage approach for electric taxis charging station site selection. *Sustainable Cities and Society*, *62*, 102396. doi:10.1016/j.scs.2020.102396

Kaya, Ö., Tortum, A., Alemdar, K. D., & Çodur, M. Y. (2020). Site selection for EVCS in Istanbul by GIS and multi-criteria decision-making. *Transportation Research Part D, Transport and Environment*, *80*, 102271. doi:10.1016/j.trd.2020.102271

Kaygusuz, K., & Sarı, A. (2003). Renewable energy potential and utilization in Turkey. *Energy Conversion and Management*, *44*(3), 459–478. doi:10.1016/S0196-8904(02)00061-4

Keçeci, F. O. (2020, June). *Türkiye'nin Enerji Görünümü ve Aralanan Fırsat Pencereleri*. Global Political Trends Center (GPoT) Policy Brief. https://www.jstor.org/stable/resrep25190

Kucukgocmen, A. (2019, April 8). Honda will stop car production in Turkey after 2021: statement. *Reuters*. Retrieved from: https://www.reuters.com/article/us-turkey-honda/honda-will-stop-car-production-in-turkey-after-2021-statement-idUSKCN1RK0H6

Liao, F., Molin, E., & van Wee, B. (2017). Consumer preferences for electric vehicles: A literature review. *Transport Reviews*, *37*(3), 252–275. doi:10.1080/01441647.2016.1230794

Meher, B. K., Hawaldar, I. T., Mohapatra, L., & Sarea, A. M. (2020). The Impact of COVID-19 on Price Volatility of Crude Oil and Natural Gas Listed on Multi Commodity Exchange of India. *International Journal of Energy Economics and Policy*, *10*(5), 422–431. doi:10.32479/ijeep.10047

Nilsson, M., & Nykvist, B. (2016). Governing the electric vehicle transition–Near term interventions to support a green energy economy. *Applied Energy*, *179*, 1360–1371. doi:10.1016/j.apenergy.2016.03.056

Önder, H., & Kaya, O. C. (2019). Elektrikli Araçların Satışı Üzerinde Sosyo-Ekonomik Faktörlerin Etkisi: Bir Panel Veri Analizi. *Anemon Muş Alparslan Üniversitesi Sosyal Bilimler Dergisi*, *7*(1), 17–21. doi:10.18506/anemon.376831

Özatağan, G. (2011). Dynamics of value chain governance: Increasing supplier competence and changing power relations in the periphery of automotive production—evidence from Bursa, Turkey. *European Planning Studies*, *19*(1), 77–95. doi:10.1080/09654313.2011.530393

Pang, C., Dutta, P., & Kezunovic, M. (2011). BEVs/PHEVs as dispersed energy storage for V2B uses in the smart grid. *IEEE Transactions on Smart Grid*, *3*(1), 473–482. doi:10.1109/TSG.2011.2172228

Paşalıoğlu, H., & Cengiz, H. (2019). Tüketici Yaşam Tarzının Elektrikli Araçları Satın Alma Niyeti Üzerine Etkisi: Tüketici Yenilikçiliği, Tüketici Benzersizliği Ve Keşifsel Davranışların Aracılık Etkisi. *Proceedings of the International Congress on Business and Marketing*. https://hdl.handle.net/20.500.12415/4676

Rauwald, C. (2020, July 1). VW Abandons Turkey Factory Plan After Car Sales Slump. *Bloomberg*. Retrieved from: https://www.bloomberg.com/news/articles/2020-07-01/vw-abandons-turkey-factory-plan-after-virus-hits-industry

Resmi Gazete. (2020, August 30). *4760 Sayılı Özel Tüketim Vergisi Kanununa Ekli (II) Sayılı Listede Yer Alan Bazı Malların Özel Tüketim Vergisi Oranlarının Yeniden Tespiti Hakkında Karar (Karar Sayısı: 2912)*. Presidency of The Turkish Republic. https://www.resmigazete.gov.tr/eskiler/2020/08/20200830-1.pdf

Revenue Administration. (2019, July 1). *Private Consumption Tax Amounts and Rates.* Presidency of The Turkish Republic. Retrieved from: https://www.gib.gov.tr/sites/default/files/fileadmin/mevzuatek/otv_oranlari_tum/ozeltuketimoranlari-OpenPage.htm

Rezvani, Z., Jansson, J., & Bodin, J. (2015). Advances in consumer electric vehicle adoption research: A review and research agenda. *Transportation Research Part D, Transport and Environment, 34*, 122–136. doi:10.1016/j.trd.2014.10.010

Richardson, D. B. (2013). Electric vehicles and the electric grid: A review of modeling approaches, Impacts, and renewable energy integration. *Renewable & Sustainable Energy Reviews, 19*, 247–254. doi:10.1016/j.rser.2012.11.042

Tascikaraoglu, A., Boynuegri, A. R., & Uzunoglu, M. (2014). A demand side management strategy based on forecasting of residential renewable sources: A smart home system in Turkey. *Energy and Building, 80*, 309–320. doi:10.1016/j.enbuild.2014.05.042

TaymazE.YilmazK. (2017, September). *Political Economy of Industrial Policy in Turkey: The Case of the Automotive Industry.* (SSRN Paper No. 3043230). doi:10.2139srn.3043230

TEHAD. (2019). *Electric & Hybrid Cars Journal.* Electric & Hybrid Cars Association. http://tehad.org/tag/electric-hybrid-cars-dergisi/

Thomas, V. J., & Maine, E. (2019). Market entry strategies for electric vehicle start-ups in the automotive industry–lessons from tesla motors. *Journal of Cleaner Production, 235*, 653–663. doi:10.1016/j.jclepro.2019.06.284

Tomo, A. (2020). "Ecobonus" and "ecotax": two recent Italian fiscal measures to promote the decarbonisation in the vehicles system. In *Economic Instruments for a Low-carbon Future.* Edward Elgar Publishing., doi:10.4337/9781839109911.00017

Tülomsaş, A. Ş. (n.d.). *İlk Yerli Türk Otomobili: Devrim.* http://devrimarabasi.com.tr/tarihce.html

TURKSTAT. (n.d.-a). *Dış Ticaret İstatistikleri* [Data Set]. https://biruni.tuik.gov.tr/disticaretapp/disticaret.zul?param1=25&param2=0&sitcrev=0&isicrev=0&sayac=5802

TURKSTAT. (n.d.-b). *Enerji Kaynaklarına Göre Elektrik Enerjisi Üretimi ve Payları* [Data Set]. https://www.tuik.gov.tr/PreIstatistikTablo.do?istab_id=1578

U.S. Department of Energy. (n.d.). *Timeline: History of the Electric Car.* https://www.energy.gov/timeline/timeline-history-electric-car

Ustabaş, A. (2014). Mikro ve Makro Etkileri Yönünden Elektrikli Otomobiller (Türkiye Ekonomisi Örneği). *Marmara Üniversitesi İktisadi ve İdari Bilimler Dergisi, 36*(1), 269-291. http://hdl.handle.net/11424/5046

Ustabaş, A., & Simav, O. (2018). Otomotiv Endüstrisindeki Dönüşüm ve Türkiye. *The International Journal of Social Sciences (Islamabad), 211*(2), 1. Advance online publication. doi:10.31454/usb.476860

World Energy Council. (2018). *The Electric Vehicle Survey.* https://www.dunyaenerji.org.tr/wp-content/uploads/2018/09/AnketSonuclari_ElektrikliAraclarAğustos2018.pdf

Yaylalı, M., & Çalmaşur, G. (2014). Türk Otomotiv Endüstrisinde Maliyet ve Toplam Faktör Verimliliği. *Atatürk Üniversitesi Sosyal Bilimler Enstitüsü Dergisi, 18*(3), 325-350. https://dergipark.org.tr/en/pub/ataunisosbil/issue/2837/38713

Yılmaz, M. (2012). Türkiye'nin enerji potansiyeli ve yenilenebilir enerji kaynaklarının elektrik enerjisi üretimi açısından önemi. *Ankara Üniversitesi Çevrebilimleri Dergisi*, *4*(2), 33–54. doi:10.1501/Csaum_0000000064

Yılmaz, S., Tastan, K., Ecek, N., & Çınar, E. (2017). Otomotiv Sektörünün Dünyadaki ve Türkiye'deki Değişimi (Change of the Automotive Sector in the World and in Turkey). *Ordu University Journal of Social Science Research, 7*(3), 685-695. https://dergipark.org.tr/en/pub/odusobiad/issue/32288/347845

Yılmaz, S., & Ustaoğlu, M. (2013). Electric Vehicles Production in Turkish Automotive Industry and Sectoral PEST Analisys. *Procedia: Social and Behavioral Sciences*, *75*, 10–17. doi:10.1016/j.sbspro.2013.04.002

ADDITIONAL READING

Gules, H. K., Burgess, T. F., & Lynch, J. E. (1997). The evolution of buyer-supplier relationships in the automotive industries of emerging European economies: The case of Turkey. *European Journal of Purchasing & Supply Management*, *3*(4), 209–219. doi:10.1016/S0969-7012(97)00015-4

Mohamad, M., & Songthaveephol, V. (2020). Clash of titans: The challenges of socio-technical transitions in the electrical vehicle technologies–the case study of Thai automotive industry. *Technological Forecasting and Social Change*, *153*, 119772. doi:10.1016/j.techfore.2019.119772

Wells, P., & Nieuwenhuis, P. (2015). EV Business models in a wider context: balancing change and continuity in the automotive industry. In *Electric Vehicle Business Models* (pp. 3–16). Springer., doi:10.1007/978-3-319-12244-1_1

Womack, J. P. (1994). The real EV challenge: Reinventing an industry. *Transport Policy*, *1*(4), 266–270. doi:10.1016/0967-070X(94)90007-8

KEY TERMS AND DEFINITIONS

Anatolian Peninsula: The larger part of the mainland of the Republic of Turkey in the western Asia, also referred to as Asia Minor.

Battery Electric Car (BEV): A passenger car run on only electric power supplied from a battery.

Carbon Emissions: The carbon dioxide equivalent of harmful gasses transferred to the atmosphere, enabling the greenhouse gas effect which is believed to cause the global warming.

Current Account Deficit: The situation observed when the imports of a country surpasses its exports in a given year.

Global Value Chain: The organization of producing a good or service of which the efforts need to be divided among people and/or places across the world.

Plug-in Hybrid Car (PHEV): A passenger car powered with both fossil fuel and electric battery.

Power Grid: Complex and multi-layered systems being used to deliver generated electricity throughout desired territories in the contemporary world.

The Jetsons: An American cartoon show aired during the 1960s which depicted an extremely advanced technological lifestyle with flying cars and robots, created by Hanna-Barbera Productions.

Thracian Peninsula: A peninsula shared by Turkey, Bulgaria, Greece in the southeastern Europe.

TOGG: Turkish acronym for Turkey's Automobile Joint Venture Group Inc.

Compilation of References

Adel El-Shahat, E. A. (2019). Electric Vehicles Wireless Power Transfer State-of-The-Art. *Energy Procedia*, *162*, 24–37. doi:10.1016/j.egypro.2019.04.004

Akçomak, I. S., & Bürken, S. (2019). *The middle-technology trap: The case of the automotive industry in Turkey* (UNU-MERIT Working Paper Series No. 2019-006). https://www.merit.unu.edu/publications/wppdf/2019/wp2019-006.pdf

Akhila, M., & Ratnan, P. (2016). Brushless DC motor drive with regenerative braking using adaptive neuro based fuzzy inference system. *2016 International Conference on Electrical, Electronics, and Optimization Techniques (ICEEOT)*, 748-751, 10.1109/ICEEOT.2016.7754784

Akram, M.A., Liu, P., Tahir, M.O., Ali, W., & Wang, Y. (2019). Article A State Optimization Model Based on Kalman Filtering and Robust Estimation Theory for Fusion of Multi-Source Information in Highly Non-linear Systems. *Sensors*, *19*(1687), 1-22.

Al Rasasi, M., & Yilmaz, M. (2016). The effects of oil shocks on Turkish macroeconomic aggregates. *International Journal of Energy Economics and Policy*, *6*(3), 471–476. https://www.econjournals.com/index.php/ijeep/article/view/2416

Alaoui, C., & Salameh, Z. M. (2005). A novel thermal management for electric and hybrid vehicles. *IEEE Transactions on Vehicular Technology*, *54*, 468–476.

Alberto Reatti, M. K. (2018). Resonant Compensations in Inductive Wireless Power System. In A. T.-C. Aguado (Ed.), Emerging Capabilities and Applications of Wireless Power Transfer (p. 383). IGI Global.

Aljanad, A., & Mohamed, A. (2015). Impact of Plug-in Hybrid Electric Vehicle on Power Distribution System Considering Vehicle to Grid Technology: A Review. *Research Journal of Applied Sciences, Engineering and Technology*, *10*, 1404–1413. doi:10.19026/rjaset.10.1841

Almazrouei, S., Hamid, A.-K., & Shamsuzzaman, M. (2018). Predictive energy management in large-scale grid connected PV-batteries system. *2018 5th International Conference on Renewable Energy: Generation and Applications (ICREGA)*, 315–318. 10.1109/ICREGA.2018.8337601

Al-Shobaki, S., & Mohsen, M. (2008). Modeling and forecasting of electrical power demands for capacity planning. *Energy Conversion and Management*, *49*(11), 3367–3375. doi:10.1016/j.enconman.2008.05.005

Amari, M., Bacha, F., & Ghouili, J. (2015). Average model of dual active bridge interfacing super-capacitor used in electrical vehicle. *International Journal of Energy Optimization and Engineering*, *4*(1), 36–54. doi:10.4018/ijeoe.2015010103

Amrhein, M., & Krein, P. T. (2005). Dynamic simulation for analysis of hybrid electric vehicle system and subsystem interactions, including power electronics. *IEEE Transactions on Vehicular Technology*, *54*, 825–836.

Anatolian Agency. (2020, January 6). *Bakan Varank: Türkiye'nin Otomobili'nin tasarımları için tescil başvuruları yapıldı.* Anatolian Agency. https://www.aa.com.tr/tr/politika/bakan-varank-turkiyenin-otomobilinin-tasarimlari-icin-tescil-basvurulari-yapildi/1693428# Automotive

Anon Namin, E. C. (2018). Solar Tricycle with Lateral Misalignment Maximum Power Point Tracking Wireless Power Transfer. In *15th International Conference on Electrical Engineering/Electronics, Computer, Telecommunications and Information Technology* (pp. 656-659). IEEE.

Armstrong, M., Moussa, C., Adnot, J., Galli, A., & Rivière, P. (2012). Optimal Recharging Strategy for Battery-Switch Stations for Electric Vehicles in France. *Energy Policy, 60.* Advance online publication. doi:10.2139srn.2169221

Ashar, M. (2016). Integration of supercapacitor with battery using DC-DC bidirectional buck boost converter in an electric vehicle. *International Research Journal of Engineering and Technology, 3,* 1513–1517.

Ates, Y., Erdinc, O., Uzunoglu, M., & Vural, B. (2010). Energy management of an FC/UC hybrid vehicular power system using a combined neural network-wavelet transform based strategy. *International Journal of Hydrogen Energy, 35*(2), 774–783. doi:10.1016/j.ijhydene.2009.11.021

Automotive Manufacturers Association. (2019). *Otomotiv Sanayii Dış Ticaret Raporu.* http://www.osd.org.tr/sites/1/upload/files/4.Ceyrek_2019_OSD_Dis_Ticaret_Raporu-5492.pdf

Automotive Manufacturers Association. (2020). *General and Statistical Information Bulletin of Automotive Manufacturers.* http://www.osd.org.tr/sites/1/upload/files/YILLIK_2020-5479.pdf

Bampoulas, A., & Karlis, A. (2017). Provision of frequency regulation by a residential microgrid integrating PVs, energy storage and electric vehicle. *2017 IEEE International Conference on Environment and Electrical Engineering and 2017 IEEE Industrial and Commercial Power Systems Europe (EEEIC / I&CPS Europe),* 1–6. 10.1109/EEEIC.2017.7977472

Ban, S., Zhang, J., Zhang, L., Tsay, K., Song, D., & Zou, X. (2013). Charging and discharging electrochemical supercapacitors in the presence of both parallel leakage process and electro chemical decomposition of solvent. *Electrochimica Acta, 90,* 542–549. doi:10.1016/j.electacta.2012.12.056

Barfoot, T. D. (2019). *State Estimation for Robotics.* Cambridge University Press.

Barrau, A. (2015). *Non-linear state error based extended Kalman filters with applications to navigation.* Automatic.

Bertoldi, B. (2018). A Non-Ideal SEPIC DCM Modeling for LED Lighting Applications. *IEEE 4th Southern Power Electronics Conference (SPEC).* 10.1109/SPEC.2018.8635896

Beşel, F. (2017). Oil prices affect current account deficit: Empirical evidence from Turkey. *Journal of Applied Research in Finance and Economics, 3*(2), 13–21. https://www.academia.edu/download/53123395/40._Besel.pdf

Bhatti, G., & Raja Singh, R. (2020). Design of a Microgrid Architecture for Rental E-Bike Charging Stations. *IOP Conference Series: Materials Science and Engineering,* Volume 906, *Virtual Conference on Advances in Electric Drives, Process Control and Automation.*

Bildirici, M. E., & Badur, M. M. (2019). The effects of oil and gasoline prices on confidence and stock return of the energy companies for Turkey and the US. *Energy, 173,* 1234–1241. doi:10.1016/j.energy.2019.02.137

Birnie, D. P. III. (2009). Solar-to-vehicle (S2V) systems for powering commuters of the future. *Journal of Power Sources, 186*(2), 539–542. doi:10.1016/j.jpowsour.2008.09.118

Blaabjerg, F., Wang, H., Vernica, I., Liu, B., & Davari, P. (2020). Reliability of Power Electronic Systems for EV/HEV Applications. *Proceedings of the IEEE.* doi: 10.1109/JPROC.2020.3031041

Bose, B. K. (2000). Energy, environment, and advances in power electronics. *Proceedings of the International Symposium on Industrial Electronics, 1*, TU1–T14. 10.1109/ISIE.2000.930467

Botterud, A. (2017). Forecasting renewable energy for grid operations. In *Renewable Energy Integration* (pp. 133–143). Academic Press., doi:10.1016/B978-0-12-809592-8.00010-X

Bottu, M., Crow, M. L., Elmore, A. C., & Atcitty, S. (2013). A Power electronic conditioner using supercapacitors to improve wind turbine power quality. *Smart Grid and Renewable Energy, 4*(01), 69–75. doi:10.4236gre.2013.41010

Brenna, M., Foiadelli, F., & Longo, M. (2014). The Exploitation of Vehicle-to-Grid Function for Power Quality Improvement in a Smart Grid. *IEEE Transactions on Intelligent Transportation Systems, 15*(5), 2169–2177. doi:10.1109/TITS.2014.2312206

Buckley, T., Gupta, A., Garg, V., & Shah, K. (2019). *Flexing India's Energy System Making the Case for the Right Price Signals through Time-Of-Day Pricing.* Institute for Energy Economics and Financial Analysis Press Release.

Bundy, K., Karlsson, M., Lindbergh, G., & Lundqvist, A. (1998). An electrochemical impedance spectroscopy method for prediction of the state of charge of a nickel-metal hydride battery at open circuit and during discharge. *Journal of Power Sources, 72*(2), 118–125. doi:10.1016/S0378-7753(97)02695-5

Canalli, V. M. (1996). Behavioral large signal averaged model for dc/dc switching power converters. *PESC Record. 27th Annual IEEE Power Electronics Specialists Conference*, 2. 10.1109/PESC.1996.548806

Cao, B., Bai, Z., & Zhang, W. (2005). Research on control for regenerative braking of electric vehicle. *IEEE International Conference on Vehicular Electronics and Safety*, 92-97. 10.1109/ICVES.2005.1563620

Cetinkaya, H. B., Kurucay, B., Stamoulis, P., Turhan, T., Erisik, O., Ozkan, M., & Ozderli, E. (2020). A Case Study on the Use of Energy Storage in Industrial Plants with a Renewable Energy Plan. *2020 7th International Conference on Electrical and Electronics Engineering (ICEEE)*, 170–175. 10.1109/ICEEE49618.2020.9102573

Chen, M., Jochem, T., & Pomerleau, D. (1995). AURORA: A vision-based roadway departure warning system. *Proc. IEEE/RSJ IROS, 1*, 243–248.

Choi, S. Y., Gu, B. W., Jeong, S. Y., & Rim, C. T. (2015, March). Advances in Wireless Power Transfer Systems for Roadway-Powered Electric Vehicles. *IEEE Journal of Emerging and Selected Topics in Power Electronics, 3*(1), 18–36. doi:10.1109/JESTPE.2014.2343674

Chowdhury, T. (2015). Design of a Micro-Grid System in Matlab/Simulink. *International Journal of Innovative Research in Science. Engineering and Technology., 04*, 5262–5269. doi:10.15680/IJIRSET.2015.0407030

Çoban, O. (2007). Türk otomotiv sanayiinde endüstriyel verimlilik ve etkinlik. *Erciyes Üniversitesi İktisadi ve İdari Bilimler Fakültesi Dergisi, 29*, 17-36. https://dergipark.org.tr/en/pub/erciyesiibd/issue/5886/77854

Çolak, I., Bayindir, R., Aksoz, A., Hossain, E., & Sayilgan, S. (2015, October). Designing a competitive electric vehicle charging station with solar PV and storage. In *2015 IEEE International Telecommunications Energy Conference (INTELEC)* (pp. 1-6). IEEE. 10.1109/INTLEC.2015.7572480

Colmenar-Santos, A., Linares-Mena, A.-R., Borge-Diez, D., & Quinto-Alemany, C.-D. (2017). Impact assessment of electric vehicles on islands grids: A case study for Tenerife (Spain). *Energy, 120*, 385–396. doi:10.1016/j.energy.2016.11.097

Corchero, C., Cruz-Zambrano, M., Heredia, F., Cairo, J., Igualada-Gonzalez, L., & Romero-Ortega, A. (2012). Optimal sizing of microgrids: A fast charging station case. *Proceedings of 9th International Conference on the European Energy Market*, 1-6. doi: 10.1109/EEM.2012.6254677

Covic, G. A., & Boys, T. (2013). Inductive Power Transfer. *Proceedings of the IEEE, 101*(6), 1276–1289. doi:10.1109/JPROC.2013.2244536

Crolla, D. A., & Cao, D. (2012). The impact of hybrid and electric powertrains on vehicle dynamics, control systems and energy regeneration. *Vehicle System Dynamics, 50*(sup1), 95–109. doi:10.1080/00423114.2012.676651

Cross, J. D., & Hartshorn, R. (2016). My Electric Avenue: Integrating Electric Vehicles into the Electrical Networks. *6th Hybrid and Electric Vehicles Conference (HEVC 2016)*, 1–6. 10.1049/cp.2016.0972

Cuk, S., & Middlebrook, R. D. (1979). A general uniðed approach to modeling switching DC- to-DC converters in discontinuous conduction mode. *Proc. IEEE PESC*, 36–57.

Cultura, A. B., & Salameh, Z. M. (2015). Modeling, evaluation and simulation of a supercapacitor module for energy storage application. *Proceedings of the International Conference on Computer Information Systems and Industrial Application*, 876–882. 10.2991/cisia-15.2015.235

Czarkowski, D., & Kazimierczuk, M. K. (1993). Energy-conservation approach to modeling PWM DC-DC converters. *IEEE Transactions on Aerospace and Electronic Systems, 29*(3), 1059–1063. doi:10.1109/7.220955

Daida, S. R. (2019). *A Design of Inductive Coupling Wireless Power Transfer System for Electric Vehicle Applications* (Master of Science thesis). Electrical/Computer Engineering, Old Dominion University.

Dai, H., Zhu, L., Zhu, J., Wei, X., & Sun, Z. (2015). Adaptive Kalman filtering based internal temperature estimation with an equivalent electrical network thermal model for hard-cased batteries. *Journal of Power Sources, 293*, 351–365. doi:10.1016/j.jpowsour.2015.05.087

Dai, J., & Ludois, D. C. (2016). Capacitive Power Transfer Through a Conformal Bumper for Electric Vehicle Charging. *IEEE Journal of Emerging and Selected Topics in Power Electronics, 4*(3), 1015–1025. doi:10.1109/JESTPE.2015.2505622

Darab, C. (2019). Electric Vehicles in Smart Grids. In H. H. Hayek (Ed.), Handbook of Research on Smart Power System Operation and Control (p. 489). IGI Global. doi:10.4018/978-1-5225-8030-0.ch009

Dharmakeerthi, C.H., Nadarajah, M., & Saha, T. (2011). Overview of the impacts of plug-in electric vehicles on the power grid. *Innovative Smart Grid Technologies Asia (ISGT), 2011 IEEE PES.* doi:10.1109/ISGT-Asia.2011.6167115

Dirkman, J. (2015). *Microgrids: Changing the Face of Energy Output and Consumption.* Retrieved from https://electricenergyonline.com/energy/magazine/884/article/Microgrids-Changing-the-Face-of-Energy-Output-and-Consumption.htm

Distributors Association. (2020, January). *ODD Genel Değerlendirme.* http://www.odd.org.tr/folders/2837/categorial1docs/2657/Sektörel%20Değerlendirme%20Ocak%202020.pdf

Dixon, J. W., & Ortuzar, M. E. (2002). Supercapacitors + DC-DC converters in regenerative braking system. *Aerospace and Electronic Systems Magazine, IEEE, 17*(8), 16–21. doi:10.1109/MAES.2002.1028079

Ehsani, M., Gao, Y., Longo, S., & Ebrahimi, K. (2018). *Modern electric, hybrid electric, and fuel cell vehicles.* CRC Press.

Ehsani, M., Rahman, K. M., & Toliyat, H. A. (1997). Propulsion system design of electric and hybrid vehicles. *IEEE Transactions on Industrial Electronics, 44*(1), 19–27. doi:10.1109/41.557495

Electric Bike Simulator. (2013). *Power Calculator.* Retrieved from http://www.electricbikesimulator.com/index,en.html

Emadi, A., Ehsani, M., & Miller, J. M. (1999) Advanced silicon-rich automotive electrical power systems. *Proceedings of the Digital Avionics Systems Conference, 2,* 8.B.1-1–8.B.1-8.

Emadi, A., Williamson, S. S., & Khaligh, A. (2006). Power electronics intensive solutions for advanced electric, hybrid electric, and fuel cell vehicular power systems. *IEEE Transactions on Power Electronics*, *21*(3), 567–577. doi:10.1109/TPEL.2006.872378

Eng, V., & Bunlaksananusorn, C. (2009). Modeling of a SEPIC converter operating in discontinuous conduction mode. *6th International Conference on Electrical Engineering/Electronics, Computer, Telecommunications and Information Technology*, 1. 10.1109/ECTICON.2009.5136983

Erbaş, M., Kabak, M., Özceylan, E., & Çetinkaya, C. (2018). Optimal siting of electric vehicle charging stations: A GIS-based fuzzy Multi-Criteria Decision Analysis. *Energy*, *163*, 1017–1031. doi:10.1016/j.energy.2018.08.140

Erdinc, O., Tascikaraoglu, A., Paterakis, N. G., Dursun, I., Sinim, M. C., & Catalão, J. P. (2017, September). Optimal sizing and siting of distributed generation and EV charging stations in distribution systems. In 2017 IEEE PES Innovative Smart Grid Technologies Conference Europe (ISGT-Europe) (pp. 1-6). IEEE. doi:10.1109/ISGTEurope.2017.8260298

Erickson, R. W., & Dragon, M. (2007). Fundamentals of power electronics. Springer Science & Business Media.

Ertuğral, S. M. (2011). Otomotiv Sektörü Ve Gümrük Birliği Sonrasi Gelişmeleri. *Sosyal Bilimler Dergisi*, (2), 75-83. https://dergipark.org.tr/tr/pub/iusosbil/issue/9499/118702

Etherden, N., & Bollen, M. H. J. (2011). Increasing the hosting capacity of distribution networks by curtailment of renewable energy resources. *2011 IEEE Trondheim PowerTech*, 1–7. doi:10.1109/PTC.2011.6019292

European Alternative Fuels Observatory. (n.d.). *NPF Highlights* [Data Set]. https://www.eafo.eu/countries/european-union/23640/summary

European Automotive Manufacturers Association. (2019). *Economic and Market Report: EU Automotive Industry First Half of 2019*. https://www.acea.be/statistics/article/economic-and-market-report-state-of-the-eu-auto-industry-first-half-of-2019

European Commission. (2011). *Energy Roadmap 2050: impact assessment and scenario analysis*. Brussels. https://ec.europa.eu/energy/sites/ener/files/documents/roadmap2050_ia_20120430_en_0.pdf

Eurostat. (n.d.). *Passenger Cars per 1,000 Inhabitants* [Data Set]. https://appsso.eurostat.ec.europa.eu/nui/show.do?dataset=road_eqs_carhab&lang=en

EU's Future Noise Policy. (2020). *Position paper on dose response relationships between transportation noise and annoyance*. European Commission.

Farrell, J. A., & Roysdon, P. F. (2017). Advanced Vehicle State Estimation: A Tutorial and Comparative Study. *IFAC-Papers Online*, *50*(1), 15971–15976.

Farzanehfard, H., Beyragh, D. S., & Adib, E. (2008). A Bidirectional soft switched supercapacitor interface circuit for hybrid electric vehicles". *Energy Conversion and Management*, *49*(12), 3578–3584. doi:10.1016/j.enconman.2008.07.004

Fawzy, T., Premm, D., Bletterie, B., & Goršek, A. (2011). Active contribution of PV inverters to voltage control – from a smart grid vision to full-scale implementation. *E & I Elektrotechnik Und Informationstechnik*, *128*(4), 110–115. doi:10.100700502-011-0820-z

Fazeli, Zeinali, & Khajepour. (2012). Application of Adaptive Sliding Mode Control for Regenerative Braking Torque Control. *IEEE/ASME Transactions on Mechatronics*, *17*(4), 745-755.

Feng, X., Sun, J., Ouyang, M., Wang, F., He, X., Lu, L., & Peng, H. (2015). Characterization of penetration induced thermal runaway propagation process within a large format lithium ion battery module. *Journal of Power Sources, 275*, 261–273. doi:10.1016/j.jpowsour.2014.11.017

Figueiredo, R., Nunes, P., & Brito, M. C. (2017). The feasibility of solar parking lots for electric vehicles. *Energy, 140*, 1182–1197. doi:10.1016/j.energy.2017.09.024

Fletcher, S., Kirkpatrick, L., Drink, R., Puttock, R., Thring, R., & Howroyd, S. (2017). The modelling of carbon-based supercapacitors: Distributions of time constants and pascal equivalent circuits. *Journal of Power Systems, 345*, 247–253. doi:10.1016/j.jpowsour.2017.02.012

Foiadelli, F., Roscia, M., & Zaninelli, D. (2006). Optimization of storage devices for regenerative braking energy in subway systems. *2006 IEEE Power Engineering Society General Meeting*, 6. 10.1109/PES.2006.1708894

Frilli, A., Meli, E., Nocciolini, D., Pugi, L., & Rindi, A. (2016). Energetic optimization of regenerative braking for high speed railway systems. *Energy Conversion and Management, 129*, 200–215. doi:10.1016/j.enconman.2016.10.011

Fuyuan, Y., Languang, L., Yuping, Y., & He, Y. (2010). Characterization, analysis and modelling of an supercapacitor. *World Electric Vehicle Journal, 4*(2), 358–369. doi:10.3390/wevj4020358

Galus, M. D., & Andersson, G. (2008). Demand Management of Grid Connected Plug-In Hybrid Electric Vehicles (PHEV). *2008 IEEE Energy 2030 Conference*, 1–8. doi:10.1109/ENERGY.2008.4781014

Gao, H., Gao, Y., & Ehsani, M. (2001). A neural network based SRM drive control strategy for regenerative braking in EV and HEV. *IEMDC 2001. IEEE International Electric Machines and Drives Conference* (Cat. No.01EX485), 571-575. 10.1109/IEMDC.2001.939368

Gao, Y., & Ehsani, M. (2014, December). Electronic Braking System of EV and HEV-Integration of Regenerative Braking. *Automatic Braking Force Control and ABS, 3*. Advance online publication. doi:10.4271/2001-01-2478

Gargies, S., Wu, H., & Mi, C. (2006) Isolated bidirectional DC/DC converter for hybrid electric vehicle applications. *6th Intelligent Vehicle Symposium*.

George, M., & Choi, S. (2003). Contrasting Hybrid Electric Bicycles and Electric Bicycle. *Inter-Professional Project, 315*(Spring), 20–25.

Goli, P., & Shireen, W. (2014, May). PV Integrated Smart Charging of PHEVs Based on DC Link Voltage Sensing. *IEEE Transactions on Smart Grid, 5*(3), 1421–1428. doi:10.1109/TSG.2013.2286745

Görener, A., & Görener, Ö. (2008). Türk otomotiv sektörünün ülke ekonomisine katkilari ve geleceğe yönelik sektörel beklentiler. *Journal of Yaşar University, 3*(10), 1213–1232. https://dergipark.org.tr/en/download/article-file/179182

Gowda, M., Kiran, Y., & Parthasarthy, S. (2014). Modelling of Buck DC-DC Converter Using Simulink. *International Journal of Innovative Research in Science, Engineering and Technology, 3*.

Gravina, R., Alinia, P., Ghasemzadeh, H., & Fortino, G. (2017). Multi-sensor fusion in body sensor networks: State-of-the-art and research challenges. *Information Fusion, 35*, 68–80. doi:10.1016/j.inffus.2016.09.005

Grazian, F., Shi, W., Dong, J., Duijsen, P. v., Soeiro, T. B., & Bauer, P. (2019). Survey on Standards and Regulations for Wireless Charging of Electric Vehicles. In *AEIT International Conference of Electrical and Electronic Technologies for Automotive (AEIT AUTOMOTIVE)* (pp. 1-6). IEEE. 10.23919/EETA.2019.8804573

Grbovic, P. J., Delarue, P., Le Moigne, P., & Bartholomeus, P. (2010, October). A bidirectional three-level dc-dc converter for the ultra-capacitor applications. *IEEE Transactions on Industrial Electronics*, *57*(10), 3415–3430. doi:10.1109/TIE.2009.2038338

Gruosso, G. (2020). A Virtual Sensor for Electric Vehicles' State of Charge Estimation. *Electronics*, *9*(2), 278.

Güneralp, B., Lwasa, S., Masundire, H., Parnell, S., & Seto, K. C. (2017). Urbanization in Africa: Challenges and opportunities for conservation. *Environmental Research Letters*, *13*(1), 1–8. doi:10.1088/1748-9326/aa94fe

Guo, Y., Xiong, J., Xu, S., & Su, W. (2016, May). Two-Stage Economic Operation of Microgrid-Like Electric Vehicle Parking Deck. *IEEE Transactions on Smart Grid*, *7*(3), 1703–1712. doi:10.1109/TSG.2015.2424912

Habib, S. (2020). Contemporary trends in power electronics converters for charging solutions of electric vehicles. *CSEE Journal of Power and Energy Systems*, *6*(4), 911–929.

Hala, S., Khalil Faris, N., & Khaled Faris, N. (2015). Supercapacitor: Modelling and characterization for electric vehicle application. *International Journal of Engine Research*, *5*, 681–686.

Hamilton, D. B. (1996). *Electric propulsion power system - overview*. Power Electronics in Transportation. doi:10.1109/PET.1996.565906

Han, D., Noppakunkajorn, J., & Sarlioglu, B. (2014). Comprehensive Efficiency, Weight, and Volume Comparison of SiC- and Si-based Bidirectional DC-DC Converters for Hybrid Electric Vehicles. *IEEE Transactions on Vehicular Technology*, *63*(7), 3001–3010.

Harighi, T., Bayindir, R., Padmanaban, S., Mihet-Popa, L., & Hossain, E. (2018). An overview of energy scenarios, storage systems and the infrastructure for vehicle-to-grid technology. *Energies*, *11*(8), 2174. doi:10.3390/en11082174

Harish, V. S. K. V., & Kumar, A. (2014). Demand side management in India: Action plan, policies and regulations. *Renewable & Sustainable Energy Reviews*, *33*, 613–624. doi:10.1016/j.rser.2014.02.021

Harnessing Renewable Energy for Sustainable Development. (2018). *International Centre for Trade and Sustainable Development*. https://ictsd.iisd.org/bridges-news/bridges-africa/news/harnessing-renewable-energy-for-sustainable-development

Hart, D. W. (2011). *Power Electronics*. Tata McGraw-Hill Education.

Hashemi, S., & Østergaard, J. (2017). Methods and strategies for overvoltage prevention in low voltage distribution systems with PV. *IET Renewable Power Generation*, *11*(2), 205–214. doi:10.1049/iet-rpg.2016.0277

Honarmand, M., Zakariazadeh, A., & Jadid, S. (2014a). Integrated scheduling of renewable generation and electric vehicles parking lot in a smart microgrid. *Energy Convers Manag*, *86*, 745-55. . doi:10.1016/j.enconman.2014.06.044

Honarmand, M., Zakariazadeh, A., & Jadid, S. (2014b). Optimal scheduling of electric vehicles in an intelligent parking lot considering vehicle-to-grid concept and battery condition. *Energy*, *65*, 572–579. doi:10.1016/j.energy.2013.11.045

Hua, C., & Kao, S. (2011). Design and implementation of a regenerative braking system for electric bicycles based on DSP. *2011 6th IEEE Conference on Industrial Electronics and Applications*, 703-707. 10.1109/ICIEA.2011.5975676

Huang, Z., Iam, I. W., Hoi, I. U., Lam, C.-S., Mak, P.-I., & Martins, R. P. (2019). Self-Contained Solar-Powered Inductive Power Transfer System for Wireless Electric Vehicle Charging. In *IEEE PES Asia-Pacific Power and Energy Engineering Conference (APPEEC)* (pp. 1-6). Macao: IEEE. 10.1109/APPEEC45492.2019.8994554

Huet, F. (1998). A review of impedance measurements for determination of the state-of-charge or state-of-health of secondary batteries. *Journal of Power Sources*, *70*(1), 59–69. doi:10.1016/S0378-7753(97)02665-7

Huh, J., Lee, S. W., Lee, W. Y., Cho, G. H., & Rim, C. T. (2011, December). Narrow-Width Inductive Power Transfer System for Online Electrical Vehicles. *IEEE Transactions on Power Electronics, 26*(12), 3666–3679. doi:10.1109/TPEL.2011.2160972

Hutchinson, N., & Bird, L. (2019). Using Renewables for Electric Vehicle Demand: A Review of Utility Program Designs & Implementation Strategies. In *Working Paper*. World Resources Institute. Available online at https://www.wri.org/publication/renewables-for-ev-demand

Igbinovia, F. O., & Krupka, J. (2018). Renewable Energy Integration in Africa: A Case Study of the Adoption of New Technology by the Electricity Transmission Company of Nigeria. *2018 International Conference on Power System Technology (POWERCON)*, 2034–2042. 10.1109/POWERCON.2018.8602197

Inácio, C., Silva, I., Faria, J., Pombo, J., Calado, M., & Mariano, S. (2020). *Controller Design and Experimental Validation of a Solar Powered E-bike Charging Station*. KnE Engineering. doi:10.18502/keg.v5i6.7104

Indian Automobile Industry Report. (2020). India Brand Equity Foundation.

International Energy Agency. (2020, June). *Global EV Outlook 2020*. https://www.iea.org/reports/global-ev-outlook-2020

Introduction to System Integration of Renewables – Analysis. (2020). *IEA*. https://www.iea.org/reports/introduction-to-system-integration-of-renewables

Jahns, T. M., & Blasko, V. (2001). Recent advances in power electronics technology for industrial and traction machine drives. *Proceedings of the IEEE, 89*(6), 963–975. doi:10.1109/5.931496

Jannesar, M. R., Sedighi, A., Savaghebi, M., & Guerrero, J. M. (2018). Optimal placement, sizing, and daily charge/discharge of battery energy storage in low voltage distribution network with high photovoltaic penetration. *Applied Energy, 226*, 957–966. doi:10.1016/j.apenergy.2018.06.036

Javier Vázquez, P. R.-S. (2018). An Inductive Power Transfer System for the Wireless Charging of Electric Vehicles: Determination of the Magnetic Coupling Factor. In A. T.-C. Aguado (Ed.), Emerging Capabilities and Applications of Wireless Power Transfer (p. 383). IGI Global.

Jetin, B. (2020). Who will control the electric vehicle market? *International Journal of Automotive Technology and Management, 20*(2), 156–177. doi:10.1504/IJATM.2020.108584

Jiaqun, X., & Haotian, C. (2015). Regenerative brake of brushless DC motor for light electric vehicle. *2015 18th International Conference on Electrical Machines and Systems (ICEMS)*, 1423-1428. 10.1109/ICEMS.2015.7385262

Jin, L.-Q., Zheng, Y., Li, J.-H., & Liu, Y.-L. (2015). A study of novel regenerative braking system based on supercapacitor for electric vehicle driven by in-wheel motors. *Advances in Mechanical Engineering, 7*(3), 1–12. doi:10.1177/1687814015573762

Johnson, R. W., Evans, J. L., & Jacobsen, P. (2004). The changing automotive environment: High-temperature electronics. *IEEE Transactions on Electronics Packaging Manufacturing, 27*, 164–176.

Jose, C. P., & Meikandasivam, S. (2020). Optimization & Validation of Intelligent Energy Management System for Pseudo Dynamic predictive regulation of Plug-in Hybrid Electric Vehicle as Donor Clients. *Transportation, 3*(1), 1–16.

Juan, J. C. A., Javier, P. F., Juan, M. V. G., & Juan, A. C. C. (2017). Regenerative Intelligent Brake Control for Electric Motorcycles. *Energies, 10*(10), 1648. doi:10.3390/en10101648

Kabalo, M., Berthold, F., Blunier, B., Bouquain, D., Williamson, S., & Miraoui, A. (2014). Efficiency comparison of wire and wireless battery charging: Based on connection probability analysis. In *IEEE Transportation Electrification Conference and Expo (ITEC)* (pp. 1-6). IEEE. 10.1109/ITEC.2014.6861783

Kanchev, H., Lu, D., Colas, F., Lazarov, V., & Francois, B. (2011). Energy Management and Operational Planning of a Microgrid With a PV-Based Active Generator for Smart Grid Applications. *IEEE Transactions on Industrial Electronics*, *58*(10), 4583–4592. doi:10.1109/TIE.2011.2119451

Kapan, K. (2018). İz Birakmadan Kaybolmuş Bir Endüstri Mirasi: Ford'un Otomotiv Montaj Fabrikasi (1929-1944). *Journal of International Social Research*, *11*(59), 384–391. doi:10.17719/jisr.2018.2646

Karunathilake, H., Perera, P., Ruparathna, R., Hewage, K., & Sadiq, R. (2018). Renewable energy integration into community energy systems: A case study of new urban residential development. *Journal of Cleaner Production*, *173*, 292–307. doi:10.1016/j.jclepro.2016.10.067

Kaya, Ö., Alemdar, K. D., & Çodur, M. Y. (2020). A novel two stage approach for electric taxis charging station site selection. *Sustainable Cities and Society*, *62*, 102396. doi:10.1016/j.scs.2020.102396

Kaya, Ö., Tortum, A., Alemdar, K. D., & Çodur, M. Y. (2020). Site selection for EVCS in Istanbul by GIS and multi-criteria decision-making. *Transportation Research Part D, Transport and Environment*, *80*, 102271. doi:10.1016/j.trd.2020.102271

Kaygusuz, K., & Sarı, A. (2003). Renewable energy potential and utilization in Turkey. *Energy Conversion and Management*, *44*(3), 459–478. doi:10.1016/S0196-8904(02)00061-4

Keçeci, F. O. (2020, June). *Türkiye'nin Enerji Görünümü ve Aralanan Fırsat Pencereleri*. Global Political Trends Center (GPoT) Policy Brief. https://www.jstor.org/stable/resrep25190

Kempton, W., Tomic, J., Letendre, S. E., Brooks, A., & Lipman, T. (2001). *Vehicle to grid power: Battery, hybrid, and fuel cell vehicles as resources for distributed electric power in California*. Working Paper Series ECD-ITS-RR-01-03.

Kempton, W., & Tomic, J. (2005). Vehicle to grid fundamentals: Calculating capacity and net revenue. *Journal of Power Sources*, *144*(1), 268–279. doi:10.1016/j.jpowsour.2004.12.025

Khan, M. Q., & Lee, S. (2019). A Comprehensive Survey of Driving Monitoring and Assistance Systems. *Sensors (Basel)*, *19*(2), 1–32. doi:10.339019112574 PMID:31174275

Khodayar, M. E., & Wu, H. (2015). Demand Forecasting in the Smart Grid Paradigm: Features and Challenges. *The Electricity Journal*, *28*(6), 51–62. doi:10.1016/j.tej.2015.06.001

Khushalani, B., Alluri, P., Kasani, V. S., & Solanki, J. (2017). Coordinated charging strategies for plug-in hybrid electric vehicles. *2017 North American Power Symposium (NAPS)*, 1–5. 10.1109/NAPS.2017.8107264

Klettke, Moser, Bossmann, Barberi, & Fournié. (2018). *Effect of Electromobility on the Power System and the Integration of the RES-S13 Report*. European Union, METIS Studies.

Koh, S. L., & Lim, Y. (2015). *Evaluating the economic benefits of peak load shifting for building owners and grid operator.* doi:10.1109/ICSGCE.2015.7454265

Ko, J., Ko, S., Son, H., Yoo, B., Cheon, J., & Kim, H. (2015, February). Development of Brake System and Regenerative Braking Cooperative Control Algorithm for Automatic-Transmission-Based Hybrid Electric Vehicles. *IEEE Transactions on Vehicular Technology*, *64*(2), 431–440. doi:10.1109/TVT.2014.2325056

Kong, L., Li, C., Jiang, J., & Pecht, M. (2018). Li-Ion Battery Fire Hazards and Safety Strategies. *Energies*, *11*(9), 2191. doi:10.3390/en11092191

Koufakis, A.-M., Rigas, E. S., Bassiliades, N., & Ramchurn, S. D. (2020). Offline and Online Electric Vehicle Charging Scheduling With V2V Energy Transfer. *IEEE Transactions on Intelligent Transportation Systems*, *21*(5), 2128–2138. doi:10.1109/TITS.2019.2914087

Kucukgocmen, A. (2019, April 8). Honda will stop car production in Turkey after 2021: statement. *Reuters.* Retrieved from: https://www.reuters.com/article/us-turkey-honda/honda-will-stop-car-production-in-turkey-after-2021-statement-idUSKCN1RK0H6

Kuperman, A., Levy, U., Goren, J., Zafransky, A., & Savernin, A. (2013). Battery Charger for Electric Vehicle Traction Battery Switch Station. *Industrial Electronics. IEEE Transactions on.*, *60*, 5391–5399. doi:10.1109/TIE.2012.2233695

Lee, C.-M., & Lai, Y.-S. (2007). Averaged switch modeling of dc/dc converters using new switch network. *7th International Conference on Power Electronics and Drive Systems*. 10.1109/PEDS.2007.4487891

Lelie, M., Braun, T., Knips, M., Nordmann, H., Ringbeck, F., Zappen, H., & Sauer, D. (2018). Battery Management System Hardware Concepts: An Overview. *Applied Sciences (Basel, Switzerland)*, *8*(4), 534. doi:10.3390/app8040534

Li, R., Wu, J. F., Wang, H. Y., & Li, G. C. (2010). Prediction of state of charge of lithium-ion rechargeable battery with electrochemical impedance spectroscopy theory. *Proceedings of the 5th IEEE Conference on Industrial Electronics and Applications (ICIEA '10)*, 684–688.

Li, Y., Chi, Y., Wang, Z., Tang, H., Li, G., Sun, S., & Cheng, Y. (2019). Study on Fault Control Requirement and Strategy of Large-Scale Renewable Energy VSC-HVDC Integration. *2019 IEEE 3rd Conference on Energy Internet and Energy System Integration (EI2)*, 1438–1441. 10.1109/EI247390.2019.9062065

Li, Y., Tian, X., Liu, C., Su, Y., Li, L., Zhang, L., Sun, Y., & Li, J. (2017). Study on voltage control in distribution network with renewable energy integration. *2017 IEEE Conference on Energy Internet and Energy System Integration (EI2)*, 1–5. 10.1109/EI2.2017.8245755

Liao, F., Molin, E., & van Wee, B. (2017). Consumer preferences for electric vehicles: A literature review. *Transport Reviews*, *37*(3), 252–275. doi:10.1080/01441647.2016.1230794

Li, L., Wen, D., Zheng, N., & Shen, L. (2015). Cognitive Cars: A New Frontier for ADAS Research. *IEEE Transactions on Intelligent Transportation Systems*, *13*(1), 395–407. doi:10.1109/TITS.2011.2159493

Li, Q., Chen, L., Li, M., Shaw, S., & Nuchter, A. (2014). A Sensor-Fusion Drivable-Region and Lane-Detection System for Autonomous Vehicle Navigation in Challenging Road Scenarios. *IEEE Transactions on Vehicular Technology*, *63*(2), 540–555. doi:10.1109/TVT.2013.2281199

Liu, X., Aichhorn, A., Liu, L., & Li, H. (2012). Coordinated control of distributed energy storage system with tap changer transformers for voltage rise mitigation under high photovoltaic penetration. *IEEE Transactions on Smart Grid*, *3*(2), 897-906. DOI: .2177501 doi:10.1109/TSG.2011

Liu, Y., Tang, Y., Shi, J., Shi, X., Deng, J., & Gong, K. (2015). Application of Small-Sized SMES in an EV Charging Station with DC Bus and PV System. *IEEE Transactions on Applied Superconductivity*, *25*(3), 1-6. doi:10.1109/TASC.2014.2374174

Liu, K., Liu, Y., Lin, D., Pei, A., & Cui, Y. (2018). Materials for lithium-ion battery safety. *Science Advances*, *4*(6), eaas9820. doi:10.1126ciadv.aas9820 PMID:29942858

Liu, L., Kong, F., Liu, X., Peng, Y., & Wang, Q. (2015). A review on electric vehicles interacting with renewable energy in smart grid. *Renewable & Sustainable Energy Reviews*, *51*, 648–661. doi:10.1016/j.rser.2015.06.036

Liu, X., Su, C., Li, Z., & Yang, F. (2018, April). Adaptive Neural-Network-Based Active Control of Regenerative Chatter in Micromilling. *IEEE Transactions on Automation Science and Engineering*, *15*(2), 628–640. doi:10.1109/TASE.2017.2667709

Liu, Y., & Xie, J. (2015). Failure Study of Commercial LiFePO4 Cells in Overcharge Conditions Using Electrochemical Impedance Spectroscopy. *J. Electro Chem. State of Charge.*, *162*, A2208–A2217.

Liu, Z., Ahmed, Q., Zhang, J., Rizzoni, G., & He, H. (2016). Structural analysis-based sensors fault detection and isolation of cylindrical lithium-ion batteries in automotive applications. *Control Engineering Practice*, *52*, 46–58. doi:10.1016/j.conengprac.2016.03.015

Locment, F., & Sechilariu, M. (2015). Modeling and Simulation of DC Microgrids for Electric Vehicle Charging Stations. *Energies*, *8*(5), 4335–4356. doi:10.3390/en8054335

Ludios, J. a. (2015). *Wireless electric vehicle charging via capacitive power transfer through a conformal bumper. In IEEE Applied Power Electronics Conference and Exposition (APEC)*. IEEE.

Lu, F., Zhang, H., & Mi, C. (2018). A Two-Plate Capacitive Wireless Power Transfer System for Electric Vehicle Charging Applications. *IEEE Transactions on Power Electronics*, *33*(2), 964–969. doi:10.1109/TPEL.2017.2735365

Lu, L., Han, X., Li, J., Hua, J., & Ouyang, M. (2003). A review on the key issues for lithium-ion battery management in electric vehicles. *J. Power Sources, 226*, 272–288.

Lu, S., Weston, P., Hillmansen, S., Gooi, H. B., & Roberts, C. (2014, December). Increasing the Regenerative Braking Energy for Railway Vehicles. *IEEE Transactions on Intelligent Transportation Systems*, *15*(6), 2506–2515. doi:10.1109/TITS.2014.2319233

Lyu, D., Ren, B., & Li, S. (2019). Failure modes and mechanisms for rechargeable Lithium-based batteries: A state-of-the-art review. *Acta Mechanica*, *230*(3), 701–727. doi:10.100700707-018-2327-8

Manna, D., & Goswami, S. (2017). Micro-Grid Operation and Control. *Proceedings of National Conference on Recent Trends in Engineering and Technology (NCRTET-2017).*

Marzougui, H., Amari, M., Kadri, A., Bacha, F., & Ghouili, J. (2016). Energy management of fuel cell/battery/supercapacitor in electrical hybrid vehicle. *International Journal of Hydrogen Energy*, *42*(13), 8857–8869. doi:10.1016/j.ijhydene.2016.09.190

Meher, B. K., Hawaldar, I. T., Mohapatra, L., & Sarea, A. M. (2020). The Impact of COVID-19 on Price Volatility of Crude Oil and Natural Gas Listed on Multi Commodity Exchange of India. *International Journal of Energy Economics and Policy*, *10*(5), 422–431. doi:10.32479/ijeep.10047

Miao, Y., Hynan, P., von Jouanne, A., & Yokochi, A. (2019). Current Li-Ion Battery Technologies in Electric Vehicles and Opportunities for Advancements. *Energies*, *12*, 1074–1094. doi:10.3390/en12061074

Mi, C. C., Buja, G., Choi, S. Y., & Rim, C. T. (2016, October). Modern Advances in Wireless Power Transfer Systems for Roadway Powered Electric Vehicles. *IEEE Transactions on Industrial Electronics*, *63*(10), 6533–6545. doi:10.1109/TIE.2016.2574993

Michalczuk, GrzesiakLech, & Ufnalski. (2012). Alithium battery and supercapacitor hybrid energy source for an urban electric vehicle. *Politechnika Warszawska*, *4*, 158–162.

Michela Longo, M. B. (2018). Focus on OLEV: On Line Electric Vehicles. In A. T.-C. Aguado (Ed.), Emerging Capabilities and Applications of Wireless Power Transfer (p. 383). IGI Global.

Miller, J. M. (2003). Power electronics in hybrid electric vehicle applications. *Proceedings of the 18th Applied Power Electronics Conference, 1,* 23–29. 10.1109/APEC.2003.1179171

Mohamed, A., Salehi, V., Ma, T., & Mohammed, O. (2014, April). Real-Time Energy Management Algorithm for Plug-In Hybrid Electric Vehicle Charging Parks Involving Sustainable Energy. *IEEE Transactions on Sustainable Energy*, *5*(2), 577–586. doi:10.1109/TSTE.2013.2278544

Mouli, C., Ram, G., Bauer, P., & Miro, Z. (2016). System design for a solar powered electric vehicle charging station for workplaces. *Applied Energy*, *168*, 434–443. doi:10.1016/j.apenergy.2016.01.110

Moussa, W. M., & Morris, J. E. (1990). Comparison between state space averaging and PWM switch for switch mode power supply analysis. In *IEEE Technical Conference on Southern Tier*. IEEE. 10.1109/STIER.1990.324626

Moustapha Elwalaty, M. J. (2020). Modeling, Analysis, and Implementation of Series-Series Compensated Inductive Coupled Power Transfer (ICPT) System for an Electric Vehicle. *Journal of Electrical and Computer Engineering*, *2020*, 1–10. doi:10.1155/2020/9561523

Mou, X., Gladwin, D. T., Zhao, R., & Sun, H. (2019). Survey on magnetic resonant coupling wireless power transfer technology for electric vehicle charging. *IET Power Electronics*, *12*(12), 3005–3020. doi:10.1049/iet-pel.2019.0529

Mou, X., Gladwin, D. T., Zhao, R., Sun, H., & Yang, Z. (2020). Coil Design for Wireless Vehicle-to-Vehicle Charging Systems. *IEEE Access: Practical Innovations, Open Solutions*, *8*, 172723–172733. doi:10.1109/ACCESS.2020.3025787

Mukhtar, A., Xia, L., & Tang, T. B. (2015). Vehicle detection techniques for collision avoidance systems: A review. *IEEE Transactions on Intelligent Transportation Systems*, *16*(5), 2318–2338. doi:10.1109/TITS.2015.2409109

Musavi, F., Edington, M., & Eberle, W. (2012). *Wireless power transfer: A survey of EV battery charging technologies. In IEEE Energy Conversion Congress and Exposition (ECCE).* IEEE.

Mwasilu, F., Justo, J. J., Kim, E.-K., Do, T. D., & Jung, J.-W. (2014). Electric vehicles and smart grid interaction: A review on vehicle to grid and renewable energy sources integration. *Renewable & Sustainable Energy Reviews*, *34*, 501–516. doi:10.1016/j.rser.2014.03.031

Nama, J. K., Verma, A., Srivastava, M., & Tomar, P. S. (2020). An Efficient Inductive Power Transfer Topology for Electric Vehicle Battery Charging. *IEEE Transactions on Industry Applications*, *56*(6), 1–8. doi:10.1109/TIA.2020.3018419

Neagu, B. C., Gavrilas, M., Grigoras, G., & Ivanov, O. (2019). Voltage Control in Microgrids in the Presence of Small-Scale Renewable Energy Source. *2019 International Conference on Electromechanical and Energy Systems (SIELMEN)*, 1–4. 10.1109/SIELMEN.2019.8905811

Ng, K. S., Moo, C. S., Chen, Y. P., & Hsieh, Y. C. (2009). Enhanced Coulomb counting method for estimating state-of-charge and state-of health of lithium-ion batteries. *Applied Energy*, *86*(9), 1506–1511. doi:10.1016/j.apenergy.2008.11.021

Niculescu, E., & Iancu, E.-P. (2000). Modeling and analysis of the DC/DC fourth-order PWM converters. *7th Workshop on Computers in Power Electronics. Proceedings (Cat. No. 00TH8535).* 10.1109/CIPE.2000.904696

Nilsson, M., & Nykvist, B. (2016). Governing the electric vehicle transition–Near term interventions to support a green energy economy. *Applied Energy*, *179*, 1360–1371. doi:10.1016/j.apenergy.2016.03.056

NITI Aayog & World Energy Council. (2018). *Zero Emission Vehicles (ZEVs): Towards a Policy Framework.* Author.

O'Connell, N., Pinson, P., Madsen, H., & O'Malley, M. (2014). Benefits and challenges of electrical demand response: A critical review. *Renewable and Sustainable Energy Reviews, 39*, 686–699. doi:10.1016/j.rser.2014.07.098

Önder, H., & Kaya, O. C. (2019). Elektrikli Araçların Satışı Üzerinde Sosyo-Ekonomik Faktörlerin Etkisi: Bir Panel Veri Analizi. *Anemon Muş Alparslan Üniversitesi Sosyal Bilimler Dergisi, 7*(1), 17–21. doi:10.18506/anemon.376831

Onoda, S., & Emadi, A. (2004). PSIM-based modelling of automotive power systems: Conventional, electric, and hybrid electric vehicles. *IEEE Transactions on Vehicular Technology, 53*, 390–400.

Özatağan, G. (2011). Dynamics of value chain governance: Increasing supplier competence and changing power relations in the periphery of automotive production—evidence from Bursa, Turkey. *European Planning Studies, 19*(1), 77–95. do i:10.1080/09654313.2011.530393

Pallas, M. A., Kennedy, F. J., Walker, I. I. (2018). *Noise emission of electric and hybrid electric vehicles*. University of Bath.

Pang, C., Dutta, P., & Kezunovic, M. (2011). BEVs/PHEVs as dispersed energy storage for V2B uses in the smart grid. *IEEE Transactions on Smart Grid, 3*(1), 473–482. doi:10.1109/TSG.2011.2172228

Pant, P., Habib, G., Marshall, J., & Peltier, R. (2017). PM 2.5 exposure in highly polluted cities: A case study from New Delhi, India. *Environmental Research, 156*, 167–174. doi:10.1016/j.envres.2017.03.024

Parise, R. (2002). Model to predict performance of all electric transportation with wireless power beams. In *37th Intersociety Energy Conversion Engineering Conference* (pp. 1-6). IEEE.

Parker. (2015). *Electric Power-Assisted Bicycles Reduce Oil Dependence and Enhance the Mobility of the Elderly electric power assisted bicycle*. Presented at 29th Australian Transport Research Forum.

Paşalıoğlu, H., & Cengiz, H. (2019). Tüketici Yaşam Tarzının Elektrikli Araçları Satın Alma Niyeti Üzerine Etkisi: Tüketici Yenilikçiliği, Tüketici Benzersizliği Ve Keşifsel Davranışların Aracılık Etkisi. *Proceedings of the International Congress on Business and Marketing*. https://hdl.handle.net/20.500.12415/4676

Patel, K., Das, N., & Khan, M. M. K. (2019). Optimization of Hybrid Solar, Wind and Diesel Energy Systems from Economic Point of View. *2019 29th Australasian Universities Power Engineering Conference (AUPEC)*, 1–6. 10.1109/AUPEC48547.2019.211861

Patterson, B. T. (2012, November-December). DC, Come Home: DC Microgrids and the Birth of the "Enernet". *IEEE Power & Energy Magazine, 10*(6), 60–69. doi:10.1109/MPE.2012.2212610

Pay, S., & Baghzouz, Y. (2003). Effectiveness of battery-supercapacitor combination in electric vehicles. *Power Tech Conference Proceedings, 2003 IEEE Bologna, 3*, 23-26. 10.1109/PTC.2003.1304472

Payman, A., Pierfederici, S., & Meibody-Tabar, F. (2008). Energy control of supercapacitor/fuel cell hybrid power source. *Energy Conversion and Management, 49*(6), 1637–1644. doi:10.1016/j.enconman.2007.11.012

Peças Lopes, J. A., Polenz, S. A., Moreira, C. L., & Cherkaoui, R. (2010). Identification of control and management strategies for LV unbalanced microgrids with plugged-in electric vehicles. *Electric Power Systems Research, 80*(8), 898–906. doi:10.1016/j.epsr.2009.12.013

Petrol Planning and Analysis Cell. (2020). *Crude Oil Price (Indian Basket)*. Retrieved from https://www.ppac.gov.in/content/149_1_PricesPetroleum.aspx

Piaggio & C. (2018). *Vespa Elettrica*. Retrieved from https://elettrica.vespa.com/_common/pdf/EN-DIGITAL-Vespa-Elettrica-10-2018.pdf

Pillai, A. M., Kanna, R. R., & Raja Singh, R. (2020). Advancement of Inventive Solar Power Based Frameworks for Rural India. *IOP Conference Series: Materials Science and Engineering*, Volume 906, *Virtual Conference on Advances in Electric Drives, Process Control and Automation.*

Pinson, P., Christensen, L.E.A., Madsen, H., Sørensen, P.E., Donovan, M.H., & Jensen, L.E. (2008). Regime switching modelling of the fluctuations of offshore wind generation. *Journal of Wind Engineering & Industrial Aerodynamics, 96*(2), 2327-2347. DOI: .03.010 doi:10.1016/j.jweia.2008

Plett, G. L. (2004). Extended Kalman filtering for battery management systems of Li PB-based HEV battery packs—Part 1. Background. *Journal of Power Sources, 134*(2), 252–261. doi:10.1016/j.jpowsour.2004.02.031

Plett, G. L. (2015). Battery management systems: Vol. I. *Battery modeling*. Artech House.

Prakash, M. M., & Paul, J. (2016, May). Control of Microgrid for Different Modes of Operation. *International Journal of Engineering Research & Technology (Ahmedabad), 5*(5). https://dx.doi.org/10.17577/IJERTV5IS051001

Prasanth, B., Sai Surya, G., Sai Vinay, G., Deepa, K., Sailaja, V., & Manitha, P. V. (2021). *Placement of Distribution generators in IEEE 14 bus system with consumer benefit maximization* (Vol. 711). Advances in Electrical and Computer Technologies.

Prasanth, B., Sanju Srivasthav, B., Srinivas Reddy, Y. S., & Deepa, K. (2020). Estimation of regenerative Braking Force in Electric Vehicles for maximum energy recovery. *17th IEEE India Council International conference*, 1-4.

Puscasiu, A., Fanca, A., Gota, D.-I., & Valean, H. (2019). Dynamic Routing of Traffic Based on Greenhouse Gases Measurements. *2019 23rd International Conference on System Theory, Control and Computing (ICSTCC),* 549–553. 10.1109/ICSTCC.2019.8886113

Qiu, C., Chau, K., Liu, C., & Chan, C. (2013). Overview of wireless power transfer for electric vehicle charging. In *2013 World Electric Vehicle Symposium and Exhibition (EVS27)* (pp. 1-9). Spain: IEEE. 10.1109/EVS.2013.6914731

Rahbari, O., Vafaeipour, M., Omar, N., Rosen, M. A., Hegazy, O., Timmermans, J.-M., Heibati, S., & Bossche, P. V. D. (2017). An optimal versatile control approach for plug-in electric vehicles to integrate renewable energy sources and smart grids. *Energy, 134*, 1053–1067. doi:10.1016/j.energy.2017.06.007

Rahbar, K., Chai, C. C., & Zhang, R. (2018). Energy Cooperation Optimization in Microgrids With Renewable Energy Integration. *IEEE Transactions on Smart Grid, 9*(2), 1482–1493. doi:10.1109/TSG.2016.2600863

Rahman, M. A., & Hoque, M. A. (1998, December). On-line adaptive artificial neural network based vector control of permanent magnet synchronous motors. *IEEE Transactions on Energy Conversion, 13*(4), 311–318. doi:10.1109/60.736315

Rahman, M. F. (2004). Power electronics and drive applications for the automotive industry, *Proceedings of the Power Electronics Systems and Applications Conference,* 156–164.

Raja Singh, R., Yash, S. M., Chavan, S. S., Indragandhi, V., Vijayakumar, V., Saravanan, P., & Subramaniyaswamy, V. (2020, November). IoT embedded cloud-based intelligent power quality monitoring system for industrial drive application. *Future Generation Computer Systems, 112*, 884–898.

Rajashekara, K. (2003). Power electronics applications in electric/hybrid vehicles. *Proceedings of IECON'03, 3*, 3029–3030.

Rauwald, C. (2020, July 1). VW Abandons Turkey Factory Plan After Car Sales Slump. *Bloomberg*. Retrieved from: https://www.bloomberg.com/news/articles/2020-07-01/vw-abandons-turkey-factory-plan-after-virus-hits-industry

Raviteja, K., Kar, P. K., & Karanki, S. B. (2018). Renewable Energy Resources Integration To Grid With Improved Power Quality Capabilities And Optimal Power Flows. *2018 IEEE International Conference on Power Electronics, Drives and Energy Systems (PEDES)*, 1–6. 10.1109/PEDES.2018.8707646

Rehmani, M. H., Reisslein, M., Rachedi, A., Erol-Kantarci, M., & Radenkovic, M. (2018). Integrating Renewable Energy Resources Into the Smart Grid: Recent Developments in Information and Communication Technologies. *IEEE Transactions on Industrial Informatics*, *14*(7), 2814–2825. doi:10.1109/TII.2018.2819169

Resmi Gazete. (2020, August 30). *4760 Sayılı Özel Tüketim Vergisi Kanununa Ekli (II) Sayılı Listede Yer Alan Bazı Malların Özel Tüketim Vergisi Oranlarının Yeniden Tespiti Hakkında Karar (Karar Sayısı: 2912)*. Presidency of The Turkish Republic. https://www.resmigazete.gov.tr/eskiler/2020/08/20200830-1.pdf

Revenue Administration. (2019, July 1). *Private Consumption Tax Amounts and Rates*. Presidency of The Turkish Republic. Retrieved from: https://www.gib.gov.tr/sites/default/files/fileadmin/mevzuatek/otv_oranlari_tum/ozeltuketimoranlari-OpenPage.htm

Rezvani, Z., Jansson, J., & Bodin, J. (2015). Advances in consumer electric vehicle adoption research: A review and research agenda. *Transportation Research Part D, Transport and Environment*, *34*, 122–136. doi:10.1016/j.trd.2014.10.010

Richardson, D. B. (2013). Electric vehicles and the electric grid: A review of modeling approaches, Impacts, and renewable energy integration. *Renewable & Sustainable Energy Reviews*, *19*, 247–254. doi:10.1016/j.rser.2012.11.042

Rizvi, S. A. A., Xin, A., Masood, A., Iqbal, S., Jan, M. U., & Rehman, H. (2018). Electric Vehicles and their Impacts on Integration into Power Grid: A Review. *2018 2nd IEEE Conference on Energy Internet and Energy System Integration (EI2)*, 1–6. 10.1109/EI2.2018.8582069

Rodney, T. (2020). *MPPT Solar Charge Controller Model*. MATLAB Central File Exchange. Retrieved June 26, 2020, https://www.mathworks.com/matlabcentral/fileexchange/73115-mppt-solar-charge-controller-model

Rodrigues, S., & Munichandraiah, N. (2000). A review of state-of-charge indication of batteries by means of A.C. impedance measurements. *Journal of Power Sources*, *87*(1-2), 12–20.

Roy, J. V., Leemput, N., Geth, F., Büscher, J., Salenbien, R., & Driesen, J. (2014, October). Electric Vehicle Charging in an Office Building Microgrid with Distributed Energy Resources. *IEEE Transactions on Sustainable Energy*, *5*(4), 1389–1396. doi:10.1109/TSTE.2014.2314754

S, P. S. (2020). Wireless Power Transfer Topologies used for Static and Dynamic Charging of EV Battery: A Review. *International Journal of Emerging Electric Power Systems*, *21*(1), 1-34.

Sandy Thomas, C. E. (2009). Transportation options in a carbon-constrained world: Hybrids, plug-in hybrids, biofuels, fuel cell electric vehicles, and battery electric vehicles. *International Journal of Hydrogen Energy*, *34*(23), 9279–9296. doi:10.1016/j.ijhydene.2009.09.058

Saransch Chhawchharia, S. K. (2018). Investigation of wireless power transfer applications with a focus on renewable energy. *Renewable & Sustainable Energy Reviews*, *91*, 888–902. doi:10.1016/j.rser.2018.04.101

Šare, A., Krajačić, G., Pukšec, T., & Duić, N. (2015). The integration of renewable energy sources and electric vehicles into the power system of the Dubrovnik region. *Energy, Sustainability and Society*, *5*(1), 27–34. doi:10.118613705-015-0055-7

Sarkar, T., Dan, A. K., & Ghosh, S. (2016). Effect of X/R ratio on low voltage distribution system connected with constant speed wind turbine. *2nd International Conference on Control, Instrumentation, Energy and Communication, 417-421.*

Sasikumar, S., & Deepa, K. (2019). *Comparative Study of LCL-S and LCC-S Topology of Wireless EV charging System. In Innovations in Power and Advanced Computing Technologies (i-PACT)*. Vellore.

Sato, S., & Kawamura, A. (2002). A new estimation method of state of charge using terminal voltage and internal resistance for lead acid battery. *Proceedings of the Power Conversion Conference*, 565–570. 10.1109/PCC.2002.997578

Schneuwly, Maher, & Auer. (n.d.). *Supercapacitors, the New Thinking in the Automotive World.* Maxwell Technologies Inc.

Schuller, A., Flath, C. M., & Gottwalt, S. (2015). Quantifying load flexibility of electric vehicles for renewable energy integration. *Applied Energy*, *151*, 335–344. doi:10.1016/j.apenergy.2015.04.004

Seddig, K., Jochem, P., & Fichtner, W. (2017). Integrating renewable energy sources by electric vehicle fleets under uncertainty. *Energy*, *141*, 2145–2153. doi:10.1016/j.energy.2017.11.140

Seki, H., Ishihara, K., & Tadakuma, S. (2009, May). Novel Regenerative Braking Control of Electric Power-Assisted Wheelchair for Safety Downhill Road Driving. *IEEE Transactions on Industrial Electronics*, *56*(5), 1393–1400. doi:10.1109/TIE.2009.2014747

Şengör, İ., Kılıçkıran, H. C., Akdemir, H., Kekezoglu, B., Erdinc, O., & Catalao, J. P. S. (2018, July). Energy Management of a Smart Railway Station Considering Regenerative Braking and Stochastic Behaviour of ESS and PV Generation. *IEEE Transactions on Sustainable Energy*, *9*(3), 1041–1050. doi:10.1109/TSTE.2017.2759105

Sensfus, F., Ragwitz, M., & Genoese, M. (2008). The merit-order effect: A detailed analysis of the price effect of renewable electricity generation on spot market prices in Germany. *Energy Policy*, *36*(8), 3086–3094. doi:10.1016/j.enpol.2008.03.035

Sergeant, P., & Bossche, A. V. (2008). Inductive coupler for contactless power transmission. *IET Electric Power Applications*, *2*(1), 1–7. doi:10.1049/iet-epa:20070059

Shabarish, P. R., Krishna, R. P., Prasanth, B., Deepa, K., Manitha, P. V., & Sailaja, V. (2020). Fuzzy based Approach to Enhance the Wireless Charging System in Electric Vehicles. *2020 4th International Conference on Electronics, Communication and Aerospace Technology (ICECA)*, 176-181.

Shetty, S. S., & Karabasoglu, O. (2014). Regenerative Braking Control Strategy for Hybrid and Electric Vehicles Using Artificial Neural Networks. In V. Mladenov, C. Jayne, & L. Iliadis (Eds.), *Engineering Applications of Neural Networks. EANN 2014. Communications in Computer and Information Science* (Vol. 459). Springer. doi:10.1007/978-3-319-11071-4_10

Shi, L., & Crow, M. L. (2008). Comparison of supercapacitor electric circuit models. *Proceedings of the IEEE Power and Energy Society 2008 General Meeting: Conversion and Delivery of Electrical Energy in the 21st Century.*

Shubhangi Das, K. P. (2018). Wireless Power Transfer in Electric Vehicles. *International Journal of Applied Environmental Sciences*, *13*(7), 643–659.

Siano, P. (2014). Demand response and smart grids—A survey. *Renewable & Sustainable Energy Reviews*, *30*, 461–478. doi:10.1016/j.rser.2013.10.022

Simaremare, A. A., Tambunan, H. B., Pramana, P. A. A., & Munir, B. S. (2018). Optimization of Renewable Energy Penetration for Microgrid System – Case Study of Tomia Island. *2018 3rd International Conference on Information Technology, Information System and Electrical Engineering (ICITISEE)*, 180–183. 10.1109/ICITISEE.2018.8720965

Simoes, M. G., Bose, B. K., & Spiegel, R. J. (1997, January). Fuzzy logic based intelligent control of a variable speed cage machine wind generation system. *IEEE Transactions on Power Electronics*, *12*(1), 87–95. doi:10.1109/63.554173

Sivarajan, N., Mathew, N.P., Gowda, N.D., & Thomas, N. (2018). Rising Fuel Prices in Bangalore - Causes and Impact. *International Journal of Research and Analytical Reviews, 5*(4).

Skunpong, R., & Plangklang, B. (2011). A Practical Method for Quickly PV Sizing. *Procedia Engineering*, *8*, 120–127. doi:10.1016/j.proeng.2011.03.022

Snehalika, & Das, S. (2019). Emerging trends in Electric Vehicle in Indian market. *2019 3rd International Conference on Recent Developments in Control, Automation & Power Engineering (RDCAPE),* 514–518. doi:10.1109/RDCAPE47089.2019.8979012

Son, H., & Kim, H. (2016). Development of Near Optimal Rule-Based Control for Plug-In Hybrid Electric Vehicles Taking into Account Drivetrain Component Losses. *Energies*, *9*(420), 1-18.

Stefan, S. (2009). *Power Sources for Hybrid Electric Vehicles* [M. thesis]. Lund University, Faculty of Engineering.

Strbac, G. (2008). Demand side management: Benefits and challenges. *Energy Policy*, *36*(12), 4419–4426. doi:10.1016/j.enpol.2008.09.030

Strunz, K., Abbasi, E., & Huu, D. N. (2014, March). DC Microgrid for Wind and Solar Power Integration. *IEEE Journal of Emerging and Selected Topics in Power Electronics*, *2*(1), 115–126. doi:10.1109/JESTPE.2013.2294738

Surya, Chennegowda, & Naraharisetti. (2020). Generalised Circuit Averaging Technique for Two Switch DC –DC Converters. In *8th International Conference on Power Electronics Systems and Applications.* PESA.

Surya, Chennegowda, Datar, Jha, & Victor. (2020, Dec.). *Accurate Battery Modeling Based on Pulse Charging using MATLAB / Simulink.* Power Electronics Drives and Energy System, PEDES.

Surya, S. (2020). Development of thermal model for estimation of core temperature of batteries. *International Journal of Emerging Electric Power Systems.*

Surya, S., & Arjun, M. N. (2020). Effect of fast discharge of a battery on its core temperature. In *International Conference on Futuristic Technologies in Control Systems & Renewable Energy (ICFCR).* IEEE. 10.1109/ICFCR50903.2020.9249999

Surya, S., Marcis, V., & Williamson, S. S. (2021). Core Temperature Estimation for a Lithium ion 18650 Cell. *Energies*, *14*(1), 87. doi:10.3390/en14010087

Surya, S., & Williamson, S. (2021). Generalized Circuit Averaging Technique for Two-Switch PWM DC-DC Converters in CCM. *Electronics (Basel)*, *10*(4), 392.

Takagi, M., Iwafune, Y., Yamaji, K., Yamamoto, H., Okano, K., Hiwatari, R., & Ikeya, T. (2013). Economic value of PV energy storage using batteries of battery-switch. *IEEE Transactions on Sustainable Energy*, *4*(1), 164–173. doi:10.1109/TSTE.2012.2210571

Takahashi, Y., & Seki, H. (2009). Fuzzy logic based regenerative braking control system of electric wheelchair for senior citizen. *2009 IEEE International Conference on Rehabilitation Robotics*, 725-730. 10.1109/ICORR.2009.5209589

Tan, K. M., Ramachandaramurthy, V. K., & Yong, J. Y. (2016). Integration of electric vehicles in smart grid: A review on vehicle to grid technologies and optimization techniques. *Renewable & Sustainable Energy Reviews*, *53*, 720–732. doi:10.1016/j.rser.2015.09.012

Tascikaraoglu, A., Boynuegri, A. R., & Uzunoglu, M. (2014). A demand side management strategy based on forecasting of residential renewable sources: A smart home system in Turkey. *Energy and Building*, *80*, 309–320. doi:10.1016/j.enbuild.2014.05.042

TaymazE.YilmazK. (2017, September). *Political Economy of Industrial Policy in Turkey: The Case of the Automotive Industry.* (SSRN Paper No. 3043230). doi:10.2139srn.3043230

TEHAD. (2019). *Electric & Hybrid Cars Journal.* Electric & Hybrid Cars Association. http://tehad.org/tag/electric-hybrid-cars-dergisi/

Thirugnanam, K., Saini, H., & Kumar, P. (2012). Mathematical modeling of Li-ion battery for charge/discharge rate and capacity fading characteristics using genetic algorithm approach. *2012 IEEE Transportation Electrification Conference and Expo. ITEC, 2012*, 1–6. doi:10.1109/ITEC.2012.6243431

Thiruvonasundari & Deepa. (2020b). Active cell balancing for electric vehicle battery management. *International Journal of Power Electronics and Drives System, 11*(2), 571-579.

Thiruvonasundari, & Deepa. (2020a). Electric vehicle battery modelling methods based on state of charge-review. *Journal of Green Engineering, 10*, 24–61.

Thomas, V. J., & Maine, E. (2019). Market entry strategies for electric vehicle start-ups in the automotive industry–lessons from tesla motors. *Journal of Cleaner Production, 235*, 653–663. doi:10.1016/j.jclepro.2019.06.284

Tomo, A. (2020). "Ecobonus" and "ecotax": two recent Italian fiscal measures to promote the decarbonisation in the vehicles system. In *Economic Instruments for a Low-carbon Future*. Edward Elgar Publishing., doi:10.4337/9781839109911.00017

Traci, R. M., Acebal, R., & Mohler, T. (1999). Integrated thermal management of a hybrid electric vehicle. *IEEE Transactions on Magnetics, 35*, 479–483.

Tuite, D. (2007). *Get the Lowdown on Supercapacitors*. Technology report electronic design online. http://electronicdesign.com/Articles/ArticleID/17465/17465.html

Tülomsaş, A. Ş. (n.d.). *İlk Yerli Türk Otomobili: Devrim.* http://devrimarabasi.com.tr/tarihce.html

TURKSTAT. (n.d.-a). *Dış Ticaret İstatistikleri* [Data Set]. https://biruni.tuik.gov.tr/disticaretapp/disticaret.zul?param1=25&param2=0&sitcrev=0&isicrev=0&sayac=5802

TURKSTAT. (n.d.-b). *Enerji Kaynaklarına Göre Elektrik Enerjisi Üretimi ve Payları* [Data Set]. https://www.tuik.gov.tr/PreIstatistikTablo.do?istab_id=1578

U.S. Department of Energy. (n.d.). *Timeline: History of the Electric Car.* https://www.energy.gov/timeline/timeline-history-electric-car

Uski, S., Rinne, E., & Sarsama, J. (2018). Microgrid as a Cost-Effective Alternative to Rural Network Underground Cabling for Adequate Reliability. *Energies, 11*, 1978. doi:10.3390/en11081978

Ustabaş, A. (2014). Mikro ve Makro Etkileri Yönünden Elektrikli Otomobiller (Türkiye Ekonomisi Örneği). *Marmara Üniversitesi İktisadi ve İdari Bilimler Dergisi, 36*(1), 269-291. http://hdl.handle.net/11424/5046

Ustabaş, A., & Simav, O. (2018). Otomotiv Endüstrisindeki Dönüşüm ve Türkiye. *The International Journal of Social Sciences (Islamabad), 211*(2), 1. Advance online publication. doi:10.31454/usb.476860

Uthaya, N., & Banu, U. A. (2018). Wireless Power Transfer in Electrical Vehicle by Using Solar Energy. *Asian Journal of Electrical Sciences, 7*(1), 6–9.

Van der Kam, M., & van Sark, W. (2015). Smart charging of electric vehicles with photovoltaic power and vehicle-to-grid technology in a microgrid; a case study. *Applied Energy, 152*, 20–30. doi:10.1016/j.apenergy.2015.04.092

VandenBos, G., Knapp, S., & Doe, J. (2001). *Primary Inductor Selection by IET Labs.* https://www.ietlabs.com/pdf/Datasheets/1482.pdf

Venkatasubramanian, VRengaswamy, RYin, KKavuri, S.N. (n.d.). A review of process fault detection and diagnosis. *Computers & Chemical Engineering*, *27*, 293–311.

Vezzini, A., & Reichert, K. (1996). *Power electronics layout in a hybrid-electric or electric vehicle drive system.* Power Electronics in Transportation. doi:10.1109/PET.1996.565910

Vishvendra Singh, A. P. (2014). Introduction to Wireless Power Transmission. *HCTL Open International Journal of Technology Innovations and Research*, *8*, 1–10.

Vita, V., & Koumides, P. (2019). Electric Vehicles and Distribution Networks: Analysis on Vehicle to Grid and Renewable Energy Sources Integration. *2019 11th Electrical Engineering Faculty Conference (BulEF)*, 1–4. 10.1109/BulEF48056.2019.9030787

Wang, H., Lou, X., Cai, Y., Li, Y., & Chen, L. (2019). Real-Time Vehicle Detection Algorithm Based on Vision and Lidar Point Cloud Fusion. *Journal of Sensors*, *1*(7), 1–9. doi:10.1155/2019/8473980

Wang, W., Zhao, D., Xi, J., & Han, W. (2018). A learning based approach for lane departure warning systems with a personalised driver model. *IEEE Transactions on Vehicular Technology*, *1*(3), 1–13.

Wang, Z., Wang, L., Dounis, A., & Yang, R. (2012). Integration of plug-in hybrid electric vehicles into energy and comfort management for smart building. *Energy and Building*, *47*, 260–266. doi:10.1016/j.enbuild.2011.11.048

Weigelt, M., Mayr, A., Masuch, M., Batz, K., Franke, J., Bican, P. M., Brem, A., Russer, J., & Russer, P. (2018). Techno-Economic Evaluation of Strategic Solutions to Extend the Range of Electric Vehicles. *2018 8th International Electric Drives Production Conference (EDPC)*, 1–7. 10.1109/EDPC.2018.8658329

Wei, J., Dong, G., & Chen, Z. (2019). Model based fault diagnosis of Lithium ion battery using strong tracking Extended Kalman Filter. *Energy Procedia*, *158*, 2500–2505. doi:10.1016/j.egypro.2019.01.391

White, S. B., Gallego, N. C., & Johnson, D. D. (2004). *Graphite foam for cooling of automotive power electronics.* Power Electronics in Transportation.

Wicked Ride Adventure Services Pvt Ltd. (2020). *Bounce™*. Retrieved from https://bounceshare.com/

Widjanarka, W., & Sutiyatno, S. (2020). Experiment electrical motorcycle used solar energy with wireless transfer energy method for transportation in isolated regions. *Journal of Physics: Conference Series*, *1517*, 012098. doi:10.1088/1742-6596/1517/1/012098

Williamson, S. S. (2013). *Energy management strategies for electric and plug-in hybrid electric vehicles.* Springer. doi:10.1007/978-1-4614-7711-2

Williamson, S., Lukic, M., & Emadi, A. (2006). Comprehensive drive train efficiency analysis of hybrid electric and fuel cell vehicles based on motor-controller efficiency modelling. *IEEE Transactions on Power Electronics*, *21*, 730–740.

Wischenewski, A., Stahl, T., Betz, J., & Lohmann, B. (2019). Vehicle Dynamics State Estimation and Localization for High Performance Race Cars. *IFAC-PapersOnLine*, *52*(8), 154–161. doi:10.1016/j.ifacol.2019.08.064

World Energy Council. (2018). *The Electric Vehicle Survey.* https://www.dunyaenerji.org.tr/wp-content/uploads/2018/09/AnketSonuclari_ElektrikliAraclarAğustos2018.pdf

World Health Organization. (2009). *Global Status Report on Road Safety: Time for action.* WHO.

Wu, C. H., & Xu, Y. X. (2019). The Optimal Control of Fuel Consumption for a Heavy-duty Motorcycle with Three Power Sources Using Hardware-in-the-Loop Simulation. *Energies*, *13*, 22. doi:10.3390/en13010022

Wu, C., Zhu, C., Ge, Y., & Zhao, Y. (2018). A Review on Fault Mechanism and Diagnosis Approach for Li-Ion Batteries. *Journal of Nanomaterials*, 1–9.

Wu, J.-F., Hu, N.-S., Hu, S., & Zhao, Y. (2002). Application of SOM neural network in fault diagnosis of the steam turbine regenerative system. *Proceedings. International Conference on Machine Learning and Cybernetics*, 184-187. 10.1109/ICMLC.2002.1176735

Wu, J., & Luo, Z. (2017). A new uncertain analysis method and its application in vehicle dynamics. *Mechanical Systems and Signal Processing*, *50–51*, 659–675.

Xiao, B., Lu, H., Wang, H., Ruan, J., & Zhang, N. (2017, November). Enhanced Regenerative Braking Strategies for Electric Vehicles: Dynamic Performance and Potential Analysis. *Energies*, *10*(11), 1875. doi:10.3390/en10111875

Xiao, B., Zeng, X., Zhang, Z., Li, B., & Yang, Y. (2016). Present status and prospect of the regenerative-braking control strategies of electric automobiles. *Auto Electr. Parts*, *12*, 1–3.

Xia, Q., Chen, L., Xu, X., Cai, Y., Jiang, H., Chen, T., & Pan, G. (2016). Running States Estimation of Autonomous Four-Wheel Independent Drive Electric Vehicle by Virtual Longitudinal Force Sensors. *Mathematical Problems in Engineering*, *1*(1), 1-17.

Xiong, R., Yu, Q., & Shen, W. (2018). Review on sensors fault diagnosis and fault-tolerant techniques for lithium ion batteries in electric vehicles. *Proceedings of the 2018 13th IEEE Conference on Industrial Electronics and Applications (ICIEA)*, 406–410. 10.1109/ICIEA.2018.8397751

Xi, Z., Dahmardeh, M., Xia, B., Fu, Y., & Mi, C. (2019). Learning of battery model bias for effective state of charge estimation of lithium-ion batteries. *IEEE Transactions on Vehicular Technology*, *68*(9), 8613–8628. doi:10.1109/TVT.2019.2929197

Xu, G., Li, W., Xu, K., & Song, Z. (2011). An Intelligent Regenerative Braking Strategy for Electric Vehicles. *Energies*, *4*(4), 1461–1477. doi:10.3390/en4091461

Xu, G., Xu, K., Zheng, C., Zhang, X., & Zahid, T. (2016, March). Fully Electrified Regenerative Braking Control for Deep Energy Recovery and Maintaining Safety of Electric Vehicles. *IEEE Transactions on Vehicular Technology*, *65*(3), 1186–1198. doi:10.1109/TVT.2015.2410694

Yadav, T., & Yadav, D. (2018). Integration of Electric Vehicle and Renewable Generation to Improve the Power Quality Problems in Distribution System. *2018 5th IEEE Uttar Pradesh Section International Conference on Electrical, Electronics and Computer Engineering (UPCON)*, 1–5. 10.1109/UPCON.2018.8596941

Yang, C.-J. (2017). Opportunities and barriers to demand response in China. *Resources, Conservation and Recycling*, *121*, 51–55. doi:10.1016/j.resconrec.2015.11.015

Yang, Y., Ali, K. A., Roeleveld, J., & Emadi, A. (2016). State-of-the-art electrified powertrains - hybrid, plug-in, and electric vehicles. *International Journal of Powertrains*, *5*(1), 1. doi:10.1504/IJPT.2016.075181

Yan, L., Yongning, C., Xinshou, T., Zhankui, Z., & Haoyong, C. (2019). The Experiences and Practices on Market with Large-Scale Renewable Energy Grid Integration. *2019 IEEE Sustainable Power and Energy Conference (ISPEC)*, 68–71. 10.1109/iSPEC48194.2019.8975174

Yao, J., Zhong, Z., & Sun, Z. (2006). A Fuzzy Logic Based Regenerative Braking Regulation For a Fuel Cell Bus. *2006 IEEE International Conference on Vehicular Electronics and Safety*, 22-25. 10.1109/ICVES.2006.371546

Yaylalı, M., & Çalmaşur, G. (2014). Türk Otomotiv Endüstrisinde Maliyet ve Toplam Faktör Verimliliği. *Atatürk Üniversitesi Sosyal Bilimler Enstitüsü Dergisi*, *18*(3), 325-350. https://dergipark.org.tr/en/pub/ataunisosbil/issue/2837/38713

Yesilbudak, M., & Colak, A. (2018). Integration Challenges and Solutions for Renewable Energy Sources, Electric Vehicles and Demand-Side Initiatives in Smart Grids. *2018 7th International Conference on Renewable Energy Research and Applications (ICRERA),* 1407–1412. 10.1109/ICRERA.2018.8567004

Yılmaz, S., Tastan, K., Ecek, N., & Çınar, E. (2017). Otomotiv Sektörünün Dünyadaki ve Türkiye'deki Değişimi (Change of the Automotive Sector in the World and in Turkey). *Ordu University Journal of Social Science Research, 7*(3), 685-695. https://dergipark.org.tr/en/pub/odusobiad/issue/32288/347845

Yılmaz, M. (2012). Türkiye'nin enerji potansiyeli ve yenilenebilir enerji kaynaklarının elektrik enerjisi üretimi açısından önemi. *Ankara Üniversitesi Çevrebilimleri Dergisi, 4*(2), 33–54. doi:10.1501/Csaum_0000000064

Yılmaz, S., & Ustaoğlu, M. (2013). Electric Vehicles Production in Turkish Automotive Industry and Sectoral PEST Analisys. *Procedia: Social and Behavioral Sciences, 75*, 10–17. doi:10.1016/j.sbspro.2013.04.002

Yoong, M. K. (2010). Studies of regenerative braking in electric vehicle. *2010 IEEE Conference on Sustainable Utilization and Development in Engineering and Technology*, 40-45. 10.1109/STUDENT.2010.5686984

Young, M. A. R., John, D. L., Kristie, Y., & John, D. L. (2019). Driver Distraction: Theory, Effects, and Mitigation. CRC Press.

Yuan, Y., Lu, Y., & Wang, Q. (2019). Adaptive Forward Vehicle Collision Warning Based on Driving Behavior. *Neurocomputing, 408*, 64–71. doi:10.1016/j.neucom.2019.11.024

Zhang, H., Xu, G., Li, W., & Zhou, M. (2012). Fuzzy logic control in regenerative braking system for electric vehicle. *2012 IEEE International Conference on Information and Automation*, 588-591. 10.1109/ICInfA.2012.6246881

Zhang, J., Kong, D., Chen, L., & Chen, X. (2012). Optimization of control strategy for regenerative braking of an electrified bus equipped with an anti-lock braking system. *Proceedings of the Institution of Mechanical Engineers. Part D, Journal of Automobile Engineering, 226*(4), 494–506. doi:10.1177/0954407011422463

Zhang, L., Wang, Z., Hu, X., Sun, F., & Dorrell, D. G. (2015). Acomparative study of equivalent circuit models of supercapacitors for electric vehicles. *Journal of Power Sources, 274*, 899–906. doi:10.1016/j.jpowsour.2014.10.170

Zhang, L., Wang, Z., Sun, F., & Dorrell, D. G. (2014). Online parameter identification of supercapacitor models using the extended Kalman filter. *Energies, 7*(5), 3204–3217. doi:10.3390/en7053204

Zhang, L., Zhenpo, W., Sun, F., & Dorrell, D. (2014). Supercapacitor modelling and parameter identification using the extended Kalman filter. *ITEC Asia Pacific, 7*, 3204–3217.

Zhu, G., & Gao, D. (2020). High-Frequency Voltage and Current Sense Circuits for Inductive Power Transfer Systems. *IEEE Transactions on Power Electronics, 35*(11), 11352–11362. doi:10.1109/TPEL.2020.2989922

About the Contributors

Umashankar Subramaniam is Associate Professor in Renewable Energy Lab, College of Engineering, Prince Sultan University, Saudi Arabia. He has 15+ years of teaching, research and industrial R&D experience. Previously, he worked as Associate Professor and Head, VIT Vellore as well as Senior R&D and Senior Application Engineer in the field of power electronics, renewable Energy and electrical drives. He is a Senior member-IEEE, and Member of IACSIT, IDES and ISTE. He has taken charge as Vice Chair - IEEE Madras Section and Chair - IEEE Student Activities from 2018. He was executive member (2014-16) and Vice Chair of IEEE MAS Young Professional from 2017 by IEEE Madras Section. He has published more than 250+ research papers in national and international journals and conferences. He has also authored/co-authored/contributed 12 books/chapters and 12 technical articles on power electronics applications in renewable energy and allied areas. He is an Editor of Heliyon, an Elsevier journal. He received Danfoss Innovator Award-Mentor during 2014-15 & 2017-18, Research Award from VIT University during 2013-18. Also he received the INAE Summer Research Fellowship for the year 2014. Under his guidance 24 P.G students and more than 25 U.G Students completed the senior design project work. Also 6 PhD scholars completed Doctoral thesis as Research Associate. He is also involved in collaborative research projects with various international and national level organizations and research institutions.

Sheldon S. Williamson, PhD, Canada Research Chair in Electric Energy Storage Systems for Transportation Electrification, and Associate Professor in the Department of Electrical, Computer and Software Engineering in the Faculty of Engineering and Applied Science, is leading groundbreaking research to extend the overall lifecycle of Lithium-ion batteries in electric vehicles using novel power electronic converter management systems. He is also focused on creating wireless and plugged fast-charging infrastructures for convenient use. Notably, he is spearheading development of the world's first method for charging electric vehicles using solar power and he aims to establish a first-of-its-kind Advanced Storage Systems and Electric Transportation (ASSET) Laboratory, featuring a solar charging station, at UOIT. Motivated to shift Canada's transportation system from fossil fuels to renewables, his research also explores electrifying mass transit using ultracapacitators. Since joining UOIT in July 2014, Dr. Williamson has been the Founder and Director of the Smart Transportation Electrification and Energy Research (STEER) group. Previously, he was an Associate Professor in the Department of Electrical and Computer Engineering at Concordia University in Montreal. He received his Bachelor of Engineering in Electrical Engineering with high distinction from the University of Mumbai in India in 1999. In 2002, he earned his Master of Science and his Doctorate in 2006, both in Electrical Engineering, specializing in Automotive Power Electronics and Motor Drives from the Illinois Institute of Technology. Noted author and co-author of over 150 papers, and several books and book chapters on electric transportation and energy storage systems, Dr. Williamson has garnered several Best Paper Awards. He is a Senior Member of IEEE and a distinguished lecturer of the IEEE Vehicular Technology Society.

Mohan Krishna. S was awarded Ph.D. in Electrical Engineering (Sensorless control of induction motor drives) from Vellore Institute of Technology (VIT), India, in March 2017. He received his B.Tech and M.Tech degrees from Amrita Vishwa Vidyapeetham, Coimbatore, India in 2009 and 2012 respectively. He also acquired a domain specific MBA (Power Management) from University of Petroleum and Energy Studies (UPES), Dehradun, India, in 2019. His research interests include, electric vehicles, smart homes and IoT based building energy management systems, state observers for induction motors, energy economics and sustainability etc. He has many research publications in SCI/SCI-E and Scopus indexed journals, book chapters and also in conference proceedings to his credit. He is Energy section advisory board member of Heliyon (Elsevier). He also serves as the Associate Editor of International Journal of Smart Vehicles and Smart Transportation (IJSVST), IGI Global, US. Additionally, he is a member of the editorial review board of International Journal of Energy Optimization and Engineering (IJEOE - Web of Science (ESCI)). He is also editing several books in the domain of electric vehicle , smart grids and energy sustainability. He also serves as a reviewer for IEEE Access, IET Control theory, IET Electric Power Applications, COMPEL, Turkish Journal of Electrical Engineering & Computer Sciences (all SCI) and several other Scopus indexed journals. He is a Life member of several professional societies like the IET, Institution of Engineers (India), ISTE, IAENG, ENFUSE and a member of IEEE. He is also a member of the IEEE YP and IEEE SIGHT (Special Interest Group on Humanitarian Technology), Bangalore section.

Febin Daya J. L. received his B.E. in Electrical and Electronics Engineering from Manonmaniam Sundarnar University, Tamilnadu, India, in 2002, M.E. in Applied Electronics from Anna University, Tamilnadu, India, in 2005 and PhD in Information andCommunication from Anna University, Tamilnadu, India in 2013. From 2005 to 2011, he was working in the Department of Electrical and Electronics Engineering at the Sri Krishna College of Engineering and Technology, Coimbatore, India. Presently he is a Professor at school of Electrical Engineering, VIT University, Chennai, India. He has published around 75 papers in International Journals and Conferences. He is involved actively in many professional bodies at the level of editor/associate editor /editorial board member/reviewer in India and abroad. He was invited as honorary program committee/technical committee/steering committee member/chair for more than 15 International conferences in India and Abroad. Febin Daya is a member of IEEE and Life Member of Indian Society of Technical Education. His current research interests include electrical drives, intelligent systems and electric vehicles.

* * *

Vishweshkumar Aithal has been working as Assistant Professor in the Department of Electronics and Communication Engineering of KLS Gogte Institute of Technology, Belagavi, an autonomous institute affiliated to Visvesvaraya Technological University, Belagavi since 3+ years. He has completed his Masters in Signal Processing. His research interests include the domain of Signal and Image Processing, Computer Networking. He has his papers published in reputed international conferences and journals. He intends to continue to do great contribution to his field of research interests mentioned above.

Bharathi Sankar Ammaiyappan is an Assistant Professor Senior Grade in the School of Electronics Engineering, VIT University Chennai Campus has 10 years of teaching and research experience including 7 years of research experience in Power Electronics, Renewable Energy Systems, Energy Storage (Li-ion Batteries and Supercapacitor) and Electric Vehicles. He has received B.E. in Electrical and Electronics Engineering from Anna University, Chennai in 2004, M.E in Power Electronics & Drives from Anna University, Chennai in 2011 and Ph.D. in Electrical Engineering from Anna University, Chennai in 2017 and Post-Doctoral Fellowship from International Advanced Research Centre (ARCI), Department of Science and Technology, Hyderabad, 2017 to 2019. During Ph.D. he has design and developed Solar Powered Electric Vehicle using Brushless DC Motor Drive (Internal Funding – 4.5 Lakhs SSN College of Engineering 2013 - 2016).During Post-Doctoral Fellowship Design and Developed Supercapacitor Powered Electric Vehicle (DST Funding – 10 Lakhs ARCI, Hyderabad 2017 - 2019).He areas of specialization are Electric Vehicle, Battery Management System, Multilevel Inverter fed BLDC drive, Renewable Energy Harvesting and Storage. He has published over 25 research publications in refereed impact factor International Journals and 35 International and National Conferences. He received around 8 best paper awards in various international conferences. He has patent published Solar Powered Battery operated Vehicle using BLDC Drive, Appl. No. 201741019394.

Yigit Aydogan was graduated from Koc University with BS degree in Economics in 2010. After working at multinational companies for a while, he has begun his academic career at Economics Department of Ege University in 2012. He received a master's degree in Economics from Ege University in 2013. He has received his PhD from Yildiz Technical University in the field of Economics, in 2019. He is a faculty member of Economics Department at Kirklareli University, since 2013.

Saravanan B. holds an ME (Power Electronics & Drives) from Government College of Technology, Coimbatore in 2005, Having overall 15 years of Experience in the field of Power Electronics & Drives, with 11+ years in the domain of Traction Power Electronics & Control which includes power converter design, development, digital control system design(DSP & FPGA), component selection, semiconductor device loss calculation, cooling system design and commissioning.

Ghanishtha Bhatti is a final year undergraduate student in the department of Electrical and Electronics Engineering at Vellore Institute of Technology (VIT), Vellore. She holds proficiency in technical writing and simulation modelling, with a keen interest in the fields of sustainable technology, robotics, and automation.

Sudhakar Hallur has been working as Assistant Professor in the Department of Electronics and Communication Engineering of KLS Gogte Institute of Technology, Belagavi, an autonomous institute, affiliated to Visvesvaraya Technological University, Belagavi since 3+ years. He has completed his Masters in Digital Communication and Networking with a Gold Medal. He also has a working experience on Software Development and Testing in Industry. His research interests include the domain of Biomedical Signal Processing, Computer Networking, Cryptography, Automobiles and Electrical Vehicles. He has his papers and book chapters published in reputed international conferences and journals. He intends to continue to do great contribution to his field of research interests mentioned above.

R. Jeyanthi graduated from Kongu Engineering College, T.N, India in 2003. She obtained M. E degree from Anna University, T.N, India in 2007 and pursuing Ph.D in Amrita Vishwa Vidyapeetham. Currently she is working as an assistant professor in Electronics & Communication Engineering Department, Amrita School of Engineering, Amrita Vishwa Vidyapeetham University, Bengaluru, Karnataka, India. She has 15+ years of teaching experience. She is a Member of IE(I). She has published 14+ research articles in international Journals and Conferences. Her research interests are Data Reconciliation and Gross Error Detection, Process control, Industrial Automation and Time Series Analysis.

Haneesh K. M. completed his B.Tech (Electrical & Electronics Engineering) from Calicut University, India and M. Tech (Electrical Machines) from Kerala University, India. Presently doing research in the control engineering area. His areas of interest include renewable energy, battery technology and hybrid vehicles.

Deepa Kaliyaperumal graduated from Alagappa Chettiar college of engineering and Technology, T.N, India in 1998. She obtained M.Tech degree from Anna University, Guindy campus, T.N, India in 2005. She received Doctoral degree from Jawaharlal Nehru Technological University, Anantapur, A.P, India in 2017. Currently she is working as Assistant professor in Electrical and Electronics Engineering Department, Amrita School of Engineering, Amrita Vishwa Vidyapeetham University, Bangalore, Karnataka, India. She has 22+ years of teaching experience. She is a life Member of IETE and ISTE, India and a senior member of IEEE. She has authored two textbooks on "Electrical Machines" and "Control Systems". She has published 27 international journal paper, 31 papers in international conference and 6 papers in national conference. 15 M.Tech Degrees were awarded under her guidance. She is the advisor for the IEEE-PES & IAS student branch joint chapter and advisor for IEEE-WIE in Amrita School of Engineering, Bengaluru from 2015. She is also joint treasurer for 2018 and treasurer for 2019 EXECOM of IEEE PES Bangalore chapter and Vicechair for IEEE PELS Bangalore chapter for 2020. Her areas of interests include Power electronics, Renewable energy technologies and Control Engineering.

Roopa Kulkarni has been working as Associate Professor with Dayanand Sagar Academy of Technology and Management, Bengaluru, affiliated to Visveshwaraya Technological University, Belagavi, Karnataka for more than 1+ years. She has previously worked as Associate Professor, KLS Gogte Institute of Technology, Belagavi for more than 15 years. She has significantly contributed to the domain of Embedded Systems, Low Power VLSI, HDL and others with a huge numbers of papers and book chapters published in national and International conferences and journals. She has guided with more than 40 projects at various levels and have secured various funds and awards.

Balamurugan M. is currently working as Assistant Professor in Department of EEE, Faculty of Engineering and Technology, JAIN (Deemed-to-be University). He received his Ph.D. degree in (Power Electronics Application in Renewable Energy Systems) from VIT University, Vellore, Tamilnadu in 2019. He received his Postgraduate Degree in Power Electronics and Drives from VIT University Vellore and Undergraduate degree in Electrical and Electronics Engineering from C. Abdul Hakeem college of Engineering and Technology affiliated to Anna University, Chennai in the year 2014 and 2012 respectively. He has 3 years of research experience as Junior Research Fellow for sponsored project funded by Department of Science and Technology (DST) in School of Electrical Engineering at VIT University, Vellore, India. He has also awarded as Best Researcher in the year 2016 by VIT University. His research interests are cascaded Multilevel Inverter, Power Electronics application in Renewable energy systems. He has published several international papers in journals, book chapters and conferences of high repute.

Kamala N.'s research interests are Artificial Intelligence and control system.Having 30 years of teaching experience. Currently working as Associate Professor in the Department of Robotics & Automation at Jain (Deemed-To-Be University), Bangalore, Karnataka.

Raghu N. has nine years of experience in Academic and Research. He is currently pursing Ph.D in Electronics Engineering, Jain University, Bangalore. His research interests were in RF Communication and Image Processing. He has published his research findings in various Scopus indexed Journals and International conferences. He is a member of various professional bodies.

Prashant Patavardhan has been serving as Professor and Head of the Department of Electronics and Communication Engineering in RVITM, Bengaluru, affiliated to Visveshwaraya Technological University, Belagavi, Karnataka for more than 1+ years. He has previously worked as Professor in Dayanand Sagar University and KLS Gogte Institute of Technology, Belagavi for more than 21 years. He has significantly contributed to the domain of Image Processing, Embedded Systems, Biometrics, Networks, Video Analytics and others with a huge numbers of papers and book chapters published in national and International conferences and journals. He has guided with more than 40 projects at various levels and have secured various funds and awards.

B. Prasanth received his B. Tech degree in Electrical and Electronics Engineering in 2020 from Amrita School of Engineering, Amrita Vishwa Vidyapeetham, Bengaluru, Karnataka, India. He has been a Student Member of IEEE PES & IAS, Bangalore Chapter 2018-19. He has published one International Journal paper, three papers in International Conferences. His area of interests includes Soft computing techniques-based control engineering, Power electronics, Power system analysis, Electric Vehicles and Machine Learning techniques.

Raja Singh R. received B. Tech. in Electrical Engineering from Pondicherry University, India, M. E in Power Electronics and Drives from Anna University, India and Ph.D. in Electrical Drives from Indian Institute of Technology (IIT) Roorkee, India in 2004, 2008 and 2017 respectively. He has more than 9 years of teaching experience in India and Tanzania, as well as, 5 years of research experience at IIT Roorkee, India. At present, he is working as an Associate Professor in the Department of Energy and Power Electronics, Vellore Institute of Technology (VIT) Vellore, India. His research includes multilevel power converter drives for renewable energy applications, power quality control, starting transient and fault analysis on industrial drives. Also, he has developed various energy-efficient control strategies for industrial drives. He has published many research papers in various journals and conferences of international repute.

Seyezhai Ramalingam is an Associate Professor in the Department of Electrical and Electronics Engineering has 21 years of teaching and research experience including 14 years of research experience in Power Electronics and Renewable Energy Systems. She received her B.E. from Manonmaniam Sundaranar University in 1996 and M.E in Power Electronics & Drives from Bharathidasan University in 1998 and Ph.D. from Anna University, Chennai. During her Ph.D. she developed a prototype of SiC-based hybrid cascaded multilevel inverter employing a novel modulation technique for fuel cell Power Conditioning System (PCS). She is the recognized supervisor of Anna University, Chennai, Satyabhama University, and Priest University, Tanjore. 9 scholars graduated with their Ph.D. under her guidance, two have submitted the thesis, and currently guiding seven Ph.D. scholars. Her areas of specialization are Electric vehicle, Battery Management system, Quasi Z-source Multilevel Inverter for photovoltaic applications, Reliability of power converters, LED Drivers, and Silicon Carbide (SiC) & GaN power devices. She has published over 250 research publications in refereed International Journals and 150 in International and National Conferences. She received around 21 best paper awards in various international conferences. She has reviewed more than 200 technical papers, served as a reviewer and editorial board member for more than 20 journals published by different bodies. She has served on roughly 40 conference and workshop program committees. She has delivered invited lectures, Keynote addresses, and Chaired technical sessions in various international conferences. She received various awards that include the Young achiever and Research Excellence Award for the Year 2017, Best Supervisor award 2017, Outstanding Faculty Award in the field of power electronics in 2016, the Best Academic Researcher Award for the year 2015 from the Association of Scientists, Developers and Faculties and the Bharat Jyothi award in 2012 for outstanding contribution to student's community. She received the best teacher award for the academic years 2017-18, 2016-2017,2016-15, 2013-14, 2012-13, 2004-2005, and the CTS Best Faculty Award for the year 2012-13. She has developed the 'Renewable Energy Conversion Laboratory' in the west wing of the EEE Department, SSNCE with the help of external and internal funding. She is in the editorial board and reviewer board of several refereed international journals. She is an active member of IEEE and a life member of ISTE. Currently, she is the Chairman, IEEE-PELS, Madras chapter.

Rakshitha Ravi currently working as Assistant Professor in the Department of Electronics and Communication Engineering at RVITM. Her specialization is in Power Electronics. She is pursuing her Ph.D. at Christ University, Bangalore, India.

Sarat Kumar Sahoo has received his Ph.D. from JNTU, Hyderabad, India in the year 2011. He has received his Master's degree in Computer Application to Industrial Drives, from Visvesvaraya Technological University, Belgaum, Karnataka, India, in the year 2002, and Bachelor's degree in Electrical Engineering, from The Institute of Engineers, Calcutta, India, October 1999. He is currently working as Professor & Head of the Department in Department of Electrical Engineering, Parala Maharaja Engineering College, Behrampur, Odisha, India. His research interests are Grid-Tie-Inverters for Solar Energy application, Distributed Generation; Renewable Integration; Rural Electrification, Micro-grid Energy System, Smart Grid and Smart meter, Electric Vehicle Controller based on Space Vector Modulation and Direct Torque Control, Multilevel inverters; Medium voltage drives. Dr. Sarat Kumar Sahoo, is also a Senior Member of IEEE, ISGF, MIE, ISTE, The Indian Science Congress Association, IISRO, UACEE, IRED, IACSIT, IDES. He has published 60 International peer-reviewed journals, about 70 conference publications, 15 Magazine articles and has published 3 books.

Usha Surendra is currently working as a Professor in the Department of Electrical and Electronics Engineering at Christ University, Bangalore, India. She has a rich experience in power systems and power electronics field.

Trupti V. N. completed B.Tech in Electrical & Electronics Engineering. Completed M.Tech in Engineering Management from Visvesvaraya Technological University. Pursuing a PhD at Jain University in the field of energy audit and management. Teaching experience of 10 years. Industrial experience of 1 year. Research publication of around 10 papers in international journals. The field of interest is energy management, renewable energy resources, hybrid energy. Published textbook on "Reduction of Cycle Time by Implementing Value Stream Mapping" and she has contributed in Book Chapter titled, "Risk, Hazard and Disaster Management about hazards from extra-high voltage lines."

Index

L

M

O

P

R

S

T

W

www.ingramcontent.com/pod-product-compliance
Lightning Source LLC
LaVergne TN
LVHW060305011026
849161LV00011B/2775